Control Systems and Applications for HVAC/R

Control Systems and Applications for HVAC/R

Thomas J. Horan

PRENTICE HALL
Upper Saddle River, New Jersey   *Columbus, Ohio*

Library of Congress Cataloging-in-Publication Data

Horan, Thomas J.
    Control systems and applications for HVAC/R/ Thomas J. Horan.
        p. cm.
    Includes index.
    ISBN 0-13-125196-1
    1. Commercial buildings—Heating and ventilation—Control.
2. Commercial buildings—Air conditioning—Control. 3. Automatic
control.   I. Title.
TH7392.C65H67   1997
697—dc20   96–23408

Editor: Ed Francis
Production Editor: Christine M. Harrington
Production Coordinator: Ruttle Graphics, Inc.
Cover Designer: Proof Positive/Farrowlyne Assoc., Inc.
Text Designer: Ruttle Graphics, Inc.
Production Manager: Deidra M. Schwartz
Marketing Manager: Danny Hoyt

This book was set in Palatino and was printed and bound by Quebecor
Printing/Book Press. The cover was printed by Phoenix Color Corp.

 © 1997 by Prentice-Hall, Inc.
Simon & Schuster/A Viacom Company
Upper Saddle River, New Jersey 07458

Printed in the United States of America

10 9 8 7 6 5 4 3 2 1

ISBN: 0-13-125196–1

Prentice-Hall International (UK) Limited, *London*
Prentice-Hall of Australia PTY. Limited, *Sydney*
Prentice-Hall Canada, Inc., *Toronto*
Prentice-Hall Hispanoamericana, S.A., *Mexico*
Prentice-Hall of India Private Limited, *New Delhi*
Prentice-Hall of Japan, Inc., *Tokyo*
Simon & Schuster Asia Pte. Ltd., *Singapore*
Editora Prentice-Hall do Brasil, Ltda., *Rio de Janeiro*

To my wife Kathy,
my best friend whose love
brings immeasurable joy and
contentment into my life.

Contents

Preface

Of all the mechanical and design tasks performed by HVAC technicians, technologists, and engineers, those dealing with control system design, calibration, and analysis tend to be the least understood and most difficult to master. To work effectively with control systems, a person must become fluent in a variety of disciplines that go beyond the boundaries found in a typical HVAC program curriculum. Today's HVAC technician must be conversant in mechanical, electrical, electronic, computer, and thermal technologies. This requirement makes the control field one of the more difficult occupations in which to excel. These requirements also make control system design, installation, and analysis one of the more challenging, exciting, and rewarding fields in which a mechanically inclined person can specialize.

The first control devices were mechanical in nature, easily understood by persons closely related to the HVAC/R field. Electromechanical relays, contactors, thermostats, and a variety of other components were easily calibrated and analyzed because their response or reaction to stimuli were easily seen. Modulating pneumatic devices used to control the flow of mass and energy were slightly more challenging to understand because their internal components consisted of small obscure springs, bleed orifices, diaphragms, levers, tubes, and adjustable dials. Although more complex, they were still based on mechanical force and balance principles that could be intuitively analyzed by the mechanically minded person.

Once electronic components evolved into a cost-effective alternative to pneumatic components, the age of "black box" control systems began. The devices operated on low voltage and current signals generated in tiny plastic coated devices with little wires protruding from their ends. There was no easy way intuitively to reference these devices back to their mechanical equivalent. An electronic control device could be opened up on the bench, its schematic reviewed, and the technician could walk away from it without knowing how it performed its function. Control companies were required to hire people with electronics backgrounds to design control systems for mechanical equipment and thermal processes for which they had little training. Mechanically based control technicians began to back away from the electronic systems, leaving them to the few technicians that had a desire to work with the new technology.

Current digital control systems have brought another technology discipline—computer science—into the HVAC control field. These new systems require computer hardware technicians to design and diagnose solid state circuit boards. Computer programmers are needed to develop the programming code that operates the control system hardware and its communication networks. Additional software must be written to operate the connected mechanical and electrical equipment that maintains a comfortable environment in the building. Consequently, operational problems are no longer confined to hardware. A bug in the software can create an intermittent problem that will be very time consuming to find. This may force the building operator to close the evolutionary path of control systems by operating the affected equipment by hand.

Today's HVAC control systems requirement for electronic and computer technology people to work on systems maintained by mechanically trained HVAC technicians created a new rift in the HVAC field. Two differently trained groups of people with different perceptions of the HVAC process must work together to maintain the operational efficiency of systems and buildings. Many HVAC technicians shy away from control-related problems that incorporate computers in favor of calling a person trained in computer science to service the system. Unfortunately, that is usually not the best alternative. The HVAC technician knows the operation and response of the building and its systems better than an outside service person and, therefore, is better qualified to tie the control system into the mechanical systems. This is the driving force behind the development of this book.

I wrote this book to compile control system information from the electronic, computer, mechanical, and thermal disciplines into one text. This book can be used by any systems-oriented person who wants or needs to know how and why a process responds the way it does to changes in a measured variable. It introduces the reader to the fundamental characteristics of all control systems and processes using HVAC systems as a basis for the applications and examples. The book is written in a manner that develops the reader's analytical skills in diagnosing and rectifying problems in any system.

Part One of this book begins with the basics. It describes the purpose, terminology, and configurations of basic control loops. The section ends with a description and applications of the basic relationship of all control systems called the linear transfer function. The transfer function is the formula used by all technicians to calibrate devices from a simple sensor to complex PID controller modes.

Part Two describes the individual components found in a control loop, their calibration parameters, and selection criteria. It also describes the construction, operation, and sizing procedures for valves and dampers. Other chapters describe how to select a controller mode based upon the characteristics of the process and how to calibrate the control loop to maximize efficiency. The last chapter gives a comprehensive overview of digital control systems and programming.

Parts Three and Four of the text describe the control strategies most often used in commercial HVAC systems. Examples include a schematic representation of the process or system, a sequence of operation, programming flowchart, and analysis of its operation. Part Four also lists troubleshooting techniques and solutions to common problems in HVAC processes.

The text can be used along with a control vendor's parts catalog to instruct new students on the different options available for a given application and how to look up items in a catalog. Laboratory assignments can be developed that analyze actual operating systems installed in the school's buildings. The catalog and associated specification sheets can be used to give specific operating parameters of the components in the system and their calibration procedures.

The book is based upon how and why a process responds in a particular way and how to adjust the parameters of the loop to improve its response. It is applicable to electromechanical, pneumatic, analog electronic, and digital electronic control systems because of its focus on the system's response, not its construction. In any control system, the adjustment of a particular control loop parameter will generate the same response. It is my hope that the book sets a firm foundation of the basics of controls so the reader may develop into a control system specialist— able to design, specify, and calibrate control systems.

Acknowledgments

I take this opportunity to thank the many people who have made this textbook possible. My wife, Kathy, who supported me when I went back to college and encouraged me to complete this project although it took considerable time away from our family; my children, Holly, Heather, Michelle, and Kelly, who couldn't understand what possessed their dad to devote countless hours in front of the computer; my in-laws, Hazel and Mike Santelli, along with Gram, who helped us out when I went back to college; and Frank and Gina, who have always made me feel special. If it weren't for the love of God and these people, I would not have been able to write this book to share with others the enthusiasm that I experience tuning and troubleshooting control systems.

I would also like to acknowledge those who sparked my interest in control systems: the faculty at Rochester Institute of Technology, in particular, Ron Amberger, who taught me how to analyze systems using the fundamental relationships that exist in science and technology; Robert Bateson, whose textbook sparked within me an interest in control systems that has developed into a rewarding career; and Dick Shaw and L. J. Bishop who have given me the opportunities to apply the knowledge found in this text.

A special thanks to members of my family who have influenced the book through their enthusiastic interest in the project and my life—my parents, Marge and Tom; George and Cindy; Kevin and Jennie; Tom and Sue; Steve and Debbie; Pat; Kathy; Matt; Jeff; Leah; Michael Jr.; Megan; Christine; Ginnette; Brittany; Katie; the rest of our clan; and Michael, Jim, Susie, and Marie who are always remembered in our thoughts.

Finally, thanks to my friends, especially Paul and Carol Lewis, Tom Page, Bob Lumsden, Ron Salik, Rob and Sue Hirsch, Bruce Weber, Richard Orlando, Larry and Denise Edmonson, and you.

PART 1

Introduction to
Control Theory

CHAPTER 1

Introduction to Control Systems

Working with control system technology is one of the most exciting and challenging aspects of the heating, ventilating, and air-conditioning field. Mastering the abilities needed to operate and analyze mechanical and computer-based control systems offers dedicated technicians the greatest potential for recognition and success in their careers. The information presented in this textbook will assist the reader in developing the skills required to analyze, modify, and calibrate any type of HVAC control system.

Before technicians can develop the skills needed to analyze the operation and response of a control system, they need to understand the *meaning* and *purpose* of a control system. Chapter 1 introduces the reader to the basic nature and purpose of any type of control system.

OBJECTIVES

Upon completion of this chapter, the reader will be able to:

1. Define the term *control* as it relates to HVAC systems.
2. Explain why control systems are required.
3. Describe the characteristics of automatic control systems.
4. Describe the different types of control systems and their advantages.

3

1.1

WHAT IS THE MEANING OF CONTROL?

Control systems are found within all aspects of our environment. The human body and other organisms contained the first control systems that were used to maintain their core temperatures, rate of food metabolism, balance, and movement. Each of these life-sustaining processes is regulated by signals generated in the organism's brain based upon input data from its environment.

Control technologies were first developed during the industrial revolution when mechanical and electrical equipment incorporated systems to automatically regulate their operation. Automating the operation of equipment improved operating efficiencies and allowed for the manufacture of standardized parts. Any mechanical or electrical system that maintains a temperature, moisture content, position, level, size, quantity, velocity, intensity, or other characteristic of a process incorporates control devices that operate collectively to maintain the process.

The first step in developing an understanding of the need for control systems is to review what it means to control a piece of equipment. A good definition of control that applies to heating, ventilating, and air-conditioning systems is "to apply a *regulating influence* upon a *device* to make it perform as required," where the *regulating influence* is a force applied by a person, electric circuit, mechanical mechanism, or machine and the device is a mechanical or electric component (valve, damper, relay, etc.) that changes the amount of mass (quantity of water, air, fuel, material, etc.) or energy (electric current, heat, etc.) that is delivered into a process. Control systems vary the strength and direction of the regulating force applied to the control device to vary the mass or energy flowing into a process in response to changes in a measured condition. By varying the regulating force, the process is maintained at a value that falls within the permissible operating range that was established when the system was designed.

Four examples of commonly used mechanical equipment that incorporate control devices are a lawn mower, radio, toilet, and sink. These are all non-HVAC systems that are used by operators who have no understanding of control systems. The throttle lever on a lawn mower, volume control of the radio, float in the toilet tank, and the faucet valves on a sink represent control devices to which a regulating force is applied to maintain one particular process condition. The lawn mower throttle lever allows the user to manually regulate the speed of the cutting blade from zero revolutions per minute (rpm) to a maximum value of 3500 rpm. This maximum speed limit is determined by the engine's design and is based on the mechanical limitations of the engine and the operator's safety. The volume control on a radio allows the listener to vary the sound intensity of the music from quiet to the maximum output capability of the unit's audio transformer. The float found inside of the toilet tank opens and closes the supply water valve to maintain the proper level within the tank. Finally, a sink's faucet valves allow the user to modulate the flow of hot and cold water from 0% to 100% to provide the desired flow rate or temperature of the water leaving the spout. Each of these examples relies upon an input signal from the user to initiate a change in the process. Control systems that require a person to apply the regulating force to the control device are called *manual control systems*.

Commercial and industrial HVAC applications use electric and mechanical control systems to position valves, dampers, and other mechanisms to maintain the desired temperature, humidity level, and static pressure within a given area or zone. These systems regulate the amount of mass (water, air, fuel) or energy (heat, electricity) transferred between a source and the process to maintain a desired condition. A window air conditioner uses a simple control system to maintain the temperature within a room at the desired level. The control system cycles the compressor on and off in response to changes in the room's temperature. When the temperature rises above a permissible level, the control system closes a relay that allows energy to flow into the compressor motor, allowing refrigerant to be circulated between the evaporator and condenser. This transfer of heat energy from the interior room to the outside prohibits the room temperature from rising above a user-defined comfortable level. As the compressor continues to operate, the room temperature decreases. As it drops below a predetermined comfort level, the control system stops the flow of energy into the compressor motor, stopping the refrigerating effect. These functions occurred without the need of a person being present to provide the force to close the control relay. Control systems that do not need human intervention to maintain a process condition are called *automatic control systems*. HVAC control systems can be designed to operate manually or automatically. The choice is based upon the equipment, process, and user requirements. Keep in mind, whether the system responds automatically or manually, its purpose—*to apply a regulating influence upon a device to make it operate as required*—remains the same.

In addition to maintaining comfort and other process conditions, mechanical and electric systems incorporate control devices that ensure that the equipment operates *safely*. These safety control devices enable and disable (turn on or off) equipment based upon a monitored equipment or process condition. When the condition exceeds a safe operating limit, the control system automatically changes the operating state of the equipment to minimize safety hazards. The compressors used in the air-conditioning and refrigeration applications incorporate safety controls that will disable (turn off) the system whenever an abnormal pressure, temperature, or electric current flow exists. Safety-oriented control systems are used to automatically protect equipment, occupants, and buildings from fires or other hazards that may result from equipment operating outside the limits of their design parameters.

The following applications describe how some typical HVAC control systems apply regulating forces upon devices to control mechanical equipment so it operates in accordance with its intended design. Application 1.1 describes the response of a control system that maintains a specified static air pressure within an HVAC duct. Application 1.2 describes a control system that maintains the temperature of the air leaving a hot water coil at a desired temperature. Each application is outlined in three sections labeled Process, Objective, and Response. The *Process* section states the commonly used description of the condition that the control loop is required to maintain at a desired value. The *Objective* section describes the specific requirements of the process. The *Response* section details how the controls operate to meet the process objective. This format is used throughout the text to introduce common HVAC processes and their typical response to changes in a measured variable.

APPLICATION 1.1

Process: Variable Air Volume System Static Pressure Control Loop
Variable Air Volume systems maintain the static pressure of the air in the duct that supplies conditioned air into the zones within a building.

Objective: Maintain the static pressure in the supply duct at 1.2 inches of water gauged by varying the amount of air (mass) entering the fan blades through the fan's inlet vane dampers.

Response: When the amount of air entering the system fan is less than the quantity of air leaving the diffusers in the zones, the static pressure within the duct will decrease. When more air enters the fan than leaves the diffusers, the duct pressure increases.

The control system measures the static pressure within the duct. As the pressure decreases, the control system opens the fan inlet vane dampers to increase the air flow into the duct and, consequently, increases the static pressure. The signal to open the dampers comes from a control device called a controller. As the static pressure increases above the design value, the controller modulates the dampers toward their closed position, reducing air flow into the duct. The controller is a control device that determines how to position the inlet vane dampers based upon the measured static pressure.

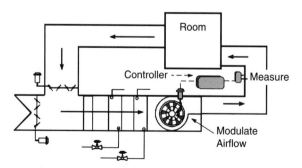

Application 1.1   Static Pressure Control System Schematic

1.2

WHAT IS THE PURPOSE OF A CONTROL SYSTEM?

The purpose of a control system is to *balance* the flow of mass or energy transferred into the process with its present *load*, where:

A. The *process* is the physical condition (temperature, humidity, level, flowrate, etc.) that is maintained by the control system at a desired value.
B. A *load* is the amount of *mass* (air, water, fuel, etc.) or *energy* (heat, electricity) needed to maintain the process at its desired condition.
C. *Balancing* the flow of mass or energy transferred into the process with the load indicates that the amount of energy entering the process *equals* the amount of energy leaving the process.

The key word in the definition of the purpose of a control system is the word *balance*. Maintaining a balance between the quantity transferred and the load in HVAC processes preserves occupant comfort and maximizes the operating efficiency of the equipment. When the mass or energy transferred into the process is greater than its present load, the efficiency of the process decreases as the process drifts outside of the limits of its desired range, producing comfort or operational problems.

A descriptive example of the definition of the purpose of a control loop is a summer temperature control application for an office space. The *process* of this application is to maintain the space temperature at a desired value of 74 °F. The load is the quantity of heat that must be removed from the space to maintain the temperature at 74 °F. The room's cooling load is comprised of a combination of solar gain, wall conductance gain, infiltration air, internal equipment heat, and occupant loads. When the quantity of energy flowing out of the conditioned space equals the total room load, its temperature remains constant. When the control system maintains the room at 74°, the process is maintained at the maximum possible efficiency for those conditions. Figure 1.1 shows a pictorial definition of the purpose of control systems, balancing the energy transfer with the load.

All control systems maintain their process condition by altering the flow of mass or energy being transferred until a balance is reached. In HVAC processes, the flow of heat, air, and water make up a majority of the mass or energy flows regulated by control systems.

Hot and chilled water are used to transfer heat, volumes of air are regulated to maintain room temperatures and movement. Moisture is transferred to affect changes in the relative humidity in zones and water flow is regulated to maintain condenser head pressures. Depending on what level a technician classifies the transfer, it will always be a mass or energy process. Which one it may be is not critical to the controls technician. Therefore, the text will use the term *heat* to classify most energy transfers that maintain temperatures and the term *mass* to describe those processes that maintain humidity levels, liquid levels and pressures.

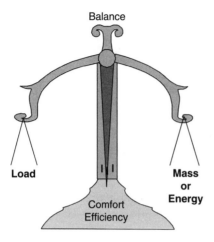

Figure 1.1   Control Loops Maintain a Balance Between the Load and the Mass or Energy
Transferred into the Process

   In actual processes, the load is *dynamic*. Dynamic means the environment is constantly changing, and consequently, the load also is varying. To maximize comfort
and operating efficiency of real systems, control systems are designed and calibrated (tuned) to balance the HVAC equipment's capacity with the current load in
order to maintain the desired conditions within the space. The following Applications describe HVAC processes in terms of the purpose of control systems.

APPLICATION 1.2

Process: Warm Air Temperature Process

Objective: Maintain the temperature of the air passing across a hot water coil at
the desired temperature of 120 °F by modulating the quantity (flowrate) of hot
water circulating through the hot water coil.

Response: Cooler air flows across the fins of a heating coil as hot water flows
through the heating coil tubes. The hot water transfers its energy to the cooler air
by conduction, raising its temperature.
   The control system measures the temperature of the air leaving the coil using a
device called a sensor. As the air temperature increases, the control system modulates
the valve toward its closed position, reducing the flow of hot water entering the coil.
As the flow of water decreases, the amount of energy transferred to the air also

decreases, reducing the air temperature back into its acceptable range. Conversely, if the air temperature is too low, the control system opens the hot water valve.

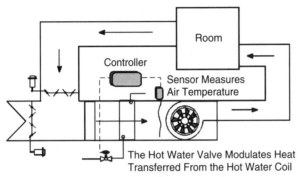

Application 1.2   Heating Control System Schematic

APPLICATION 1.3

Process: Residential Heating System—Maintain the temperature of the room with the thermostat at 72°.

Objective: The control system, consisting of a thermostat, fuel valve, fan limit switch, and required safety devices, maintains a balance between the heat flowing into the room through the registers with the energy leaving the room through the walls, ceiling, and floor.

Response: Air from the room is transported across a gas-fired furnace heat exchanger by the furnace fan. The air is returned to the occupied space to offset the present heating load. As the room temperature decreases below the comfort

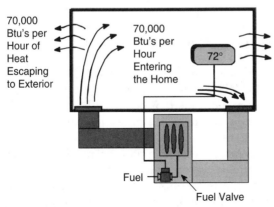

Application 1.3   Home Temperature Control

limit established by the occupants, the control system starts the furnace. As the cooler air passes across the fins of a heat exchanger, heat from the combustion of the fuel is transferred to the air and transported back to the room. When the temperature increases above the high comfort limit, the furnace is cycled off. The cycling of the furnace maintains an average balance between the amount of heat entering the room and the amount transferred to the colder outside ambient. In this application, the controller and sensor are combined into one device.

1.2.1 Design Load Conditions

Before control systems can be described, a brief introduction to the design strategy used to select the fan, heating and cooling equipment will be presented. This information is required because control systems can only maintain a balance between the energy flow and the process load when the electric or mechanical systems that provide the energy are sufficiently large, without being oversized. If the equipment is improperly sized or applied, there is no control system available that will be able to maintain the process condition and maximize its operating efficiency. The sizing and application of equipment is usually the responsibility of the system's design engineer.

In HVAC applications, the equipment used to maintain comfort processes is selected based upon the construction characteristics of the building along with the maximum ambient conditions that are likely to occur over the course of a typical year. The temperature and humidity of the outside air along with the wind's speed and the position of the sun have a profound effect on the thermal load occurring within a conditioned space. These conditions must be taken under consideration when determining the size of the equipment needed to serve the needs of a space. Internally generated sensible and latent heat from the space occupants along with the operation of lights, office and mechanical equipment also affect the load.

To adequately size a mechanical system, the period of time when the external load variables plus the internal load variables equal the greatest load is used to select the mechanical system's equipment. This will be the largest expected load that is likely to occur under expected operating conditions. Since this load is used to design the mechanical system, it is called the *design load*. The application engineer or technician selects the HVAC equipment's capacity (btu's/hour) so the control system is able to balance the load of the building throughout the year.

The outside air temperature and humidity levels have the greatest effect on maximum load experienced by the HVAC equipment in most process designs. The dry bulb and wet bulb temperatures that occur with enough frequency to create the maximum or design load are called the *design temperatures*. Summer design temperatures are used to establish the maximum cooling load of a system. Likewise, winter design temperatures establish the system's heating load. These temperatures are used to determine the maximum heating and cooling load that can be expected in the building a few hours each year. The HVAC mechanical equipment is considered to be fully (100%) loaded during design temperature conditions. Summer design temperature values indicate the maximum temperatures within a particular geographical area will exceed 2.5% (or 1% or 5%) of the total daily hours during the months of June through September. Winter design temperatures indicate the total hours the temperature can be expected to be below the design values during the months of December through February.

The value 2.5% has been standardized by the American Society of Heating, Refrigeration and Air Conditioning Engineers (ASHRAE). ASHRAE is an organization comprised of people who work in HVAC-related fields. ASHRAE publishes four volumes of information—*Fundamentals, Systems & Equipment, Refrigeration and Applications*—which describe the design, operation, and maintenance of HVAC equipment and systems. These manuals are updated every four years to reflect advances and changes in the field. ASHRAE also funds and publishes research papers and operational standards that are used by designers, operators, and students as reference material. Students in the HVAC field are advised to join student chapters and attend meetings, thereby establishing themselves in the field as they share information with the experts in their field.

Table 1.1 shows a partial table that lists some of the ASHRAE design temperature data for selected cities in the state of Arizona. Note that the table lists values for 1%, 2.5%, and 5%. A cooling system for a building located in Yuma would be sized based upon an outside air dry-bulb temperature of 109°F coupled with a wet bulb temperature of 72°F. These temperatures are expected to be exceeded 2.5% of the time. In other words, designing to meet the thermal load during periods with these temperatures ensures that the system will be able to handle the cooling load for 97.5% (100% − 2.5%) of the year. For any periods when the outside temperature exceeds these values, the temperature within the building may rise beyond the desired value for the duration of this interval.

1.2.2 Partial Load Operation

As a consequence of sizing HVAC equipment to handle the load conditions that only occur 2.5% (or 1% or 5%) of the year, when the outside air is at its design temperature and humidity conditions, comfort HVAC equipment must operate with partial loads during most of its operating life. A partial load describes the operation of equipment under conditions where it is transferring less energy than is

Table 1.1 CLIMATIC CONDITIONS FOR THE UNITED STATES

State and Station	Winter °F Design Dry-Bulb 99%	97.5%	Summer °F Design Dry Bulb and Mean 1%	Coincident Wet-Bulb 2.5%	5%	Mean Daily Range	Design Wet-Bulb 1%	2.5%	5%	Temp. °F Median of Annual Extr. Min	Max
ARIZONA											
Douglas	27	31	98/63	95/63	93/63	31	70	69	68	104.4	14
Phoenix	31	34	109/71	107/71	105/71	27	76	75	75	112.8	26.7
Yuma	36	39	111/72	109/72	107/71	27	79	78	77	114.8	30.8

The complete listing can be found in the ASHRAE Fundamentals Volume, in the Load Calculation Section.

stated on its capacity nameplate. Using the 97.5% ASHRAE design criteria, equipment will be expected to operate at less than full load 97.5% of each year.

There is a penalty incurred when equipment is operated at partial load. The operating efficiency of most mechanical equipment decreases as the load decreases. Cooling equipment, boilers, chillers, fans, and pumps experience decreases in the amount of energy they can transfer for each unit of energy they consume while operating under partial load. Control systems are designed to maximize operational efficiency during partial load operation. This makes it all the more important that the control system is designed, installed, and calibrated correctly. The controls must regulate the equipment in a manner that results in the efficient use of energy resources, while still maintaining occupant comfort, safety, and health.

1.3

CONTROL SYSTEMS CLASSIFICATIONS

Control systems are broadly classified as either *automatic* control or *manual* control. In Section 1.1, automatic control systems were defined as those systems that use devices to vary the flow of mass or energy into the process without human intervention. Once these systems are installed and calibrated, they operate continuously in accordance with their design intent. The only human interaction occurs when a technician periodically evaluates, calibrates, or makes adjustments to the control system. The heating process in Application 1.2 depicts an automatic control system that maintains the temperature of the air without the need of a person to position the hot water valve in response to changes in the temperature of the air leaving the coil.

Manual control systems require people to adjust the position of the control devices based upon some observable process characteristic. A person must constantly monitor the process and adjust the regulating force applied to the control device to maintain the process within its design limits. If the control system described in Application 1.2 was designed as a manual system, the hot water valve would have to be closed manually to reduce the heat transferred into the airstream as the air temperature became too warm. A kitchen faucet is a common manual control device that uses an operator to manually adjust the temperature of the water leaving the spout.

Figure 1.2 depicts an automatic valve controlled by an automatic control loop. Any changes measured by the control system automatically produce changes in the valve's position. The manual control system that performs the same function uses a hand valve and gauge to monitor the process. A person must be present to complete the connection between the process condition and the valve. Any change in the process requires the operator to adjust the hand wheel on the valve in order to maintain the process balance.

To maintain comfort, safety, and efficiency, most of the control systems used in HVAC applications are designed to operate automatically. Some of the advantages of automatic control systems are:

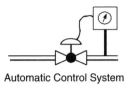

Automatic Control System

Note: There is no connection between the gauge and
the valve—a person must supply that connection to
maintain a balance within the process.

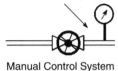

Manual Control System

Figure 1.2   Major Categories of Control Systems

1. By removing the human element from a process, the equipment operates
   more efficiently and with a higher degree of safety than it would under manual control. Automatic systems do not take coffee breaks, read books, or get
   bored watching the process.
2. An operator need not be present if an emergency occurs. Automatic controls
   can initiate a safety control sequence that will safely shut down the apparatus
   to provide for the protection of people and equipment.

1.3.1 Modulating (Analog) and Two-Position (Binary) System Classifications

Automatic and manual control systems are further categorized as being either
modulating or *two-position* systems. Modulating control systems position their control device at any point within its operating range as shown by the dashed line in
Figure 1.3. The dashed line represents the position of a valve over a period of time
when the process is undergoing a change. It shows that the valve can be opened to
any position between of 0% to 100% of full flow. The valve will stay at a given
position for as long as the process load remains constant.

Modulating control systems are used wherever it is necessary to maintain a
process variable at a desired value as the environmental conditions change. The
lawn mower throttle, volume potentiometer, and faucet valves described in Section 1.1 are modulating controls that allow the mower engine speed, volume, or
flowrate to be set at any point between 0% and 100% of position. The ability to
position the device at any value allows the control system to more efficiently balance the load. Most of the comfort related loops in commercial and industrial
HVAC systems use modulating control devices. Modulating control systems are
also called *analog* control systems.

Two-position control devices can only be commanded to one of two possible
positions, 0% or 100%. There are no intermediate positions available. These

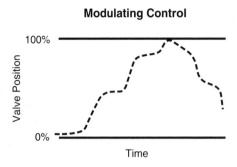

Figure 1.3   Modulating Control

locations are commonly labeled on/off, open/closed, or 0%/100%. The heavy solid line in Figure 1.4 shows the response of the binary control device. It is positioned either 100% open or closed. It cannot be commanded to any position within the center portion of its operating range. The vertical dotted line indicates the device quickly changes position without stopping at an intermediate position.

Two-position control system designs are used most often to prevent processes from exceeding a predetermined operating limit. The limit can be selected to maintain a minimum or maximum process value. Switches, relays, contactors, and motor starters are common two-position devices used to cycle HVAC equipment on and off. Circuit breakers are two-position safety devices that are used to prevent excessive current flow from being drawn by an electrical device. Two-position control devices used in commercial control systems are more commonly referred to as *binary* or *digital* devices.

1.3.2 Types of Control Systems

Electromechanical

There are four common types of control systems used in the HVAC industry. The oldest type of control systems are called *electromechanical* systems. They are made up of two-position or *binary* operating devices consisting of relays, contactors, mechanical thermostats, switches, etc., that are wired together into a system. They operate by opening and closing a pair of contacts in response to forces generated by their internal springs, bellows, diaphragms, and/or levers. Opening and closing contacts cycles other control devices and equipment on and off in response to changes in a measured process condition.

Electromechanical controls are generally less expensive to purchase because of their two-position nature. These systems also are easier to analyze and maintain than modulating control systems. They were commonly used for applications where the processes can be maintained at their desired conditions by cycling the equipment on and off.

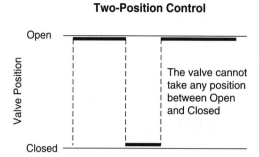

Figure 1.4 Binary or Two-Position Control

The control system used in a basic residential refrigerator or air conditioner is commonly designed with electromechanical devices. Changes in the temperature inside the box alter forces within the thermostat that caused the contacts to snap open or closed, cycling the condensing unit. Simpler home furnace thermostats also incorporated a bimetal coil that expands and contracts in response to changes in the room's temperature. In this design, the movement of the bimetal spring tips a glass bulb, allowing a bead of mercury to slide from one tip of the capsule to the other. When the mercury flows between two electrodes, an electrical circuit is completed cycling the furnace on. Advances in electronic control systems are reducing the use of electromechanical systems. Electronic systems and devices offer more features in a smaller package that maintain better control of equipment with minimum maintenance.

Pneumatic

Commercial buildings require control systems to maintain the temperature, humidity, and static pressure within many different areas simultaneously. In these applications, electromechanical systems tend to become very complex and large. *Pneumatic control systems* were developed to replace the electromechanical systems in larger applications.

Pneumatic systems use devices that operate using air pressure. An instrument air compressor supplies the control systems with clean, dry, oil-less air within a pressure range of three to thirty pounds per square inch (3–30 psi). The air's pressure supplies the forces used to position control devices so they can establish a balance between the energy transfer and the load.

Internally, pneumatic devices are constructed with a system of levers and diaphragms that vary the magnitude of the control signals based upon a measured variable. When the process changes, it disrupts a mechanical balance existing within the device, causing the control signals to change. The signals continue to change until a new balance occurs between the forces within the control device. Pneumatic devices are more complex than electromechanical devices. They are also more sensitive to dust and dirt in the environment, increasing the need for periodic maintenance and calibration. Pneumatic devices can be designed to pro-duce analog or binary (modulating or two-position) signals. The choice is based on the intended operational characteristics of the process.

Analog Electronic

With the development of the solid state transistor in the 1960s, *analog electronic control systems* emerged to compete against pneumatic control systems for new construction applications. Analog electronic devices generate continuous direct current signals (modulating) within a range of 0–15 volts. They permit electrical control systems to modulate mass or energy flow into a process to maintain a bal-anced condition. This response was impossible to generate with electromechanical systems.

Analog electronic systems have some advantages over pneumatic systems. They are less susceptible to the effects of accumulated dust and mechanical wear of moving parts, the two largest maintenance problems associated with pneu-matic and electromechanical systems. Electronic devices also can operate without the need for an instrument quality air compressor to operate the controls. A major disadvantage of these systems is the difficulty of analyzing the operation of the electronic circuit boards by technicians who are familiar with mechanical (pneu-matic) systems.

Digital Electronic

Digital electronic control systems for HVAC applications evolved with the expansion of the electronic computer industry. *Digital* systems differ from the *analog* elec-tronic systems by the type of signals they produce. Digital systems represent all data with strings or groups of logic 1s and 0s. A logic 1 is typically represented by the presence of a 5-volt signal and a logic 0 is represented by a signal that is less than 1 volt. These 1s and 0s are used to represent words, numbers, or signals that can be stored in memory for future use and analysis. The binary information also can be converted into analog signals to modulate systems, thereby taking the place of analog electronic systems. *Personal computers and calculators are digital systems that operate using binary strings of 1s and 0s to perform math functions, generate graphic drawings, control communications, and display data.*

Digital control systems have become the industry standard due to the capabili-ties they add to a control system. These systems are in reality small computers packaged in a control panel that is mounted in the field, near the equipment it controls. These microcomputers can effectively schedule the operation of the equipment, signal the operator if a process operates outside of its intended limits, automatically calibrate its control loops, turn off loads to save energy, and store operational histories that technicians use to analyze the operation of the system. The only disadvantage of this type of system is the need for the technician to become familiar with computer operation and programming so they can apply this technology to its greatest potential.

1.4

SUMMARY

The purpose of a control system is to regulate the flow of mass or energy into a process in response to some changing condition. Automatic HVAC control systems are used to safely maintain the requirements of a process in an energy-efficient manner. Electromechanical, pneumatic, analog electronic, and digital electronic systems are used to provide the regulating forces used to balance energy transfer with the load in HVAC/R applications. Digital electronic systems provide the greatest potential for improving building operational efficiency through the use of custom-developed programs that can integrate all of the building equipment into a single system.

Analog or modulating control systems generate continuous signals within the operating range of the system. Two-position or binary control systems generate pulsing signals of 0% or 100%. Analog signals commonly are used in commercial applications because they can regulate the energy transferred into the process at a level that matches the load. Binary signals are used to cycle equipment whenever the process exceeds its design limits. A control system takes advantage of the characteristics of binary and analog signals to maximize the operating efficiency of the process.

1.5

EXERCISES

Determine if the following statements are true or false. If any part of the statement is false, the entire statement is false. Explain your answers.

1. To control means to balance the energy transfer with the process load.
2. HVAC control systems typically are used to maintain comfort and safety.
3. The purpose of a control loop is to apply a regulating influence upon a device to make it operate as required.
4. Automatic and man-made are two classifications of control systems.
5. Binary and analog are terms used to categorize control systems.
6. Binary and two-position are terms that describe the same type of control response.
7. Analog control systems are also known as on/off systems.
8. Analog systems are typically used to prevent the process from exceeding a predetermined limit.
9. Pneumatic systems operate using instrument air pressures within a range of 3 to 30 pounds per square inch.
10. Digital control systems are the type of control system most common installed today.

Respond to the following statements and questions:

1. Define automatic control.
2. What are some advantages that automatic control has over manual control systems regarding safety and efficiency?
3. Describe an example of a control loop found in the locations listed below. List the process (purpose) and the load (mass or energy transferred) for each loop described.
   a. automobile
   b. kitchen
   c. basement
   d. bedroom
   e. family room
4. What is the primary purpose of modulating control loops?
5. For what applications are binary controls commonly used?
6. List the different types of control systems used today, and summarize their characteristics.
7. Why must equipment be sized for the warmest and coldest temperatures expected in a particular locale?
8. What is a major disadvantage of sizing equipment for design load conditions?
9. List three control systems found in your classroom and classify them using the information provided in Section 1.3.
10. Why are control systems so important in building operations?

Control System Terminology

People in the control field use somewhat proprietary (exclusive) terms to write and converse about controls. The terms used to describe control systems, devices, designs, and events have special meanings and must be used correctly to effectively present required information. Control equipment specification sheets, calibration instructions, and trade and association publications use these terms to communicate vital information to control technicians. Understanding this terminology allows technicians to perform daily tasks and to remain current with advancements being made throughout the control industry.

Chapter 2 presents some of the common terms used within the HVAC control industry. These terms are presented here and used throughout the remainder of the text. They also appear in the glossary at the back of the book. This chapter lays the foundation of knowledge required for advancement within the control field.

OBJECTIVES

Upon completion of this chapter, the reader will be able to:

1. Accurately define the various control terms presented.
2. Use correct terminology to describe basic processes.
3. Read and understand published articles describing fundamental control applications.

2.1

AIR HANDLING SYSTEM TERMINOLOGY

Much of the equipment that incorporates commercial control devices is associated with processes that condition air streams for comfort and environmental conditioning. This equipment heats, cools, humidifies, filters, ventilates, and pressurizes different areas within a building to meet the requirements of each room. Changing the characteristics of air flowing through mechanical equipment to meet the needs of a room is called *conditioning* the airstream. Once conditioned, the air is sent to the individual areas within a building to offset their thermal, moisture, pressurization, and ventilation loads. By balancing the heat and quantity of conditioned air entering a space with the load of that area, the desired comfort or process conditions within that area are maintained.

Air handling units are pieces of equipment used to condition air within a building to maintain the desired environment. The following terms are commonly used to describe the components and processes associated with commercial air handling equipment. Keep in mind that different manufacturers may describe the same component or process with slightly different terminology. These differences will be noted within the definition.

2.1.1 Air Handling Unit

An air handling unit is a collection of mechanical and electrical components whose operations are integrated to condition the air that flows through it's filters, coils, and fan. The conditioned air is supplied to various zones within a building to maintain the desired environment within those areas.

A building may have one or more units, depending upon the loads of the conditioned areas. Air handling units (AHUs) have fans that move air coming back from the rooms and from the outside, through various heat transfer coils, dampers, and other equipment to produce changes in air temperature, relative humidity, cleanliness, and pressure. An AHU can be designed to perform some or all of the following processes associated with conditioning air:

- Clean the airstream that is to be delivered back to the space using mechanical, chemical, or electronic filters.
- Remove odors, fumes, and carbon dioxide concentrations by diluting the room air with ventilation (outdoor) air.
- Adjust the air temperature with heating and cooling heat exchangers to maintain comfort or process requirements.
- Alter the humidity by using cooling coils to dehumidify or steam injection to raise humidity levels to maintain comfort or process requirements.
- Vary static pressures within rooms or the entire building to reduce or prevent infiltration or exfiltration.

Minimally, all AHUs have mechanical filters, heat exchangers, a fan, air flow dampers, hot and chilled water, steam or refrigerant valves, and control devices. An AHU is specified (sized and selected) by the design engineer to include the components and processes (heating, cooling, humidifying, pressurizing, etc.) that are necessary to maintain the requirements of the processes that it serves. Figure 2.1 shows a photograph of a multiple zone AHU.

2.1.2 Return or Recirculated Air

Return air is the volume of air that is returned to the AHU from the conditioned areas within the building served by that AHU, so it can be reconditioned and sent back to the spaces to balance the loads within those areas.

Air leaves a conditioned space through its return air grilles and travels through the return duct system back to the air handling unit because of the pressure differential created by the fan. The fan generates a low pressure zone in the air handling unit whenever its blades are turning. The air that flows back to the AHU from the conditioned spaces is called the return air because it is previously conditioned air that is *returned to the AHU* from the connected spaces. Return air is reconditioned as it passes through the AHU and is sent back to connected zones through the supply air duct. Return air is also called *recirculated* air because it is constantly returned to the building spaces after it is reconditioned. Figure 2.2 is a schematic representation of a return air path.

Return air dampers are used to vary the amount of air that will be reconditioned by the AHU. A damper is a mechanical device that modulates open and closed to change the resistance in the return air duct. When the return air dampers are commanded toward their closed position, the resistance of the return path increases. As the resistance to flow increases, the quantity of flow decreases, reducing the amount of air returned to the AHU. The opposite response occurs when the dampers open — air flow increases.

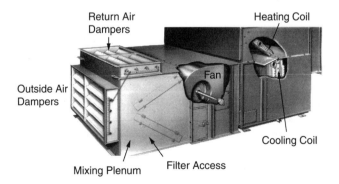

Figure 2.1 Air Handling Unit

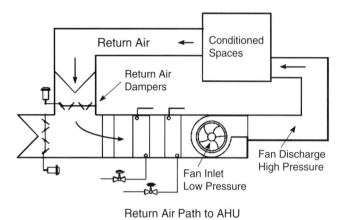

Return Air Path to AHU

Figure 2.2 Return Air System

2.1.3 Outside or Ventilation Air

Outside air is the volume of air drawn into an AHU from the out-of-doors to provide air used to dilute contaminants in the conditioned spaces served by the AHU.

Outside air enters an AHU through a set of louvers located in the building's exterior wall. These louvers are slotted openings that prevent rain from entering the building through the outside air opening. Fastened to the louvers is a screen used to prevent birds, leaves, papers and other large foreign matter from entering the air handling unit. The differential pressure produced by the fan allows the outside air to enter the system so it may be conditioned and distributed to each area the AHU serves. Figure 2.3 schematically shows the outside air path into an air handling unit.

Some outside air must always be supplied to the rooms when occupants are present. The volume of outside air brought into a building is used to dilute the carbon dioxide (CO_2), harmful vapors, and other contaminants that build up within the areas served by the unit. Building codes and industry standards specify the quantity of outside air to be introduced into any space that is occupied by people. This requirement is driven by the need to maintain acceptable levels of indoor air quality (IAQ) within occupied areas.

The amount of outside air required by a room is a function of the number of occupants in the area, their activity characteristics, and the presence of any other processes that may be taking place in the zone. As one of the leaders in HVAC research, ASHRAE has been instrumental in developing ventilation standards for the HVAC industry. ASHRAE Standard 62–89 advises a quantity of 15 to 20 cfm of ventilation air is required for each person occupying an area where there is no smoking or no other harmful contaminants are being generated. The ventilation air requirements increase wherever smoking is permitted or where there is a process

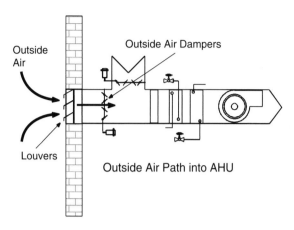

Outside Air Dampers

Outside Air

Louvers

Outside Air Path into AHU

Figure 2.3 Outside Air Intake

occurring that gives off annoying or hazardous fumes. The volume of outside air brought into a system is modulated by racks of outside air dampers, similar in construction and appearance to those used to regulate the return air volume. *Ventilation air* and *make-up air* are also used to describe the amount of air being brought into the AHU from outside of the building envelope. Ventilation air is used because the outside air being brought into the system is used to provide the ventilation requirements of the conditioned spaces. Make-up air is used to describe air that is brought into a building to replace air exhausted from the building.

2.1.4 Mixed Air

Mixed air is the volume of air that is produced when the system's return air stream is blended with its outside air stream within the AHU's mixing plenum.

During AHU operation when return air is being brought back to be reconditioned, the outside air volume is *mixed* with return air before the entire air mass is reconditioned. The resulting combination of return and outside air is called mixed air. Figure 2.4 shows that the mixed air passes through the filters and heat exchangers to be conditioned and sent to the rooms connected to the air handling unit.

The temperature of the mixed air can be regulated by modulating the return air and outside air dampers. Changing the percentage of outside air being added to the mixture alters its temperature and humidity properties. When the outside air is colder than the return air, modulating the outside air dampers open while proportionately closing the return air dampers will reduce the mixed air temperature. The opposite response occurs when the outside air is colder than the return air temperature and the outside air dampers are modulated toward their closed position — the mixed air temperature increases.

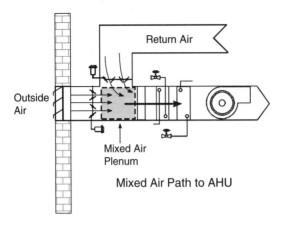

Figure 2.4   Mixed Airstream

2.1.5 Exhaust Air

Exhaust air is the portion of return air that does not pass through the return air dampers into the mixing plenum. Exhaust air is released outside of the building through exhaust dampers and louvers.

Exhaust air is not reconditioned by the AHU. It leaves the building and is replaced by an equal amount of outside air. Whenever outside air is brought into a building for ventilation or temperature control, a corresponding quantity of return air must be released or exhausted from the building. This strategy prevents excessively high static pressures from building up within the building envelope. These high pressures can prohibit doors from opening or closing properly, creating a serious safety hazard. Excessive pressures also increases the amount of conditioned air that is expelled (exfiltration) from cracks around exterior doors and windows. Exfiltration is the uncontrolled loss of conditioned air outside of a building, increasing the cost of system operation.

Exhaust air is sometimes called *relief* air. There is a slight difference in the proper use of these two terms. Exhaust air correctly describes AHU systems that have a return or exhaust fan. This fan is required to generate the high pressure needed to force the air out of the building through the exhaust dampers and louvers. If the fan is located upstream to (before) the exhaust air dampers, the fan is a return air fan. Figure 2.5 shows a return fan system. If the fan is located downstream of the exhaust air dampers, the fan is called an exhaust fan. Both of these fans permit some of the return air to leave the building as exhaust air.

Relief air refers to conditioned air that exits the building as a result of a difference in static pressure across gravity or motor operated relief air dampers. When more outside air is brought into the building than exits through openings, the static pressure in the building increases. When the relief dampers are opened, some conditioned air leaves the building, reducing the building's pressure. A return fan is not required in relief air applications.

2.1.6 Supply Air

Supply air is the fully conditioned air that exits the AHU. It is transported through the mechanical system supply ducts and is used to offset the loads within the areas it serves.

Once the mixed air passes through the air handling unit's ventilating, filtering, heating, humidifying, and cooling processes, it is fully conditioned and ready to be sent to its connected zones. This conditioned air is called the supply air because it is *supplied* to the areas in need of conditioning. The supply air is transported through the supply duct by the pressure difference generated by the unit's fan(s). Supply air enters the rooms through supply air registers or diffusers and mixes with the room air to offset its thermal, humidification, and pressure loads, thereby maintaining the desired process conditions. Supply air is also called *discharge* air because it is discharged from the supply fan in the AHU. Figure 2.6 depicts the supply air system.

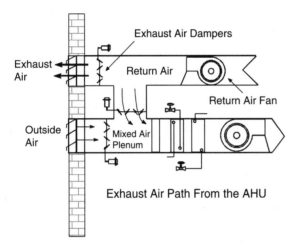

Figure 2.5   Exhaust Airstream

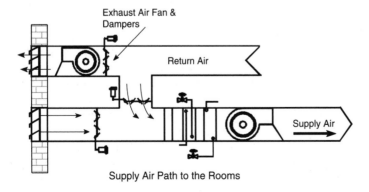

Figure 2.6   Supply Air Path

2.2

CONTROL LOOP TERMINOLOGY

The previous sections presented an overview of the basic elements associated with an air handling unit. The following terms are associated with the control systems that are used to operate the air handling unit. Note: the words *control* and *controlled* are used in many of the definitions. Be sure to use the correct form of the word when applying these terms.

2.2.1 Control Loop

A control loop is a group of interconnected control devices that operate collectively to maintain a single process at its desired condition.

When controllers, sensors, valve actuators, damper actuators, signal converters, motor starters or other control devices are connected together with signal lines, they create a control loop. All the components in a control loop work together to maintain a *single* process at its required condition. The process may be comfort, manufacturing, or safety related. Control loops maintain temperatures, humidities, pressure, and levels in HVAC processes. In other systems, control loops maintain speed, light levels, direction, intensity, flowrates, etc. No matter what the purpose or function of the group of interconnect control devices, it is called a control loop. Modulating control loops are designed to maintain a process at its desired condition throughout all foreseen load changes. The modulating control devices alter the flow of mass or energy into the process until a balanced condition exists between the amount of mass or energy entering and leaving the process. Binary control loops prevent a process from exceeding a predefined limit or limits. These loops are most commonly found performing a safety-related function although they are also used to maintain processes within an acceptable range of its design value.

Figure 2.7 shows two separate modulating control loops. One loop maintains the mixed air temperature at 60 °F and the other maintains the supply air temperature at 60 °F. Although both processes are designed to maintain the same temperature of 60 °F, two separate loops are required, one for each process. One loop modulates the outside and return air dampers while the other modulates the hot and chilled water valves. Each loop has its own sensor to measure the process and controller to modulate the dampers or valves. The control devices that make up the loop are shaded grey.

Generally, individual control loops are combined with several other related control loops to form a *control system*. A control system includes all the necessary components and logic to regulate the condition of several process characteristics while monitoring the system to ensure safe operation. The phrase *control loop* is often shortened to the term *loop*.

The key to understanding a control loop is to realize it can only maintain one temperature, pressure, level, humidity, or flowrate value. If a space requires spe-

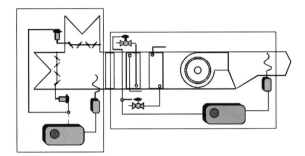

Two Separate Control Loops

Figure 2.7   Individual Control Loops

cific temperature, pressure and humidity values, it will need a minimum of three control loops. Application 2.1 outlines the control loops found in a typical residential window air conditioning unit.

APPLICATION 2.1

Process: Room Temperature Control Using a Window Air Conditioner

Objective: Maintain the temperature in a room at a desired value of 74 °F by circulating room air across a direct expansion cooling coil located in the air conditioner.

Operation: Return air from the room is drawn through a filter and a direct expansion (DX) heat exchanger by the unit's evaporator fan. Sensible and latent heat are removed from the room (return) air and transferred outside the home by the condensing unit. The conditioned air is supplied back to the occupied space to offset the space cooling load.

Response: Various control loops are incorporated into a system to maintain the comfort of the occupants of the room. The control system balances the energy transferred into the room from warmer surrounding areas with the energy extracted from the room by the evaporator. The system also monitors the operation of mechanical components to ensure they are operating safely.

Control Loops:
1.   The temperature in the room is monitored by the air conditioner's thermostat. The sensing bulb is located in front of the evaporator coil so it measures the return air temperature. The thermostat's electrical contacts open and close

based upon the temperature of the room, cycling the compressor as required. The thermostat, the sensor and the compressor's starting relay contacts make up one control loop that regulates the operation of the condensing unit. This is an automatic, two-position control loop.

2. A second control loop is required to regulate the evaporator fan's speed. A push button switch regulates the speed of the fan (slow, medium, high) based upon occupant preference. This is a manual, two-position control loop.

3. Another control loop monitors the current draw of the compressor. If the current drawn by the compressor motor exceeds a safe limit, the control loop opens the circuit, disabling the system. This is an automatic, two-position control loop.

4. There is also a manual, modulating control loop that regulates the quantity of ventilation air drawn into the unit. Based upon occupant preference, a manually controlled lever opens a small access port to the outside air.

2.2.2 Controlled Medium

The controlled medium is the fluid that absorbs or releases the mass or energy transferred to or from the process.

The most common controlled mediums (fluids) found in HVAC processes are air and water. Air is the controlled medium that passes through an air handling unit to absorb energy from the heating coil or release energy to the cooling coil. The air stream also absorbs water vapor (mass) in humidifying processes or gives up water through condensation or adsorption in dehumidifying processes. Air, as the controlled medium, can be transported to the conditioned spaces by a fan or natural circulation.

Water is another controlled medium used to absorb or release thermal energy with chillers, boilers, and other similar mechanical systems. Hot water can transport heat to a zone to offset the losses through the building envelope when it is cold outside. Chilled water is used to absorb heat from the air passing through the heat exchanger fins and release it in the chiller evaporator. Water is transported to locations that are remote from the mechanical rooms using pumps.

A controlled medium has several different properties, including temperature, pressure, moisture content, enthalpy, specific heat, and specific volume. Each property of the controlled medium that must be maintained at a desired value constitutes a separate process and, therefore, must be maintained by its own control loop. Each loop is designed to maintain one property of a controlled medium at its desired value. A zone needing both temperature and humidity control requires an air handling unit with a temperature control loop and a humidity control loop to maintain both properties of the same controlled medium (air).

2.2.3 Controlled Variable

The controlled variable is the property of a controlled medium that a control loop is maintaining at a desired value.

As stated in the previous section, a controlled medium may have more than one property that must be maintained at a desired condition in order to meet process requirements. The controlled variable is *one* of the properties of the controlled medium. The controlled variable is maintained at its required value by its own control loop. The word variable is used by the control industry to indicate that its value changes or *varies* in response to changes that affect the process. The controlled variable may be easier to remember if it were called the controlled *property*. The most common controlled variables of HVAC processes are temperature, static pressure, and relative humidity.

2.2.4 Set Point

The set point is the desired (required) value at which the controlled variable is maintained by a control loop to satisfy the requirements of the process.

Set points are *set* by the installing technician or zone occupant. Its value or magnitude is selected based upon a processes' design specifications or the occupants level of comfort. When an air handling unit is designed to satisfy simultaneous loads in several conditioned spaces, the supply air temperature set point is set to a fixed value to offset the highest expected load that may occur in any one of those spaces. In this application, the process design specifications are used to initially determine the set point, that typically falls within the range of 52 to 58 °F.

In larger system applications, the set point is adjusted to match the actual response of the process after the system has operated under normal conditions for a few months. In applications where a thermostat is mounted on a wall, the set point is a function of the occupants' comfort level. If the occupants in the room feel cool, they can raise the thermostat's set point a few degrees. The set point can be easily altered to account for changes in the occupants' activity level or changing exterior and solar loads.

2.2.5 Control Point

The control point is the actual value of the controlled variable at a particular point in time.

The control point indicates the instantaneous or present condition of the controlled variable. Whenever a controlled variable is measured by an instrument, the value indicated on the instrument's display is the control point of the process. When a person walks by a room thermostat that has a set point of 71°F and glances at the temperature indicator, it shows the control point of the process. If the indicating thermometer showed 70 °F, the control point of the room is 70 °F. The set point is still 71°F because that is the desired value of the controlled variable (room air temperature). The control point can take on any value within the operating range of the process.

Ideally, the control loop should maintain the controlled variable equal to its set point under all expected load changes. In actual operation, the control point is usually close to the set point but not equal to it. The difference between the two readings is due to the dynamic characteristics of the process, load and control system. Time delays that occur naturally allow the control point to deviate slightly from the set point. Application 2.2 applies the terms defined above using a central air-conditioning application.

APPLICATION 2.2

Process: Maintain the temperature in a home at a desired value using a central air-conditioner.

Objective: An occupant in the home sets the thermostat to 75 °F. This value is the set point of the temperature control loop. It is the objective of the air conditioning system and its controls to maintain the room temperature (control point) near 75 °F.

Operation: A thermometer in the room with the thermostat indicates the present temperature is 78 °F (control point). Air from the room (return air) is drawn to the furnace whenever the system fan is operating. The controlled medium (room air) passes across a direct expansion (DX) heat exchanger ("A" coil) located in the furnace plenum. The controlled medium transfers the heat it absorbed from the process to the cold refrigerated coil. The air leaving the coil is conditioned to a level that will maintain the process within a few degrees of the 75 °F set point. The property of the controlled medium that is altered to offset the load is the temperature of the supply air.

2.2.6 Error

The error is the mathematical difference between the control loop set point and the process control point.

As mentioned previously, the control point is usually a different value than the set point due to the dynamic nature of the process. The mathematical difference between the set point and control point is called the loop or process error. The error is calculated by subtracting the control point from the set point.

error = set point − control point

There is always a mathematical sign, either a (+) or a (−), associated with the error's value. When a process set point is 55°F and its control point is 56°F, the error is equal to −1° (55 − 56). If the set point is 140° and the control point is 138°, the error is +2°. Note, a negative error indicates the set point is *below* the control point and a positive error signifies the set point is *above* the control point. The error is used by the controller to generate a response that maintains the desired process conditions by reducing the size of its error.

The error can be incorrectly referred to as the loop *offset* or *drift*. Error, offset and drift cannot be used interchangeably. As stated previously, an error is a *signed*

(+/−) value. Conversely, offset and drift are *absolute* values. Absolute values do not have a (+) or a (−) sign associated with their numbers. Offset and drift are typically used to indicate the magnitude of difference that exists between the set point and the control point. The direction of the offset, negative or positive, is of no interest. If the set point of a process is 74 °F and the control point is 72 °F, the error is +2 °F and the offset (drift) is 2 °F.

2.2.7 Control Signal

A control signal is the entity by which information is exchanged between the control devices within a loop or system.

All control devices generate or receive signals that are *related* to the magnitude of the process control point. Whenever a measurable change occurs within a process, it generates a corresponding change in the control point which in turn produces a proportional changes in the loop's control signals. These changes in a signal's magnitude transfer information between the interconnected control devices. Through control signals, the control loop can respond in a manner that will maintain the required conditions of the process. When a signal arrives at a control device, it is called an *input signal*. The signal a control device generates and sends to other devices is called an *output signal*.

2.2.8 Signal Types

The signal type describes the physical medium used by control devices to transport the information contained in the control signal.

The most popular signal types used in the HVAC industry are voltage, current, resistance, and pneumatic (air pressure). The type of signal selected for a control system is based upon the signal type that may already exist in the building or the preferences of the building owner, operator, and designer.

Standard ranges also differentiate the signal type. Typical ranges of modulating or analog signals of electronic-based control systems are 0–5 v dc, 0–10 v dc, 4–20 mA and 0–20 mA along with changes in electrical resistance. Commonly used two-position or binary signals used in electronic-based control systems include 0 and 10 v dc, 0 and 5 v dc, 4 and 20 mA and 0 and 24 v ac. Pneumatic signal ranges of 3–13 psi, 3–15 psi and 3–18 psi are also commonly used in HVAC control systems.

Signals are further classified by their dynamic nature. Analog signals are *continuous* or modulating by nature. A continuous signal is always present and generates changes in magnitude in response to changes in the process. An analog signal can be any value between its design minimum and maximum limits, with the magnitude of the signal containing the information that is being exchanged between loop devices. An alternating current electric signal is a common example of an analog signal.

Binary signals are *discontinuous* in nature. They relay information using two values, 1/0, on/off, high/low, max/min, etc. The discontinuous signal indicates that the signal changes abruptly from its minimum to its maximum value. At the instant of change, there is no signal present. Information is exchanged by noting the present state of the signal or by responding to a change in the signal state. Binary signals are used in computer control systems and safety strategies. Figure 2.8 shows analog and binary signals.

2.2.9 Signal Paths

Signal paths are the interconnecting wires or tubing that physically connect the control loop devices together, allowing them to exchange the information contained in their signals.

Signal paths are the hardware connections that exist between devices in a control loop. Control signals travel through these interconnecting paths as they deliver data to the other components within the system. Pneumatic control loops use signal paths fabricated of copper or plastic tubing that contain the instrument air pressure signal. Changes in the air pressure indicate a change in the control point or set point has occurred. Electrical control systems use wires for their signal paths, providing a conductive path between control devices.

2.2.10 Zone

A zone is an area within a building where a controlled variable is maintained at the same set point.

A zone is a geographical area within a building whose limits are defined by the characteristics of its process load. Each zone has different load characteristics and, therefore, requires its own control loops to maintain the set point within that area. A single control loop cannot maintain the process conditions within zones that are experiencing different loads at the same time.

A large zone may be divided into smaller sections; offices, work areas, etc. Each subsection within a zone will have the same temperature, humidity, and pressure requirements because they are part of a single zone. The control loops respond to changes that occur within the area of the zone where the sensor or thermostat is located. When each area within a section of the building experiences similar internal and external load changes they can be combined into one zone.

If the areas within a section of a building do not share the same load profile, each different profile represents a zone of control. Each will have its own control

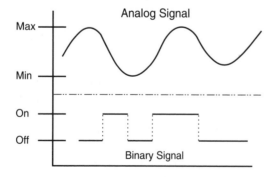

Figure 2.8   Signal Classifications

loops to maintain process set points. Larger buildings are typically broken down into north, south, east, west, and internal (no outside walls) zones for each floor. An AHU that only serves one zone is called a single zone air handling unit. An air handling unit that is required to maintain the desired conditions for many separate zones at the same time is called a multiple zone AHU.

2.3

SUMMARY

A control loop is a group of interconnected control devices that maintain the controlled variable at its desired set point. Every control loop is designed to maintain the set point of the process as long as the load varies within its design limits. A control loop modulates the flow of mass or energy into the process to reduce the error between the control point and the set point. Energy and mass are transported and stored in the controlled medium. The controlled medium is usually air or water.

The set point is the desired value of the controlled variable and the control point is the actual value of the controlled medium. Any difference between the set point and the control point is called the error. If the error is positive, the set point is greater than the control point. If the error is negative, the set point is less than the control point. Offset and drift also describe the difference between the set point and control point without indicating the direction of the error.

The controlled variable is a property of the controlled medium that is maintained at a value that satisfies the process requirements. A controlled medium may have more than one property that must be maintained at a given set point. Each separate controlled variable requires its own control loop.

Information is transmitted to control devices by control signals. Control signals can be analog or binary in nature. Analog control signals experience changes in their magnitude while binary signals produce changes in their state. The signal types used in HVAC applications may be electrical or pneumatic. Electrical signals require wires to connect the control devices, pneumatic signals use plastic or copper tubing.

2.4

EXERCISES

Determine if the following statements are true or false. If any part of the statement is false, the entire statement is false. Explain your answers.

1. Exhaust air is another name for the return air that leaves a building instead of being reconditioned by the AHU.
2. Conditioning air creates changes its temperature, pressure, and moisture level.

3. Air handling units are used to move conditioned air to building zones to create a balance between their load and the energy contained in the conditioned air.
4. Supply and discharge air are terms that describe the same air stream entering the air handling unit.
5. All the return air that enters the mixing plenum is reconditioned.
6. The set point is the desired value of the process.
7. The controlled medium of a process is maintained at a desired set point.
8. The process error is calculated by subtracting the set point from the control point.
9. The controlled medium transports energy from its source to the process.
10. Offset and drift describe the same process condition.

Respond to the following statements and questions completely and accurately, using the material found in this chapter.

1. Describe an air handling unit.
2. What is supply air and what is it used for?
3. List three properties or condition of air that are changed when air is conditioned in an AHU.
4. What is mixed air?
5. What device is used to regulate the quantities of outside and return air that make up mixed air?
6. Why is outside air required in commercial air handling systems?
7. What organization establishes standards for the amount of outside air required in HVAC applications?
8. When is exhaust air required in a commercial HVAC application?
9. Explain the differences between the terms control point and set point.
10. Explain the difference between the controlled medium and the controlled variable.
11. Explain the difference between an analog and a digital control signal.
12. What are the most common signal types and their ranges used in the HVAC industry?
13. Using your home heating or cooling system, record the set point, control point, and calculate the offset and error.
14. What is the difference between a zone and a room?
15. What are the differences among the terms offset, error, and drift?

CHAPTER 3

Basic Control Loop Components

As stated in Chapter 2, a collection of control devices that maintain a controlled variable at its set point is called a control loop. Control loops are configured in a multitude of combinations to meet the needs of countless processes. Every loop, no matter how simple or complex, requires a minimum of three control devices in order to fulfill its design function: a sensor, a controller, and a final controlled device. These devices permit the control loop to measure, evaluate, and respond to changes occurring in the process control point. Chapter 3 describes the purpose and operating characteristics of sensors, controllers, and final controlled devices.

OBJECTIVES

Upon completion of this chapter, the reader will be able to:

1. List the components required in a basic control loop.
2. Explain the purpose of the three required components in a loop.
3. Describe the flow of information through the control loop.

3.1

CONTROL LOOPS

Control systems range from simple single loop configurations, as in a light circuit, to systems consisting of thousands of loops, similar to those in the large office complexes and corporate centers. No matter what the size of the system, they all began with individual devices that are collected into loops and various control systems. Figure 3.1 depicts the development of a building automation control system. It begins with individual control devices. These components are linked together with signal paths into specific control *loops*. Each individual loop maintains one process set point. The related control loops are combined into control *systems*. Each control system includes all the loops necessary to maintain a zone at its required condition along with the operation and safety controls for their mechanical systems.

In some buildings, control systems are left in a stand-alone mode. They do not communicate with each other. To improve the operational efficiency of the building or complex, the HVAC comfort control systems can be integrated with each other and with control systems that maintain lighting, fire, safety, security, and personnel access functions. These large supervisory systems are called *Building Automation Systems* (BAS). They are enormous control systems that combine the information and actions of all of the separate systems into one platform to permit sharing of information between the different subsystems. Building automation systems allow a building operator to monitor and manage all the integrated control systems from one computer.

Large HVAC control and BAS systems present complexities that often make troubleshooting an operational problem with a piece of connected equipment initially appear overwhelming. A typical commercial system, with all its compo-

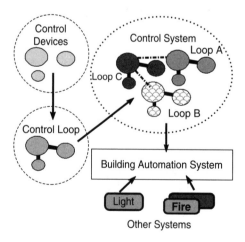

Figure 3.1 The Composition of a Building Automation System

nents, wires and tubes gives little indication as to a logical starting point to begin analyzing its operation. Through field experience and training, a control system technician learns to approach a troubleshooting exercise by dividing a large system into its individual control loops. After the individual loops are identified, their control devices can be analyzed to determine their operational status. Therefore, to build proficiency in analyzing systems, a technician learns to identify the components that make up an individual control loop.

The operational function and field location of the three required control devices found in every control loop are introduced in the following sections. The information presented makes it possible for the technician to identify the individual control loops within a larger system.

3.1.1 Control Loops Measure, Evaluate, and Respond

To accurately maintain a process at its set point, a control loop must be able to perform the following operations:

1. Measure the control point of the process.
2. Calculate the size and direction of every change in the error.
3. Generate a response that reduces the size of the error.
4. Transfer the correct quantity of mass or energy into the controlled medium of the process thereby reducing the error to acceptable levels.

These four operations are performed by the three essential components of a control loop — the sensor, the controller, and the final controlled device.

1. *Sensors* are installed in the system to measure the control point of a process.
2. *Controllers* calculate the process error (set point − control point) and generate a control signal that will reduce the error.
3. *Final controlled devices* have the responsibility of varying the quantity of mass or energy transferred into the controlled medium of the process. The following subsections describe these devices in more detail.

3.1.2 Control Loop Sensors

Sensors are control devices that measure the control point of a process. They generate an output signal related to the magnitude of the control point.

Sensors provide the ability for a control loop to *monitor* changes that occur in the condition of the *controlled variable*. These process changes produce corresponding changes in the sensor's output signal. Variations in the sensor's output signal indicate a loop response is required to maintain the balance between the mass or energy being transferred and the process load. If a response is not initiated, the process will exceed its design limits and the loop is described as operating *out of control*.

A sensor performs two related tasks: measuring the control point and generating a related output signal. The measurement function is performed with the portion of the sensor called the sensing element. The sensing element is made of a material that changes its properties in response to changes in the magnitude or size of the measured variable. Examples include temperature, pressure, and humidity sensors. Temperature sensors are made of materials that change their electrical resistance or length in response to changes in the intensity of the heat in the measured variable. Humidity sensors can be constructed to measure changes in the moisture level of the controlled medium using materials that change their size in response to changes in their adsorbed moisture. Pressure sensors are made that alter the position of a diaphragm in response to minute changes in the measured pressure. In each example the sensor's measuring element altered its characteristics in response to changes in the measured variable. Figure 3.2 shows a pneumatic control system sensor. The coil of copper capillary tubing is the sensing element. It is filled with a fluid that alters its pressure in response to changes in the measured temperature. These pressure changes are converted into a control signal by the sensor's signal generator. An overview of pneumatic device operation is presented in Chapter 7.

The second task of a sensor is to generate an output signal that *accurately* represents the measured value of the control point. The output signal contains important information describing the current state of the process. Other devices in the loop need that information in order to perform correctly. If the sensor is generating an inaccurate signal, all of the devices using this information will also produce incorrect responses to the actual process conditions. The sensor's signal is sent to all of the components connected to its output signal path. Figure 3.2 shows the pneumatic signal port that connects the sensor's output signal to other devices in the loop. Figure 3.3 shows a variable resistance sensor used in electronic control systems.

Pneumatic Signal
Connection Port

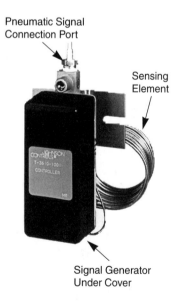

Sensing
Element

Signal Generator
Under Cover

Figure 3.2   Pneumatic Sensor

Sensing
Element

Fitting used to
connect signal
wires to the
sensing
element

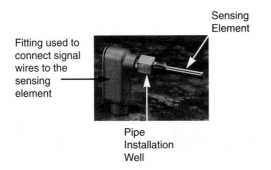

Pipe
Installation
Well

Figure 3.3   Resistance Water Temperature Sensor

3.2

CONTROLLERS

*A controller is the component in the control loop that generates an output signal to regu-
late the flow of mass or energy into the process, thereby reducing the error.*

The controller is the most complex component found in a control loop. Its purpose is to respond to changes in the control point in a manner that reduces the
process error. Controllers receive information regarding the state of the process by
means of the sensor's output signal. It compares the control point signal from the
sensor with the set point signal which is usually stored in the controller and generates an output that is *mathematically* related to the change that occurred in the
process.

The controller sends its output signal to a valve, damper, relay, or other device
in the system that controls the energy or mass flow into the process. As the controller alters its output signal, it produces changes in the amount of mass or
energy being transferred into the process, as it works to reestablish a balance
between the new load and the energy transfer.

A simple controller has one input signal and one output signal connection.
Pneumatic controllers use barbed fittings as connection or termination points,
electronic systems use screw type connections on termination strips. A controller's
input connection is connected to the sensor's output signal connection. Its output
connection is connected to the valve, damper, or relay's input connection. There
are more complex controllers available, called dual input controllers, that have
two input connections and one output connection. These controllers accept two
sensor signals, one of which measures the control point similar to a single input
controller. The second input connection is connected to a sensor that automatically
adjusts the set point of the process with respect to outside air or some other related
temperature. These controllers are covered in more depth in Chapter 8. Figure 3.4
shows a dual input pneumatic controller.

Dual Input Pnuematic Controller

Input   Supply Air   Output
Ports   Port   Port

Figure 3.4   Pneumatic Controller

3.3

FINAL CONTROLLED DEVICES

The final control device is a component in a control loop that performs the actual modulation of the flow of energy into the process, thereby establishing a balance between the load and the energy transfer.

The final controlled device is the third component found in all control loops. It is responsible for varying the flow of mass or energy into a process. Hot and chilled water valves, steam valves, return air dampers, outside air dampers, exhaust air dampers, electrical relays, and contactors are examples of final controlled devices commonly used in HVAC systems. Valves are used to control the flow of fluids into the process, dampers control the flow of lower pressure air. while relays and contactors switch the flow of electric current to control the operation of electrical loads.

A final controlled device is constructed of two dependent sections: *the flow control apparatus*, and the *actuator*. The flow control apparatus opens and closes the flow path, allowing mass or energy to be transferred to the process. Valves and dampers are the most common modulating flow control apparatuses used in commercial HVAC applications. Control relays and contactors are the most common two-position flow control devices.

The *actuator* provides the connection between the controller's output connection and flow control apparatus. It converts changes in the controller's output signal into movement that repositions the flow control apparatus. This motion is transferred to the flow apparatus directly or through a mechanical linkage to produce changes in the flow area of the apparatus. The flow control apparatus is mechanically coupled to the actuator to create a single component called a *final controlled device*. Figure 3.5 shows a pneumatic actuated water valve.

Actuators receive the output signal from the controller and convert it into rotational or linear motion. Pneumatic actuators are the most common actuator type used in larger commercial HVAC modulating control systems They produce a linear motion in response to changes in the controller output signal. Linear motion is movement that occurs in a straight line. Figure 3.6 shows how a diaphragm/piston assembly extends (or contracts) in a straight (linear) line as the input signal to the actuator changes. Pneumatic actuators are more commonly used because they are relatively inexpensive when compared to motorized actuators, they have a long trouble-free operating life, and they are easy to repair.

Final Controlled Device

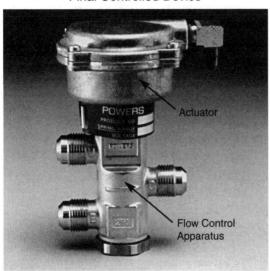

Figure 3.5   Final Controlled Device

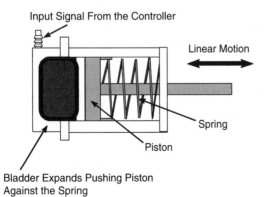

Figure 3.6   Pneumatic Linear Actuator

Electric coil actuators are also available as linear actuators although with less frequency than pneumatic devices. Electric coil actuators vary the electric current within the coil and therefore, the magnetic force it creates, to move the shaft that is located in the center of the coil.

Electrical relays and contactors also produce a linear motion when their coils are energized. The creation of an electromagnetic field draws the armature (contacts) into the stator (coil), against the force produced by an integral spring. When the current is interrupted, the armature is forced away from the stator by the spring, returning the contacts to their normal position.

Rotational actuators are primarily electric motor-driven devices. These actuators produce a circular (rotational) movement by the turning of an armature-driven gear box whenever their input signal changes. These actuators may be set up to produce a two position or a modulating response. Lower torque rotational actuators are available which can be directly driven by signals from digital control systems. These devices are used to position smaller valves and dampers. A low-torque electric motor actuator is shown in Figure 3.7.

High-torque electric motor actuators are primarily used in commercial packaged air conditioning equipment. In these applications, power is applied across one set of terminals to drive the flow control device open. Power is applied to another set of terminals to rotate the motor in opposite direction, causing the flow control apparatus to close. Common, clockwise, and counter-clockwise are typically used to label these actuator terminals. The method by which the motor is

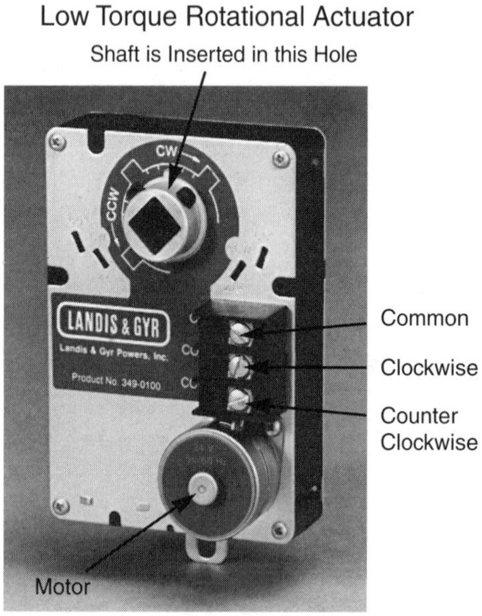

Figure 3.7 Rotational Actuator

connected to the flow control apparatus determines whether the flow path opens or closes when the signal is applied to the clockwise and common terminals. Final controlled devices are also known as FCDs or simply, controlled devices.

3.4

THE CONTROL AGENT

The control agent is the fluid that transfers the required mass or energy to the process controlled medium. The control agent flows through the final controlled device.

The control agent is not a loop component like a sensor, controller, or final controlled device although it is still essential to the operation of the control loop. Without the control agent, there can be no transfer of mass or energy into the process to maintain a balance with its load. Hot and chilled water, steam, refrigerant, air, and electric current are the most common control agents used in HVAC processes.

The control agent can easily be mistaken for the controlled medium. To correctly differentiate between the two, remember the control agent's flowrate varies with changes in the position or the control loop's actuator. The controlled medium of a process is identified as the fluid that is in physical contact with the loop's sensor.

3.5

BASIC CONTROL LOOP CONSTRUCTION

The previous sections have described a sensor, a controller, a final controlled device, and the control agent, which are the essential elements of all control loops. More complex loops, applications, and strategies may require additional control devices to condition (modify) control signals, select one signal from many inputs and perform logic functions, but the same basic four elements are still found within every control loop.

Although more than three components are typically found in a control loop, the basic configuration of the loop's design is still determined by the configuration of these three required devices. The flow of information contained in the signal paths is depicted in Figure 3.8 and is described as follows:

1. The control loop's sensor measures the control point of the controlled variable and generates an output signal that is sent to the controller.
2. The controller inputs the signal from the sensor and calculates the process error. The controller generates a change in its output signal that reduces the magnitude of the error. This output signal is sent to the actuator of the final controlled device.
3. The final controlled device's actuator inputs the signal from the controller and converts the change in signal magnitude into a corresponding change in

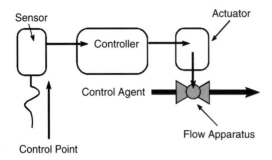

Figure 3.8   Flow of Information in a Loop

linear or rotational motion. The motion of the actuator changes the flow area of flow control device's orifice, modulating the flow of the control agent.

4. The change in the flowrate of the control agent produces a corresponding change in the mass or energy transferred into the process, reducing the process error.

   This diagram of the flow of information in a control loop is helpful when a technician is isolating the control devices that make up one particular loop that is part of a larger system. The first step is to locate the sensor within the process that is experiencing problems. Once the sensor is located, find the controller that receives the output signal from that sensor. Trace the output of the controller to the final controlled device and the loop has now been identified. Application 3.1 examines an air handling unit's discharge air temperature control loop using the terminology presented above.

APPLICATION 3.1

Process: Discharge Air Cooling Process

Objective: Maintain a set point of 56 °F in the air handling unit's discharge duct.

Control Loop Elements:
1. a sensor located in the discharge air stream
2. a single input controller
3. a final controlled device, chilled water valve with a pneumatic actuator
4. chilled water as the control agent to transfer energy from the air stream

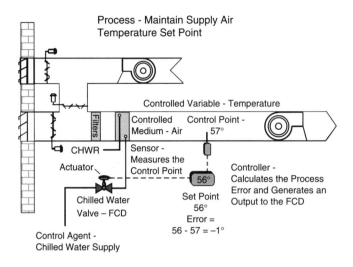

Application 3.1   Control Loop Construction and Terminology

Response: The sensor is located in the controlled medium. It measures the control point and generates an output signal that is sent to the controller. The controller compares the sensor's output signal with the set point signal and calculates the loop error. The error is used to generate a change in the controller's output signal. This signal is sent to the actuator of the final controlled device. The actuator repositions the chilled water valve's plug modulating the flow of energy from the process.

3.6

SUMMARY

Every functional control loop requires a minimum of four elements to maintain a process set point: a sensor, a controller, a final controlled device, and a control agent. The sensor monitors the magnitude of the controlled variable and sends a related signal to the controller. The controller compares the control point with the set point and generates an output signal based upon the error. The controller's output signal is sent to the final control device via the signal path between the two devices. The final controlled device actuator receives the signal from the controller and converts the change in the controller's output signal into linear or rotational motion. The movement of the actuator changes the flow area of the orifice in the flow control apparatus, varying the quantity of the control agent entering the process. Differences in the flowrate of the control agent produce associated changes in the magnitude of the controlled variable. Changes in the controlled variable will reduce the error that exists between the control point and set point. When a control loop is designed and calibrated correctly, it will optimize the use of energy as it maintains the process at its set point.

3.7

EXERCISES

Determine if the following statements are true or false. If any part of the statement is false, the entire statement is false. Explain your answers.

1. Control systems can have more than one control loop.
2. A sensor measures a change in the magnitude of the controlled medium.
3. Sensor input signals are sent to the controller which monitors the control point.
4. Controllers can have more than one output signal connection.
5. The final controlled device generates a signal that is sent to the controller.
6. Information in a control loop flows in the following order: process to the sensor to the controller to the actuator to the flow control apparatus and back to the process.
7. Energy or mass is modulated into the process by the final controlled device.
8. Controllers calculate the error and generate an output signal that reduces the magnitude of the offset.
9. The change in the sensor's output signal is related to the change in the controlled variable.
10. The control agent is an element found in every control loop.

Respond to the following statements and questions:

1. List the four elements required in all control loops.
2. List the two parts of a final controlled device and describe the purpose of each part.
3. What is the purpose of the controller in a control loop?
4. Describe the function of a sensor.
5. What is the purpose of the control agent?
6. Describe a non-HVAC control loop found at home. Determine the loop sensor, controller, final controlled device, control agent, process, set point, control point, and error.
7. Why is a controller more complicated than a sensor or final controlled device?
8. How can the inability of a commercial boiler to control its water temperature affect the operation of temperature control loops in the building?
9. Explain why a control loop cannot operate without a controller.

10. List 3 sensors found in residential applications that are not air conditioning related.
11. List 3 final controlled devices found in non-HVAC residential applications.
12. What is the difference between an energy transfer and mass transfer process?
13. How can a technician determine the difference between a controlled medium and a controlled agent?
14. What is the meaning of the control agent's quality and why is it important for the correct operation of a control loop?
15. Describe a building automation system.

CHAPTER 4

Control Loop Configurations

There are two possible configurations used to interface a control loop with its process. One installation method creates an open loop configuration and the other creates a closed loop configuration. Both configurations have different operating characteristics and therefore, different applications. Chapter 4 describes the characteristics and system layout of open and closed loop configurations. To simplify the representation of each configuration, block diagrams and component transfer functions are also introduced in this chapter.

OBJECTIVES

Upon completion of this chapter, the reader will be able to:

1. Describe an open loop configuration and its applications.
2. Describe a closed loop configuration and its applications.
3. Draw a block diagram of an open control loop.
4. Draw the block diagram of a closed control loop.
5. Define a transfer function and its purpose.

4.1

OPEN AND CLOSED LOOP CONFIGURATIONS

As stated in the previous chapter, all information in a control loop flows in the same sequence. It enters the loop at the sensor location via the sensing element. Information is passed on to the controller which in turn alters the information and sends it on to the final controlled device. Loops that are more complex, having additional control devices within the signal path, still pass information using the same sequence, starting at the sensor and ending at the final controlled device. Keeping this in mind, it is correct to state that open and closed loop configurations do not describe how a loop's control devices are interconnected. Open and closed loop configurations describe how the control loop is *connected* or coupled to the process. More specifically, the phrases describe the sensor's location within a process. Its location determines how the control loop responds to changes in the measured variable. The following sections describe the responses of open and closed loop control configurations.

4.1.1   Open Loop Response Characteristics

All open loop configurations generate a binary or two-position response to changes in the measured variable. This configuration is used in applications to prevent the control point from exceeding a predetermined design limit.

As stated in the previous section, the configuration of a control loop is a function of the location of its sensor within the process. To create an open loop control configuration, the sensor is placed *upstream* of the location where the control agent effects a change in the controlled medium. An upstream location is any position in the process preceding the point where mass or energy is transferred into the process. Figure 4.1 depicts areas upstream and downstream of the location where heat transfer takes place. Figure 4.2 depicts the sensor in the upstream or open loop location.

A careful examination of Figures 4.1 and 4.2 show the major differences between open and closed loop configurations. In Figure 4.1, the control loop maintains the supply air temperature at 102° in the area where the sensor is located. By repositioning the sensor, the entire function of the control loop changes. Figure 4.2 shows the supply air temperature is now 120°. The controller will not close the valve even though the temperature is above the process set point because the sensor is located upstream of the heating coil. Consequently, the loop never measures the supply air temperature. In the upstream location, the only function the control loop can perform is to prevent the supply air temperature from getting too cold when the outside air temperature decreases below its set point of 42°. It cannot prevent the supply air from becoming too warm. When the outside air temperature exceeds the set point (42°) the controller commands the hot water valve closed and the supply air cools down. This response indicates that an open loop configuration cannot be used to maintain a balance between the energy transfer and the process load

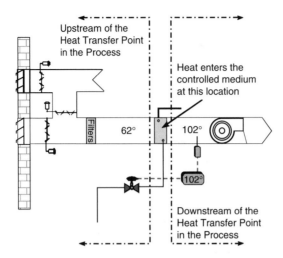

Figure 4.1   Upstream and Downstream Locations

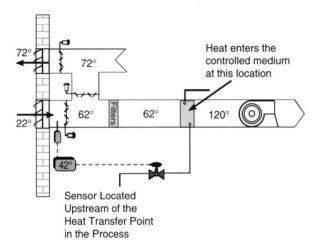

Figure 4.2   Open Loop Configuration

because it cannot measure the load when its sensor is located upstream of the point where the control agent transfers mass or energy into the process.

Another difference between open and closed loop configurations is terminology associated with the sensor's function. In previous chapters, it was stated that the sensors purpose was to measure the control point of the process and that a change in the quantity of mass or energy into the process altered the magnitude of the control point. These statements are only true in closed loop configurations where the sensor can measure the effect of the transfer into the controlled medium. In open loop configurations, the sensor is located before the point of transfer so it does not actually measure the process control point. To be correct, it is stated that open loop sensors monitor the measured variable. When the measured variable exceeds the process set point, the controller responds.

Open loop controllers can only produce a two-position or binary signal response to changes in the measured variable because modulating signals require a sensor positioned downstream of the point of energy transfer in the process. The output signals of open loop controllers are typically equal to the minimum and maximum values of their output signal range. Therefore, a open loop controller with a 3 to 15 psi output signal range can only send a 15 psi or a 3 psi output signal to the final controlled device. A controller with an output signal range of 4 to 20 mA will generate a 4 mA or a 20 mA signal based upon its set point and the magnitude of the measured variable. Open loop controllers are incapable of generating any signal between their minimum and maximum output signal range values. Figure 4.3 depicts the signal change of an open loop controller in response to changes in the process control point.

Open loop controllers cycle the final controlled device between its fully open and fully closed position, producing pulses in the flowrate of the control agent. These pulses of energy generate pulses or cycles in the measured variable *downstream* of the sensor's location. Consequently, it cannot measure changes in the controlled medium that result from changes in the flowrate of the control agent. This response is the identifying characteristic of an open control loop.

Open loop configurations are used in applications where the process cannot be permitted to exceed a predetermined design *limit*. The magnitude of the limit is equal to the minimum or maximum allowable value of the measured variable. Whenever the measured variable exceeds this limit, the controller commands its

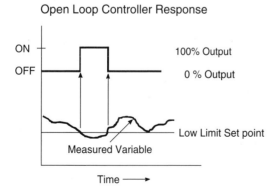

Figure 4.3   Open Loop Controller Response

output signal to its complementary or opposite value. This response forces the final controlled device to its opposite position, varying the flowrate of the control agent 100% of its previous value. This produces a corresponding change in the process. The process is forced back within its design limits.

The set point of an open loop process may be either a high or a low limit value. The choice is based upon the operating requirements of the process. In low limit applications, whenever the measured variable falls below the *low limit set point*, the controller changes the state of its output signal. Conversely, in high limit applications, whenever the measured variable exceeds the *high limit set point*, the controller output signal changes state. These responses occur quickly, as soon as the control point exceeds the limit. This immediately alters the magnitude of the flow of the control agent entering the process. The control loop's response may produce a change in the temperature, pressure, or humidity level of the controlled medium or the response may cycle equipment on or off. Refrigeration systems use open control loops to prevent the compressor from operating if the refrigerant pressure becomes either too high or too low or if its current draw exceeds a safe level. Air handling units have open loops that measure high and low air temperature limits, humidity limits, high static pressure limits, the presence of smoke limits, and high motor currents.

Not all open loop applications have limits that are associated with a safety condition. Time scheduling is a common nonsafety-related open loop application. In these strategies, the clock's output contacts change state after a time limit has been reached. Once the present time exceeds a predetermined start time, the output contacts of the timer are commanded to start the equipment. After the time exceeds the shutdown set point, the equipment is commanded off. The sensor or real time counter never monitors whether its change in contact position started the equipment, therefore, it is an open loop application. Application 4.1 summarizes the operation of an open loop preheat coil that prevents the temperature of the air in the mixing box of the AHU from dropping below freezing.

APPLICATION 4.1

Process: Maintain a Minimum Temperature of the Ventilation Air Entering the AHU

Objective: When the outside air approaches temperatures that are cold enough to freeze coils (typically below 38 °F), the steam valve is commanded to 100% open, preheating the outside air entering the AHU to approximately 60 °F. Warming the ventilation air as it passes between the fins of the coil minimizes freezing any coils located downstream of the preheat coil.

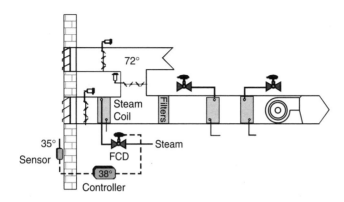

Application 4.1   Open Loop Preheat Coil

Response: As the outside air temperature decreases below the loop's set point of 38 °F, the controller's output changes state to its minimum value allowing the normally open steam valve to open 100%. When the outside air temperature rises above 38 °F, the controller's output signal changes state to its maximum value, closing the steam valve.

4.1.2 Feed-Forward Control

As a consequence of the location of the sensor in an open loop configuration, all the information from the control signals flows from the measured variable, through the sensor, controller, and final controlled device. From the final controlled device, the information continues *forward*, out of the process. It never returns back to the sensor to indicate whether or not the last change in the controller's output signal had the desired effect on the process. Since all process information continues onward, out of the process, open loop configurations are called *feed-forward* control loops. The information of the present state of the process never returns to the controller via the sensor.

   Feed-forward loop configurations cannot be used to maintain a process at its set point. Maintaining a set point requires a balance to be maintained between the energy flow and the process load. When a sensor is located upstream of the effect of the control agent on the process, it is unable to measure the load after the mass or energy has been transferred to the controlled medium. Without this information, modulation of the control agent is not possible. Consequently, open loop or feed-forward configurations can only be used to maintain a process from exceeding a design limit using a binary or two-position controller response. Figure 4.4 shows the flow of data in an open (feed-forward) loop.

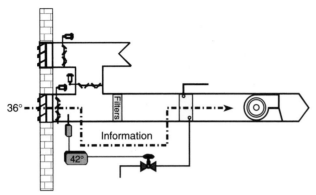

Open Loop - Feedback Information Flow

Figure 4.4 Information Flow in a Feed-Forward Loop

4.2

CLOSED LOOP CONFIGURATIONS

Closed loop control configurations generate a response that can maintain the controlled variable at its set point.

Closed loop control configurations respond in a manner that open loops cannot. Closed loops can *modulate* (infinitely vary) the flow of the control agent into the process to maintain a balanced condition thereby maintaining the control point at set point. Closed loop configurations can also pulse energy into the process using a two-position controller to maintain the control point within a few units of the set point. As stated in the previous sections, two-position open loop controls can only prevent a process from exceeding a design limit, they cannot maintain a process at or near its set point.

To layout control devices in a closed loop configuration, the loop's sensor must be located in the controlled medium, downstream of the position where the control agent transfers mass or energy into the process. In this location, the sensor can measure the effect of the control agent on maintaining the control point near its set point. If too much energy is being transferred, the sensor signals the controller to reduce the flowrate of the control agent. Conversely, if the flowrate of the control agent is too small, the sensor measures the deficiency and signals the controller to increase the flowrate. Keep in mind, these changes in flowrate can be accomplished using a modulating or two-position controller response. Figure 4.5 shows the differences between open and closed loop controller response intervals. Open loops become responsive whenever the measured variable exceeds a low or high process limit. Closed loops respond whenever the control point is between the high and low limits of the process operating range.

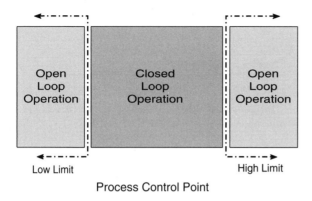

Figure 4.5   Differences Between the Operating Ranges of Open and Closed Loop Configurations

Closed loop control configurations maintain the controlled variable at set point by measuring the effect the control agent is having on the process control point. The amount of mass or energy transferred is related to the size of the difference that exists between the control point and the set point of the process. A modulating closed loop response varies the flow of control agent in increments that match the changes in the control point. The control point changes whenever a change in the load occurs.

To generate a varying signal response, closed loop controllers produce a modulating or analog signal. This signal varies the position of the final controlled actuator and consequently, the flowrate of the control agent to any value between 0% and 100% of its maximum flowrate. Modulating the control agent allows a balance to be maintained between the mass or energy transferred and the process load. Figure 4.6 shows the way a modulating closed loop controller response tracks the changes in the process control point. Compare this response to the two-position response shown in Figure 4.7. Note: an advantage of closed loop modulating configurations over closed loop two-position configurations is the tighter control provided by the modulating loop. The control point does not drift as far away from the set point before the analog controller responds, maintaining the control point closer to the set point through all load changes.

Most closed loop applications use a modulating controller over a two-position controller. A modulating closed loop maintains the control point near the set point using a smooth response that goes unnoticed by the occupants. A controller with a two-position response maintains the control point within an allowable range of the set point, but the occupants may become uncomfortable at either end of the control point's operating range.

Closed loop configurations are used in applications where the process must be maintained within a predetermined range of the set point. All the HVAC control loops used to maintain a temperature, pressure, level, or humidity set point are

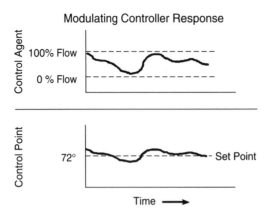

Figure 4.6   Closed Loop, Modulating Controller Response

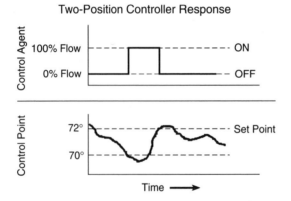

Figure 4.7   Two-Position Closed-Loop Response

configured as closed loops. They all have their sensors placed in the controlled medium, downstream of the heat or fluid transfer location.

4.2.1 Closed Loop Feedback

Information in a closed loop configuration travels in a circular path. It enters the loop at the sensor's location in the controlled medium, travels through the sensor, controller, and final controlled device, back into the process via the controlled agent and finally, back to the sensor as a change in the process control point. The sensor's downstream location closes the information path between the process and the controller. This data route is called the *feedback* path. It derives its name from the characteristic that information on the current condition of the process is *fed back* to the controller via this path. Figure 4.8 depicts the feedback or circular

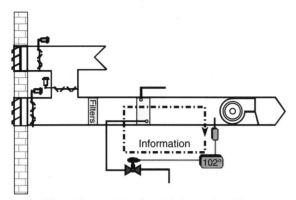

Closed Loop - Feedback Information Flow

Figure 4.8 Closed Loop Information Path

flow of information in a closed loop configuration. Compare this figure with the open loop information path in Figure 4.4.

Feedback describes a condition where a device uses the results of its previous response to generate a new change in output signal. In control loops, the feedback loop provides a path for the essential process information to be sent back to the controller if it is going to vary the flow of the control agent into the process as a function of the previous change in the control point. In other terms, the feedback path establishes a method for the controller to respond in a self-correcting manner. This self-correcting action allows the controller to constantly adjust its output signal until the control point falls within acceptable range of the set point. Without a feedback path, the controller could not effectively maintain the control point of the process. Consequently, the intent of the process could not be maintained. Application 4.2 describes a closed loop heating process designed to maintain the discharge air temperature of an AHU at 140º.

There is one negative aspect associated with a closed loop configuration. The presence of the feedback path increases the possibility of the control loop becoming unstable during a response to a change in the control point. Unstable loops oscillate and, therefore, cannot maintain the requirements of the process. Open loop configurations cannot become unstable. Stability is covered in depth in subsequent chapers.

APPLICATION 4.2

Process: Modulating, Closed Loop Temperature Control

Objective: Maintain the temperature of the air in the hot deck at 140 ºF by modulating a hot-water valve.

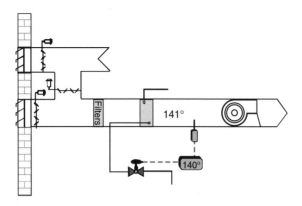

Application 4.2   Modulating, Closed Loop Heating Process

Response: In a hot water coil heating application, the hot water (control agent) is modulated into the coil by varying the position of the hot water coil's valve. The valve is positioned by the controller's output signal which changes in response to changes in the control point. If the last response of the controller allowed too much energy to be transferred to the air, the sensor is positioned so it can feed this information back to the controller. The controller can now modify its output signal (self-correct) to command the controlled device to a new position. In this instance, the valve will be modulated toward closed to reduce the amount of energy transferred, maintaining the control point within its acceptable range.

In this application, the sensor can feed back the controllers effect on the controlled variable, signifying the loop is configured as a closed loop. If the sensor were located before the coil, it could not measure the effect the control agent has on the controlled medium and the control point could not be maintained within its design range.

4.3

TRANSFER FUNCTIONS

A transfer function is the mathematical equation that describes the relationship that exists between changes in the input signal of a device and the corresponding changes in its output signal.

Every control device receives at least one input signal and generates one output signal. A mathematical relationship exists between these two signals that describes how the output signal changes in response to changes in the input signal. This mathematical relationship is called a *transfer function*. Without getting too involved in the actual mathematics of the transfer function, remember that it is mathematically and consequently, scientifically based.

Transfer functions describe the changes that occur in the size, phase and timing characteristics of an output signal of a control device in response to a change in its input signal. The *size* portion of the transfer function regulates the change in the magnitude of the output signal. The *phase* characteristic of the transfer function indicates which direction (increase or decrease) the output signal changes in response to a change in the input signal. The *timing* portion of the transfer function defines any *delay* that occurs in the generation of the output response after the input signal changed. These three signal characteristics are defined in the mathematical formula that is used by the control technician to calibrate the devices after they are installed in a system. These characteristics are covered in greater detail in the following chapters.

Figure 4.9 depicts each of the three changes in the output that are described in the transfer function. Changes in timing are shown as a delay between the point where the input signal began to change and the corresponding change in the output signal. The phase change is depicted by the signals changing in opposite directions. Other transfer functions have the signals change in the same direction. The third change depicted is the change in the size of the output signal with respect to the corresponding change in the input signal. The size parameter of the transfer function is the most commonly analyzed variable of the transfer function and is called the *gain* of the device. The transfer function was introduced at this time because of their requirement in the following section on block diagrams. It is described in greater detail in Chapter 5.

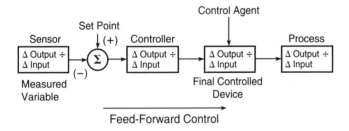

Figure 4.9   Elements of a Transfer Function

4.4

BLOCK DIAGRAMS

A block diagram is a graphical representation of the flow of information between components in a control loop.

Block diagrams are used to simplify the visualization of the interconnections that exist between the control devices in a loop, the flow of the information in their signals, and their relationship to the process. They are also used to show the input/output signal relationship or transfer function of each device and process. It quickly points out the configuration of the loop and all the input and output signals that are part of the control loop. This presentation format makes it easier to analyze the entire loop response for changes in the input conditions.

In a block diagram each control device is represented by a rectangle or block. The blocks are connected with lines that represent the signal paths. The lines also show the direction that information flows between components. Each block has sufficient lines to represent all signals that enter and leave each component. Along with the signal paths, the block contains the equation that represents the transfer function that depicts the mathematical relationship between the input and output signals of each device. Although the generic formula is shown in these figures, it will be presented in more depth in Chapter 5.

The summing junction is a circle with the mathematical symbol for summation (Σ) centered in the circle. The summing junction represents the procedure of subtracting the control point from the set point to calculate the error.

The block diagram contains all the information required to represent the process making it easy to visualize the relationships between the loop components. Note the flow of information in the open loop and the lack of a feedback path between the sensor and the process in Figure 4.10. Block diagrams can be used to represent any control loop, from a simple light switch to a complex HVAC strategy. If constructed correctly, they present all the information required to analyze a control loop. Figures 4.11 shows block diagrams of a closed loop configuration.

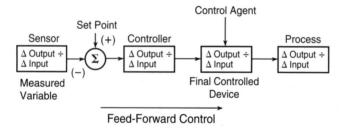

Figure 4.10   Block Diagram of an Open Loop

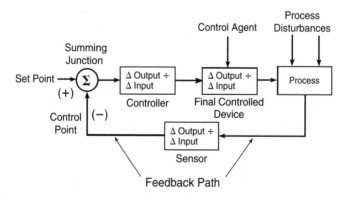

Figure 4.11   Block Diagram of a Closed Loop

4.5

SUMMARY

Open loop controllers always generate a two-position response to prevent the process from exceeding a predetermined design limit. The sensor is positioned upstream of the location where the mass or energy of the control agent enters the process. Consequently, there is no feedback path in this configuration and the controller never knows whether the desired response has occurred. The advantage of an open control loop configuration is that it cannot become unstable during normal operation. The disadvantage of this configuration is that it cannot maintain a process at a given set point. Open loop configurations are primarily used in safety-related applications.

Closed loop configurations locate their sensor within the controlled medium, downstream of the location where the control agent changes the process. This location allows the controller to receive feedback concerning its previous response to the change in process conditions. The feedback path creates a self-correcting loop configuration where each subsequent change in the controller output signal is based on the effect that its previous signal change had on the control point. The major disadvantage associated with closed loop configurations is their increased susceptibility of becoming unstable when responding to unusual load changes. Closed loop configurations are used in applications that maintain the process control point with a desired range of the set point.

A transfer function is a mathematical equation that describes the relationship between a change in the input signal of a control device and the corresponding change in its output signal. It describes the changes in magnitude, phase, and timing that the output signal experiences with each change in the input signal. Every sensor, controller, final controlled device, signal converter, etc., has its own transfer function. Block diagrams are used to simplify the representation and analysis of a control loop. The structure of the diagram shows the flow of information between the components along with the mathematical relationships that govern their response.

4.6

EXERCISES

Determine if the following statements are true or false. If any part of the statement is false, the entire statement is false. Explain your answers.

1. Open loop configurations have a feedback path.
2. The sensor position within a process determines if the loop is an open or closed configuration.

3. Feed-forward and open loop are two phrases that describe loops where the sensor is located upstream of the effect of the control agent has on the control point.
4. All two-position controllers are open loop controllers.
5. The feedback path can introduce instability into a control loop.
6. Oscillations in the final controlled device are a sign of instability.
7. The transfer function is a mathematical formula that describes the relationship between the change in the input and output signals of a control device.
8. A block diagram is a representation of the flow of information in a control loop.
9. A process also has a transfer function.
10. A safety control loop is typically a feed-forward configured loop.

Respond to the following statements and questions:

1. Describe the operating characteristics of an open loop control configuration.
2. Describe the operating characteristics of a closed loop control configuration.
3. What is the fundamental difference between open and closed loop control?
4. List three examples of an open loop configuration found in a home.
5. List three examples of a closed loop configuration found in a home.
6. What is a summing junction?
7. Is the burner control on a stove an open or closed loop control strategy? Explain your answer.
8. Is the temperature control loop of a residential refrigerator an open loop or closed loop configuration? Explain your answer.
9. Draw a block diagram of a refrigerator temperature control loop.
10. Draw a block diagram of a stove burner control loop.

CHAPTER 5

Elements of a Transfer Function

A transfer function is the equation that relates changes in a control device's output signal to the changes that occurred in its input signal. Technicians use the transfer function to calibrate and analyze the operation of control loops. Chapter 5 describes the characteristics of a typical transfer function for a control device.

OBJECTIVES

Upon completion of this chapter, the reader will be able to:

1. Describe the purpose of a transfer function.
2. Describe the purpose and characteristics of the various mathematical elements that make up a transfer function.
3. Describe the effects of a transfer function on the output signal of a device.

5.1
■■■■

CONTROL DEVICE TRANSFER FUNCTIONS

A transfer function is a mathematical expression that defines the response of a control device to a change in its input signal.

The word *transfer* signifies that any change in the input signal is *transferred* through the device and appears as a change in its output signal. *Function* is a term that describes a mathematical relationship that exists between two variables. The mathematical relationship exists between changes in the input and output signals of a control device. All control devices have a transfer function that instructs the device on how to change its output signal in response to the change that is occurring in its input signal. The equation looks at the size, direction, and timing characteristics of the input signal's change to generate a related change in its output signal.

The transfer function of a control device is modified by adjusting a dial, slide mechanism, software database, or some other procedure. During the calibration procedure, a technician adjusts the operating parameters of a control device's transfer function so it generates the proper output signal for a known input condition. For example, a sensor may be adjusted so it generates an 11.4 psi output signal when the sensing element is at 70 °F. Likewise, a controller may be calibrated so it positions a valve 100% open whenever the loop error exceeds + 2°.

A control device may have up to six separate adjustments that can be made to their transfer function during the calibration process. Sensors have the simplest transfer function, having only one field adjustment. Controllers have the most complex transfer function, having up to six adjustments that must be made by a technician during the calibration procedure. These adjustments make it possible for the controller to respond correctly to the size, phase, and timing characteristics of the process changes. The following sections describe the elements of a transfer function.

5.2
■■■■

THE PROPORTIONAL GAIN OF A TRANSFER FUNCTION

The proportional gain of a control device defines the change in the size of the output signal that will occur in response to the change in size of the input signal.

The proportional gain element of a transfer function is responsible for generating the initial change in the size (magnitude) of the output signal in response to a change in the input signal. The proportional gain defines the amount of change in the output signal that occurs in response to a unitary (single unit) change in the input signal. A change of one whole unit is used to standardize the way the proportional gain of a device is written. Applying this definition to a sensor, the proportional gain element of a relative humidity sensor defines the amount of change in its output signal that occurs for each 1% change in the measured relative humidity. Similarly, a temperature sensor's proportional gain indicates the amount of change in its output signal that is generated for each 1° change in the measured variable. The proportional gain parameter of a transfer function is equal to:

$\Delta Output / \Delta Input$

where:

1. Δ symbolizes the phrase "change in"
2. Δ Input represents a unitary change in the input variable.

If a pneumatic sensor generates 0.12 psi for each 1° change in the measured temperature, its proportional gain is written as:

$0.12\ psi / 1\ °F\ or\ simply\ 0.12\ psi / °F$

Similarly, if an electronic sensor generates 0.16 milliamps for each 1° change in the measured temperature, its proportional gain is written as 0.16 mA/°F.

Proportional gain is used to describe the size responsive element of the transfer function. The word *proportional* signifies that the change in the output signal is related to the change in the input signal by a mathematical ratio. A ratio is a term in an equation that is represented by a fraction (x/y). As previously shown, the gain of a transfer function is indeed represented by a fraction, namely $\Delta Output / \Delta Input$. The term *proportional* is used to differentiate between three different gain elements that can be used in the transfer function of a controller: proportional, integral, and derivative gain. Integral gain and derivative gain are presented in subsequent sections of this chapter.

The total change in an output signal is a function of the amount of change that occurred in its input signal. Using the temperature sensor that has a proportional gain of 0.16 mA/°F as an example, if the measured temperature changes 2°, the output changes a value equal to twice ($2x$) its gain value or 0.32 mA. Similarly, a change of 4° in the measured temperature generates a change in the output signal that is equal to four times ($4x$) the sensor's gain or 0.64 mA. Figure 5.1 shows a block diagram of a controller having a gain of 2 psi/°F. As the input temperature increases 1°, the output will increase 2 psi. Conversely, if the temperature decreases 1°, the output signal will decrease 2 psi.

The magnitude of the proportional gain has a direct effect on the response of the device to changes in its input signal. Control devices with large values for their

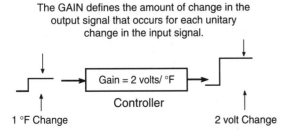

Figure 5.1   The Response of the Proportional Gain Element of a Transfer Function

proportional gain generate large changes in the size of their output signal in response to a unitary change in the size of their input signal. Large proportional gain values create a control device that responds to *small* changes in its input signal. These devices are described as being *sensitive* to changes in their input signals. They are used to control process that must react to small changes in the control point to maintain the process offset within an acceptable range.

Control devices can also be calibrated with small values for their proportional gain. Consequently, they will require a larger change in their input signal before they generate a measurable change in their output signal. These low gain devices are tuned so they are **not** sensitive to changes in their input signals. They can be used to create a stable control loop that maintains the control point within a larger but still acceptable range of the set point. Due to the relationship between the size of the proportional gain and the responsiveness of the device, the term *sensitivity* is often used synonymously with the term gain. Some control companies use the term sensitivity exclusively while others use the term *proportional gain* or simply, *gain.* In each case, the term describes the amount of change in the output signal that occurs for each unitary change in the input signal. Application 5.1 introduces the derivation of the proportional gain of a typical HVAC sensor.

APPLICATION 5.1

Device: Temperature Sensor

Purpose: Measure changes in the controlled variable and generate a related change in the sensor output signal.

Operating Ranges: Input Range: 0 to 100 °F
Output Range: 3 to 15 psi
Proportional Gain: 0.12 psi/°F

Response: The proportional gain of this sensor indicates the output signal changes 12 hundredths of a psi in magnitude for each 1 °F change in the measured temperature (input signal). When the input signal changes 2 °F, the output signal *changes* a magnitude of $2° \times 0.12$ psi/°F $= +0.24$ psi. Similarly, if the input decreased 0.3 °F, the output would change $-0.3° \times 0.12$ psi/°F $= -0.036$ psi.

Change in Input Signal (Δ Input)	Change in Output Signal (Δ Output)
+ 8	+ 0.96
+ 2	+ 0.24
0	0
− 3	−0.36
− 5	−0.6
−10	−1.2

The proportional gain element of sensor and final controlled device transfer functions is typically set to a fixed value during the device's manufacture and cannot be field adjusted by a control technician during the calibration process. Consequently, these components maintain the same input/output relationship that they had when they left the factory. Controllers are constructed so their proportional gain can be adjusted after they are installed. This allows the technician to tune the control loop so its response matches the desired response of the process.

5.3

THE PHASE ELEMENT OF A TRANSFER FUNCTION

The phase element of a transfer function defines the direction that the output signal changes in response to a change in the input signal.

Phase is a term that describes the relative position or direction of one signal in reference to the position or direction of another signal, at the same point in time. In control technology, the phase defines the direction that the output signal changes (increase or decrease) as the input signal increases. The phase element of the transfer function of a control device is commonly referred to as the *action* of a device.

Since a signal can only increase or decrease as it changes, there can only be two possible control actions, *direct acting* and *reverse acting*. A *direct acting* device generates changes in its output signal that occur in the *same direction* as the change in the input signal. Therefore, when the input signal increases, the output signal will also increase. Conversely, when the input signal decreases, the output signal will also decrease. Figure 5.2 shows the responses of a direct acting and a reverse acting control device.

Sensors and final controlled devices have *fixed* actions that cannot be reversed after the component leaves the factory. Most pneumatic and electronic sensors are direct acting devices. Their output signals increase as the magnitude of the

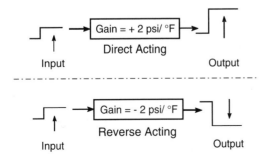

The Phase or Action element of a transfer function describes the direction of change in the output signal with respect to the change in the input signal

Figure 5.2   Phase or Action Element Response

measured variable increases. However, there is a popular electronic temperature sensor called a thermistor, commonly used in digital control systems, that has a reverse acting response. Its output resistance decreases as the measured temperature increases. These reverse acting devices are described as having a negative temperature coefficient (NTC) and are described in greater detail in Chapter 7.

Most controllers have the capability of being configured to respond as a direct or reverse acting device. Their action is field selectable and based upon the response characteristics of the final controlled device along with the operational and safety requirements of the process. The system designer or control technician analyzes the process and determines whether the final controlled device should fully open or completely close in response to an increase in the control point. If the controller's output signal must change in the same direction as the change in the control point to correctly position the final controlled device, a direct acting controller is required. Conversely, if the controller's output signal must decrease when the controlled variable increases, the required controller phase is reverse acting. Applications 5.2 and 5.3 describe the analysis used to determine if the controller's phase element should be set as direct or reverse acting.

APPLICATION 5.2

Process: Zone Temperature Control

Objective: Maintain the set point of the zone by modulating a normally open hot water valve.

Components:
1. Fin tube convection heat exchanger
2. A normally open (NO) hot water valve (FCD)
3. A direct acting temperature sensor
4. A controller

Response: As the room temperature increases, the heat transferred by the fin tube must be decreased to maintain the zone's set point. The valve was selected so it opens as its input signal decreases. This maintains a safe operating response if the control system malfunctions or the signal to the actuator is interrupted. If either fault occurs, the valve moves to its wide open position, maintaining the transfer of heat into the zone. Therefore, the valve modulates toward its closed position as the room temperature increases.

Controller Action: The signal to the control valve must increase to close the valve, thereby reducing the heat transfer into the zone. Therefore, as zone temperature increases, the controller's output signal must also increase, requiring a direct acting controller action.

APPLICATION 5.3

Process: Variable Air Volume, Supply Duct Static Pressure Control.

Objective: Maintain the static pressure in the discharge air duct at a set point of 1.1 inches of water.

Components: 1. Normally closed (NC) inlet vane dampers (FCD)
2. A direct acting differential pressure sensor
3. A controller

Response: As the static pressure in the duct increases, the inlet vane dampers modulate toward their closed position to reduce the static pressure at the sensor's location. The inlet vane damper is configured so its blades close upon loss of signal to their actuator. This causes the pressure in the duct to decrease to zero if the control signal to the actuator is interrupted, prohibiting the duct from becoming over-pressurized.

Controller Action: As the static pressure in the duct decreases, the signal to the inlet vane dampers must increase to open the damper blades slightly, allowing more air into the duct to raise its static pressure. Therefore, as the static pressure decreases, the controller output must increase. A controller with a reverse action is required.

5.4

THE TIMING ELEMENT OF A CONTROLLER TRANSFER FUNCTION

As stated in the previous chapter, the time element of a transfer function describes the delay that occurs between a change in an input signal and the associated change in the output signal. This time delay is of little concern to HVAC control technicians because HVAC processes respond relatively slowly when compared with the magnitude of the signal delay. Consequently, the delay between the input and output signal change has no measurable effect on the response of the process. In this text, the timing element of a transfer function is used to describe changes in the controller's output signal that occur in response to the timing characteristics of the changes that occur in the process error.

There are two remaining transfer function elements that alter an output signal in response to the time related characteristics of a change in the process control point. These transfer function elements change the output signal in proportion to the speed at which the control point changes and the duration which an offset exists between the control point and the set point. These two timing elements are called the *derivative gain* and the *integral gain*. Since these elements respond to the timing characteristics of process changes, they are considered within the scope of this text to be the timing characteristics of a controller's transfer function.

Controller transfer functions may contain one or both of these *timing related elements* to alter the output signal in response to the speed and/or duration of changes in the process control point. The integral gain determines the relationship between the length of time a process error exists and the change in the controller's output signal. The derivative gain determines the relationship between the speed at which the process control point changes and the corresponding change in the controller's output signal. Sensors and final controlled devices do not have timing elements in their transfer functions. Sections 5.4.1 and 5.4.2 describe the effect that the integral and derivative elements responses have on the controller's output signal.

5.4.1 The Integral Gain of a Controller Transfer Function

Integration is a mathematical procedure that is used to calculate the area under any curve generated by a variable that changes with respect to time. In HVAC transfer functions, integration is used to calculate the *amount of error* that exists over a specific time interval. This value is called the *time weighted error*. Simply stated, the time weighted error is calculated by multiplying the magnitude of the process error by the length of time it exists and summing these products over the length of a time interval. The calculation of the time weighted error occurs over a period called the element's *reset* time. Therefore, as the magnitude of the process error or the length of time it exists increases, the time weighted error increases. Conversely, the time weighted error decreases as the magnitude of the error and/or length of time it exists decreases. The time weighted error equals zero whenever the process error (set point − control point) equals zero.

Controllers can be configured to respond to the time weighted error by incorporating the transfer function element called the *integral gain*. Controller transfer functions that incorporate the integration timing element are stated as being configured with an *integral mode* or *integral gain*. The purpose of the integral element is to reduce any offset within the process to zero after a change has occurred. It accomplishes this function by continuously adjusting the controller's output signal in proportion to the size of the time weighted error until the process error is forced to zero. The integral mode of the transfer function is also called *reset mode* because this transfer function element causes the control point to be *reset* back to the value of the set point whenever a process change occurs.

The integral gain element modifies the output signal of the controller in *proportion* to the magnitude of the *time weighted error*. The time weighted error is multiplied by the integral gain to determine the amount of change in the output signal that occurs due to the integral element of the transfer function. As the size of the time weighted error increases, the change in the controller's output signal also increases. Conversely, as the size of the time weighted error decreases, the change in the controller's output signal also decreases. The change in the controller's output signal repositions the final controlled device so it transfers *more mass or energy* than is required to create a balance with the load. The additional transfer forces the control point back toward the set point, reducing the process error. Once the error is driven to zero, the time weighted error (error x length of time) also equals zero and the controller's output signal stabilizes at a fixed value. The controller's output signal will remain at that value until the process load or set point changes. Application 5.4 describes the response of a process incorporating integral gain.

APPLICATION 5.4   INTEGRAL CONTROLLER MODE

Process: Variable Air Volume AHU, Static Pressure Control

Components: 1. Normally closed inlet vane dampers
2. A direct acting differential pressure sensor
3. A reverse acting controller incorporating the integral mode

Process Requirements: As the static pressure in the duct increases, the inlet vane damper blades must modulate toward their closed position to reduce the mass transferred into the duct and consequently, reduce the static pressure at the sensor's location. After the response is completed, the process error will equal zero and the control point will equal the set point of 1.4 inches of water.

Controller Response: As the static pressure in the duct increases, the signal from the direct acting sensor also increases. The controller's integral element computes the time weight error, multiplies it by the integral gain and uses the resulting product to modify the controller's output signal. As its output signal changes, the time weight error decreases, reducing the size of the subsequent changes that are occurring in the output signal. This response continues until the error is reduced to zero. The accompanying figure depicts the error being driven to zero.

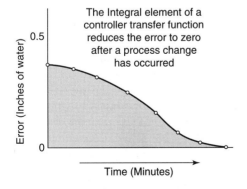

Application 5.4   Supply Air Static Pressure Controller Response

There are several differences between the intent, function and response of proportional and integral elements of a controller's transfer function. The following summation outlines the significant differences:

1a. Proportional gain is used to establish a balance between the mass or energy being transferred within a process and its present load based upon a fixed ratio.

1b. Integral gain is used to eliminate the error between the control point and the set point after a change in the process has occurred.

2a. The proportional gain modifies the output signal in proportion to the magnitude of the change in the process error.

2b. The integral gain modifies the output signal in proportion to the magnitude of the time weighted error (error x length of time).

3a. The proportional gain element moves the final controlled device to a position where the mass or energy transferred within a process balances the process load. Consequently, a small offset typically exists between the process control point and its set point.

3b. The integral gain element moves the final controlled device to a position where the mass or energy transferred within a process exceeds the quantity required to balance the process load until the process control point and its set point are equal (error equals zero).

4a. Proportional gain generates one change in the output signal in proportion to the size of the error. The output signal stays at that value until another load change occurs.

4b. Integral gain continuously modifies the output signal in proportion to the size of the time weighted error. The output signal continues to change as the error is driven to zero. Once the error equals zero, the controller's output signal stays at its last commanded value until another load change occurs.

5.4.2 The Differential Gain of a Controller Transfer Function

Differentiation is another mathematical procedure that is used to calculate the instantaneous rate of change of a variable. In HVAC controller applications, differentiation is used to determine the instantaneous *rate of change* of the error with respect to time. The *derivative* of the error indicates how fast the process error is changing. A large value indicates the error is changing rapidly. Conversely, a small rate of change indicates the process is changing very slowly or is relatively stable.

Transfer functions incorporating a derivative element require an additional gain parameter called derivative gain. The *derivative gain* adjusts the magnitude of the output signal in *proportion* to the rate of change of the process error. The rate of change of the error is multiplied by the derivative gain to determine the amount of change in the controller's output signal. Using the derivative element in a controller transfer function commands the final controlled device to move more quickly, thereby reducing the effect the load change has on the control point.

Differentiation is incorporated into a controller transfer function in order to adjust its output signal in response to the speed at which the error is changing. In these applications, the rate of change of the error is used to predict the total change in the control point based upon its present rate of change. When the load is changing quickly, the change in the output signal produced by the derivative element of the transfer function is greater than when the error is changing at a slower rate. When the error is changing slowly, its rate of change is small and consequently, the contribution of the derivative element of the transfer function to the change in the controller' output signal is also small. Finally, if the error is not changing (constant) the contribution of the derivative element to the change in the controller's output signal is zero.

Differentiating the rate of change and using the result to adjust the controller's output signal allows the loop to anticipate the effect of a load change before the entire change occurs. Therefore, the derivative element of the transfer function has the effect of limiting the growth of the error that results from a fast load change. The derivative element is also called *anticipatory* or *rate* mode. Application 5.5 provides an example of a process incorporating a derivative timing element in its controller transfer function.

APPLICATION 5.5

Process: Zone Temperature Control Loop that Modulates Heat Transfer Using a Chilled Water and a Hot Water Coil to Prevent the Control Point From Deviating More than 0.5 °F

Components: 1. One hot and one chilled water coil
2. One normally open hot water valve and one normally closed chilled water valve
3. One direct acting discharge air temperature sensor
4. One controller with derivative mode in its transfer function

Operation: As the zone temperature increases, the heat transferred by the hot water coil must be decreased to maintain the control point within acceptable range of the set point. If the hot water valve is closed and the temperature continues to increase, the chilled water valve begins to modulate open. If the load change occurs at a fast rate, the hot and chilled water valves must respond quickly to prohibit the room temperature from drifting beyond the $+/-0.5$ °F limits as the loop responds.

Controller Response: As the zone temperature increases, the sensor's signal to the controller also increases. The controller's output signal must respond quickly to close the hot water valve and, if necessary, begin to open the chilled water valve. A direct acting controller with derivative gain is required to speed up response during intervals of fast load changes. When the load changes quickly, the controller's output signal increases a larger amount to reposition the valves quickly. Responses due to slow load changes will be very similar to the response of a transfer function without derivative gain because the contribution of the derivative term is very small.

The derivative gain element of a controller's transfer function is the least used of the three controller gains because its response is based upon a fast load change. In HVAC processes, most changes occur at a rate that is slow compared to the mass or energy transfer rate. Therefore, these changes are easily managed by a controller transfer function that does not incorporate a derivative element. In those applications where the process tolerances are small and the control point must be kept close to the set point during any load change, the derivative element can be used.

5.5

SUMMARY

Transfer functions are equations that describe how the output signal of a control device changes in its size, phases and timing in response to a change in its input signal. The size element of the transfer function produces changes in the magnitude of the output signal based upon the size of the change in its input signal is called the *proportional gain* or *sensitivity*. The phase component describes the change in the *direction* of the output signal with respect to the change in the direction of the measured variable and is called the *action* of the device. Timing elements only apply to controller transfer functions. They modify the controller's output signal as a function of the time-based characteristics of the change in the process error. The integral element responds to the time weighted error, modifying the controller's output signal until the process error is reduced to zero. The derivative element modifies the controller's output signal as a function of the rate of change of the process error. It acts as an anticipatory element, reducing its maximum size of fast changing process errors.

Each element of a controller's transfer function, proportional gain, integral gain, and derivative gain modifies the output signal. A technician adjusts these gain values so the control loop's response to changes in the process matches the requirements of the process. As the value of a gain element increases, its contribution to the change in the output signal also increases. Understanding how each element in the transfer function alters the output signal allows the technician to make the necessary adjustments to a loop to optimize comfort, safety, and the efficiency of the process.

5.6

EXERCISES

Determine if the following statements are true or false. If any part of the statement is false, the entire statement is false. Explain your answers.

1. Transfer functions mathematically describe how an output signal changes in response to a change in the input signal.
2. Size, magnitude, and phase are the three elements of a transfer function.
3. Sensor and final controlled devices have fixed transfer functions.
4. The proportional gain of a transfer function generates changes in the magnitude of the output signal.
5. The action of a controller is determined by the phase element of its transfer function.
6. The integral gain changes the output signal based upon how large the error is and how long it exists.
7. As the size of the error increases, the contribution of the reset element of the controller's transfer function also increases.
8. The derivative gain changes the output signal of a controller based upon the rate of change of the process error.
9. Small changes in the error produce large changes in the output signal of controllers using proportional, integral, and derivative gains in their transfer function.
10. Integral and derivative gains are also known as reset and rate gains.

Respond to the following statements and questions completely and accurately, using the material found in this chapter.

1. What is a transfer function?
2. What purpose does the proportional gain element of a transfer function serve?
3. What information does the phase component of a transfer function provide to the technician?
4. Name two elements of a transfer function that respond to the dynamic or changing characteristics of the input signal.
5. What applications require the integration element in their controller's transfer function?
6. What applications require the derivative element in their controller's transfer function?
7. What application requirements would incorporate both the integral and derivative elements in their controller transfer function?
8. If the integral gain of a controller's transfer function is increased, will the length of time required to reduce the error to zero increase or decrease? Explain your answer.

9. If the loop response becomes unstable, are the gains in the transfer function most likely too large or too small? Explain your answer.

10. What does the proportional gain of a device indicate?

11. What effect does changes in the value of the gain have on the sensitivity of a device?

12. Will the contribution of the integral element of a controller transfer function increase or decrease as the size of the error increases? Explain your answer.

13. Which gain element is used to minimize the loop offset during fast changing errors?

14. What is the reset interval?

15. What effect does the derivative element of a transfer function have on the change in the output signal when the error is changing slowly? Explain your answer.

16. If the reset interval is increased, does the controller take a longer or shorter time interval to reset the control point back to set point?

17. If the derivative gain is decreased, will the error be reduced to a smaller value? Explain your answer.

18. If the sensitivity of a device is high, does it produce a large or small change in the output signal in response to a $\frac{3}{4}^{\circ}$ change in the measured variable?

19. Why do sensors and final controlled devices omit a timing element from their transfer functions?

20. What controller transfer function reacts quickly to changes in the error and reaches a steady state condition without a process error remaining?

CHAPTER 6

Applying Transfer Functions

HVAC control technicians spend a fair amount of their time installing and calibrating control loops. The calibration procedures often require a technician to tune the parameters of a component's transfer function so the loop responds correctly to process changes. To become proficient in performing calibration tasks, a control technician must be able to derive and apply the transfer function of each control device in a loop. Without the use of transfer function equations, system troubleshooting and analysis becomes a trial and error procedure. Generally, the transfer function of a device is developed using the information found on the component's nameplate along with the operating parameters of the process. This chapter details the development of the equation of a typical control transfer function. These procedures form the foundation upon which all calibration and analysis of modulating control loops is built.

OBJECTIVES

Upon completion of this chapter, the reader will be able to:

1. Describe the parameters and variables that make up the transfer function.
2. Develop a transfer function for a sensor.
3. Develop a transfer function for a controller.

4. Develop a transfer function for a final controlled device.
5. Find the unknown values in a transfer function using a calculator.

6.1

LINEAR CONTROL SYSTEMS

Linear control components and systems produce changes in their output signals that are proportional to the changes that occurred in their input signals.

The term *linear* is often used in the description of the response of control devices, loops and processes. When the response of a linear control device, loop or process is plotted on a graph, it appears as a straight line segment. When the transfer function of a linear device is written, it has the same general format as the formula used to define a straight line segment. These are the reasons control devices and processes are described using the term *linear.*

A linear response is synonymous with the phrase *proportional response.* As stated in Chapter 5, the term proportional indicates that there is a fixed relationship between the input and output signals of a control device. This relationship is used to determine if the output signal of a sensor or controller is correct for a given input signal. It is also used to quantify the flow through a final controlled device and to develop set point reset schedules. Processes also can be described as linear if a change in the flow rate of the control agent produces a proportional change in the measured variable of the process. The linear characteristic of a process is used to predict a processes' response to a change in its load.

6.1.1 The Equal Percentage of Change Characteristic of Linear Systems

The proportionality relationship between a change in the input of a linear system and the corresponding change in its output response yields another serviceable relationship. When a device or process is linear, the *percent of change* in its output is equal to the percent of change that occurred in its input. All linear devices and processes experience this relationship of equivalent *percentage* changes between their input and output signals. Therefore, when a linear process experiences a 10% drop in the control point, the loop's linear steam valve will open an additional 10%, increasing the heat transferred into the process by 10%. This response occurs because both the process and loop's component responses are linear. Carrying this relationship one step further, when a linear device or process experiences a small change in its input, the corresponding change in the output is also small. Conversely, when a large input change is introduced into the process, a large response follows. This relationship exists throughout the operating range of the process and is used by control technicians to quickly analyze operating control systems.

6.2

THE GENERAL FORM OF THE LINEAR TRANSFER FUNCTION EQUATION

The general form of a transfer function is an equation that is the basis for developing transfer functions for all linear control devices, reset schedules, and processes. Most of the equations used by a control technician are based upon the equation:

Output = (Δ Input) $\times$ Proportional Gain + Bias

The first step in developing any transfer function is to understand the meaning of each of the terms found in the general equation. The following subsections detail this information, describing the characteristics of each of the *parameters* and *variables* found in the general form equation. Keep in mind, a *variable* is a term in the equation that changes in response to changes occurring in the process. These changes include variations in temperature, pressure, humidity, level, and other measured variables. A *parameter* is a term in the equation that remains constant (unchanging) throughout the operation of a device. In the general form of a transfer function equation, the proportional gain and the bias are the unchanging parameters and Output and Δ Input are the variables.

6.2.1 The Transfer Function Variables

The variable *Δ Input* represents the amount of change that occurred in the measured variable. In sensor transfer functions, this variable may be a process temperature, pressure, level, or relative humidity. In controller transfer functions, Δ Input represents the change in the process error. The Greek letter Δ (delta) is used to symbolize the phrase "change in." Therefore, *Δ Input* denotes the *change in the input variable.*

To calculate the magnitude of Δ Input, the minimum value of the control devices' input range is subtracted from the current value of the measured variable (see Example 6.1). Therefore, in temperature sensor applications, the minimum value of the sensor's input temperature range is subtracted from the process control point; in controller equations the change in the input signal is equal to the process error and in final controlled devices, the minimum value of the actuator's input signal is subtracted from the controller's current output signal.

Example 6.1

An outside air temperature sensor has an operating range of -20 to 120 °F. The present temperature at the sensor's location is 67 °F. Calculate the value of Δ Input.

Solution:

Δ *Input = (Sensor Input − Minimum Measurable Value)*
Δ *Input = [67 − (−20)]*
Δ *Input = 87 °F*

6.2.2 The Proportional Gain

The *proportional gain* is the parameter in a transfer function that defines the amount of change in the output signal that occurs for each unitary change in the process error. The gain is equal to the ratio:

$\dfrac{\Delta \text{ Output Signal}}{\Delta \text{ Input Signal}}$

where:
   1. Δ Output is the output signal span of a control device.
   2. Δ Input is equal to its input signal span.

Taking the output signal span and dividing it by the input signal span equals the device's proportional gain (see Example 6.2).

Example 6.2

> The outside air temperature sensor used in Example 6.1 has an input operating range of −20 to 120 °F and an output signal range is 3 to 18 psi. Calculate the value of the sensor proportional gain.

Solution:

$Proportional\ gain = \dfrac{Sensor\ output\ span}{Sensor\ input\ span}$

$Proportional\ gain = \dfrac{(18 - 3)}{[120 - (-20)]} = \dfrac{(15)}{(140)} = 0.107\ \dfrac{psi}{°F}$

6.2.3 The Output Signal Bias

The *bias* is a parameter found in the transfer function of sensors, controllers, and final controlled devices. It indicates the initial value or starting point of an operating signal or range. The bias is equal to the value of the output signal of a control device whenever the *change in its input signal is equal to zero* (see Example 6.3).

Example 6.3

> The outside air temperature sensor used in the previous examples has an operating range of −20 to 120 °F and generates a 3 to 18 psi output signal. Determine the value of the sensor bias.

Solution:

The bias of the sensor is equal to the value of the output signal when the change in the input signal is zero. Therefore, the sensor generates an output signal equivalent to the bias whenever the input value is −20°F. The output signal when the temperature is −20° is equal to 3 psi.

6.3

DEVELOPING THE TRANSFER FUNCTION FOR A SENSOR

The transfer function for a linear sensor is the simplest linear control equation to develop. Its has two parameters, proportional gain and bias, and two variables, Δ Input and Output. The values of the parameters are calculated using the information contained within the input and output signal ranges of the device as outlined in the previous examples. The input and output ranges of a sensor are typically found on its packaging or dust cover label, or noted on the control prints. These ranges can be found by referencing the sensor's part number with the manufacturer's parts catalog. Once the ranges are found, the proportional gain and the bias are determined using the methods shown in the following subsections.

6.3.1 Sensor Gain

The first parameter to calculate is the sensor's proportional gain. The ratio is equal to the maximum *design* change in the output signal divided by the maximum *design* change in the input signal. The design values are those numbers written on the sensors name plate. They indicate the range of the device where a linear response is guaranteed. If a sensor is installed in the field and exposed to conditions that exceed its design operating range, the output signals it generates will not be guaranteed to remain accurate.

The maximum design change in a signal is called its *span*. It is equal to the difference between the maximum and minimum values of a signal's operating range. For example, a sensor that is designed to measure temperatures within a range of 40 to 240 °F and generate a corresponding output signal of 3 to 15 psi, has an input span of 200 °F and an output span of 12 psi. Since the gain is equal to the change in the output signal divided by the change in the input signal, the formula for proportional gain can be written in terms of the sensor's spans:

$$\text{Gain} = \text{output span} \div \text{input span} = \frac{\text{output span}}{\text{input span}}$$

Example 6.4

An outside air temperature sensor has an operating range of 20 to 120 °F and has an output signal range of 4 to 20 mA. Calculate this sensor's proportional gain.

Solution:

$$Proportional\ gain = \frac{\Delta Output}{\Delta Input}$$

$$Proportional\ gain = \frac{\Delta\ output\ span}{\Delta\ input\ span} = \frac{16}{100} = 0.16 \frac{mA}{°F}$$

In Example 6.4, the minimum value of a range was subtracted from its maximum value to calculate the span. Next, the output span is divided by the input span to determine the device's proportional gain. The sensor in this example has a proportional gain of 0.16 mA per °F. This indicates that each one degree change in temperature produces a corresponding 0.16 mA change in the sensor's output signal. The gain should always be labeled using output units over input units.

6.3.2 Sensor Bias

The parameter *bias* must be determined to complete the sensor's transfer function. The output signal bias indicates the amount of signal that must be added to the *change* produced by the elements in the transfer function to determine the *exact* value of the output signal as opposed to the *amount* of change that occurred in the output signal in response to the change in the input signal. The sensor's output signal bias always equals the minimum value of its output signal (see Example 6.5).

Example 6.5

An mixed air temperature sensor has an operating range of 20 to 120 °F and generates a 4 to 20 mA output signal. Determine the sensors output bias.

Solution:

Bias = minimum value of the sensors output range, therefore,
Bias = 4 milliamps

6.3.3 Step-By-Step Development of a Sensor Transfer Function

The following data was listed on a sensors label. This data will be used to develop the transfer function of the sensor.

Sensor Data:

Input Range: 40 to 150 °F
Output Range: 3 to 15 psi

Task: Develop the transfer function of this sensor.

Solution:

Step 1 Calculate the sensors input and output spans.

$$input\ span = 150 - 40 = 110\ °F$$
$$output\ span = 15 - 3 = 12\ psi$$

Step 2 Calculate the sensors gain.

$$Proportional\ gain = \frac{output\ span}{input\ span}$$

$$Proportional\ gain = \frac{12\ psi}{110\ °F} = 0.109\ \frac{psi}{°F}$$

The proportional gain indicates the output signal changes approximately one-tenth of a pound of pressure for each one degree change in the measured temperature. Therefore, if the temperature increases 10 °F the output will increase 1.09 psi (10 × 0.109). Conversely, if the temperature decreases 20 °F, the output decreases 2.18 psi (−20 × 0.109).

Step 3 Determine the sensors bias.
The bias is equal to the minimum value of the sensors output signal range. Therefore,

$$Bias = 3\ psi$$

Step 4 Write the transfer function using the values from steps 2 and 3.

$$Output = Proportional\ gain \times (\Delta Input\ °F) + bias$$

$$Output = 0.109\ \frac{psi}{°F} \times (\Delta\ Input\ °F) + 3\ psi$$

6.3.4 Applying the Transfer Function to Calculate the Output Signal

After a component's transfer function has been developed, it can be used to calculate the output signal for any input condition that falls within the sensor's linear operating range. Technicians use this transfer function to analyze the operation of a loop sensor. Remember, if the measured variable exceeds the design operating range of the sensor, the output cannot be accurately determined so it is important that the sensor remain calibrated.

To check a sensor for proper operation, an instrument of known accuracy is placed next to the sensing element of the sensor that is being evaluated. This location permits the calibration instrument to measure the same conditions as the sensor. To determine the value of Δ Input, the value of the measured variable (as indicated by the calibration instrument) is subtracted from the minimum value of the sensor's input range. Next, the value of Δ Input is entered into the transfer function to calculate the magnitude of the corresponding output signal. Once the output signal is calculated (using the measured value from the instrument), it is compared to the actual signal being generated by the sensor. If both values are within reasonable range of each other, the sensor is considered to be calibrated and no further adjustments are necessary (see Example 6.6). If the values of the calculated and actual output signals differ significantly, the technician has the option of calibrating the sensor or replacing it. Remember, a process cannot be maintained at its set point if its sensor is out of calibration.

Example 6.6

A direct acting temperature sensor measures conditions within a linear range of 20 to 120°F and generates a proportional output signal of 3 to 15 psi.

Sensor Data:

Input Range: 20 to 120 °F
Output Range: 3 to 15 psi

Task: Calculate the sensor's output signal at 60 °F.

Solution:

Step 1 Write the transfer function.
$$Output = Proportional\ gain \times (\Delta Input\ °F) + bias$$
$$Output = 0.12\ \frac{psi}{°F} \times (\Delta Input\ °F) + 3\ psi$$

Step 2 Calculate the change in the input signal for 60°F.
To calculate Δ Input, subtract the measured value (control point) from the minimum value of the sensors design input range.
$$\Delta Input = 60 - 20$$
$$\Delta Input = 40\ °F$$

Step 3 Enter the value of Δ Input into the transfer function and solve for the output signal.
$$Output = Proportional\ gain \times (\Delta Input) + bias$$
$$Output = 0.12\ \frac{psi}{°F} \times (40\ °F) + 3\ psi = 7.8\ psi$$

Step 4 Check the answer for accuracy.

No problem is completely solved until the answer has been checked for accuracy. To determine if the output for a linear device is correct, a quick check can be performed. The check is based on the proportional relationship that exists between the change in the input signal and the corresponding change in the output signal of linear devices. To perform the check, calculate the percentage of change that occurred in the input and output signals and compare them to each other. In linear devices, when the math is performed correctly, the percentage of change for the input and output signals will be equal.

For this example, the input changed 40°F of the sensor's 100°F input span. The percentage change in input signal is equal to the amount of change divided by the span:

40°/100° = 0.40 or 40%

The output signal changed 4.8 psi (7.8 − 3) out of the psi output signal span. The percentage of change in the output signal is:

(7.8 − 3)/12 psi = 0.40 or 40%

Since both signals changed 40% and the associated mathematical computations were performed correctly, the output signal of 7.8 psi is correct. To complete the calculation, be sure all units are shown in the equations. By canceling the units in the equation, the technician can be sure the variables and parameters were correctly placed in the formula.

6.3.5 Sensor Action

In the previous examples, the action of the sensor was noted as direct acting. Recall from Chapter 5 that the action of a control device defines the relationship between the direction of change in the input signal and the corresponding direction of change in the output signal. The action of a sensor is determined during the manufacturing process and is either *direct acting* or *reverse acting*.

A direct acting sensor has signals that change in the same direction. When the input signal increases, the output signal also increases. As the input signal decreases, the output signal also decreases. Reverse acting sensors have signals that change in opposite directions. As their input signals increase, their output signals decrease. Conversely, as an input signal decreases, the output signal increases.

The output signal of a direct acting sensor is equal to the minimum value of the sensors output range when the sensor is measuring a process variable equal to the minimum value of its input range. The output signal of a reverse acting sensor is equal to the maximum value of the sensors output range when the sensor is measuring a process variable equal to the minimum value of its input range. This relationship is the result of the signals changing in opposite directions. As the measured value of a direct acting sensor increases, the output increases. When the measured value of a reverse acting sensor increases, the output decreases. Keeping these relationships in mind allows the technician to quickly evaluate the operation of a sensor based on the process control point. Figure 6.1 shows a graphic representation of the signal relationships for direct and reverse acting sensors.

The action of the sensor is indicated by the values used for the gain and bias parameters in the sensor's transfer function. If a sensor is direct acting, as in all the previous examples, the gain is a positive number. If a sensor is reverse acting, the gain becomes a negative number.

The action of the sensor also determines the value used as the sensor's bias. Recall, the bias is equal to the output signal when the change in the input signal is equal to zero. In the transfer function for a direct acting sensor, the bias is equal to the minimum value of the sensor's output signal range when the change in the input signal is zero. This relationship is not true for a reverse acting device. The

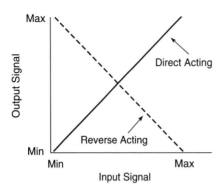

Figure 6.1   Sensor Action

bias of a reverse acting sensor is equal to the maximum value of the sensor's output range when the change in the input signal is equal to zero. This is easy to see when viewing Figure 6.1. Note the output signal equals the maximum value of the output signal range when the input signal is at its minimum value (see Example 6.7).

Example 6.7

A reverse acting relative humidity sensor measures the relative moisture content of the air within a linear range of 20% to 80%. It generates a proportional output signal of 0 to 10 volts. As the relative humidity increases, the output signal generated by the sensor decreases.

Sensor Data:

Input Range: 20% to 80%
Output Range: 0 to 10 volts

Task: Calculate the sensors output signal at 65% rh.

Solution:

Step 1   Write the transfer function.

$$Output = Proportional\ gain \times (\Delta\,Input) + bias$$

$$Output = (-0.167)\,\frac{v}{\%\ rh} \times (\Delta\,Input\ \%\ rh) + 10\ volts$$

Step 2   Enter the value of Δ Input into the transfer function and solve for the output signal.

$$Output = Proportional\ gain \times (\Delta\,Input) + bias$$

$$Output = -0.167\,\frac{volts}{\%\ rh} \times (45\%\ rh) + 10\ volts = 2.49\ volts$$

Step 3. Check the answer for accuracy.

For this example, the input changed 45% rh of the possible 60% input span. The percentage change in input signal is:
45%/60% = 0.75 or 75%
The output signal changed 7.5 volts (10 − 2.49) out of the 10 volt output signal span. The percentage of change in the output signal is:
7.5 volts/10 volts = 0.75 or 75%

Since both signals changed 75% and the associated mathematical computations were performed correctly, the output signal of 2.49 volts is correct.

6.4

THE CONTROLLER TRANSFER FUNCTION

The procedure used in the development of the sensor transfer function applies to all linear devices. Slight differences exist between the transfer functions of different control loop components but the procedure is fundamentally the same. Any differences are associated with the *definition* of the parameters that make up the transfer function of a controller and final controlled device. The following sections outline the development and use of a controller transfer function.

6.4.1 The Process Throttling Range

The first difference between a sensor and controller transfer function is the method used to calculate the proportional gain. Although the nomenclature Δ *Output/Δ Input* remains the same, the procedure used to determine the value of Δ Input is different from that used for a sensor transfer function. Δ Input for a controller still defines the maximum allowable (linear) range of the input signal, but the input signal's range is not a characteristic of the construction of the controller as it is for a sensor. Consequently, Δ Input cannot be found on the label of the controller as it can with a sensor.

When calculating the controller's gain, the value of Δ Input is based on the process requirements. It is equal to the *desired operating range* of the process. The calibrating technician selects a value for Δ Input that defines the linear operating range of the process. This range of operation sets the allowable minimum and maximum values of the control point. Under normal operation, the control point will not exceed these limits. This desired range of linear control is called the processes' *throttling range.*

The throttling range defines the acceptable operating range of the process. It indicates the full range of input conditions that will modulate the final controlled device from its minimum (closed) position to its maximum (open) position. At one limit of the throttling range, the controller output signal is equal to its minimum value. As the control point moves toward the other limit of the throttling range, the controller's output signal modulates toward its maximum value.

The units used to describe the throttling range are the same as the units of the controlled variable because the throttling range is the linear range of the control point of the process. As depicted in Figure 6.2, the throttling range is usually divided by two, with one half added to the set point to obtain the maximum limit of the process' linear range. The remaining one half is subtracted from the set point, yielding the low limit of the linear range of the process. The control point can be equal to any value within this range and the control loop is considered to be *"in control."*

If the control point moves beyond the limits of the throttling range, the final controlled device will be unable to respond to the additional change in load. This occurs because the final controlled device is already at the limit of its operating range when the control point equals a limit of the throttling range. Consequently, the control loop will no longer be able to balance the energy transfer with the load and the loop begins to operate *out of control.* It will not return to its linear control range until a change in the process load drives the control point back within the throttling range.

A typical throttling range for a supply air temperature control loop is 4 °F. This value indicates that the control point can equal within the range of the set point $+/-2°$. Figure 6.3 depicts a graph of a 4° throttling range. Whenever the control point is less than 53° or greater than 57°, the control loop is operating out of control and the final controlled device is positioned either 100% open or 100% closed. Application 6.1 describes the throttling range of a mixed air temperature process.

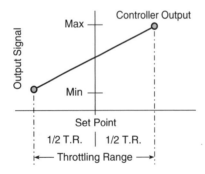

Figure 6.2 Process Throttling Range

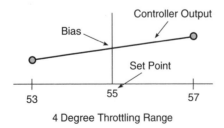

Figure 6.3 Throttling Range

APPLICATION 6.1

Process: Mixed Air Temperature Control Loop

Objective: To maintain the temperature of the mixed air at the sensor location within the throttling range of the process. The throttling range is 6° F and the set point is 62° F.

Response: The direct acting controller generates a linear output signal of 0 to 10 volts. The signal is sent to normally closed (NC) outside air and normally open (NO) return air damper actuators. As the mixed air temperature increases, the controller's output signal increases, opening the outside air dampers more while proportionately closing the return air dampers. The increase in the volume of cooler outside air reduces the mixed air temperature.

Throttling Range: The throttling range of 6 °F defines the linear limits of the process to a minimum value of 59 °F and the maximum value of 65 °F. At 59 °F, the controller's output signal is 0 volts, the outside air dampers are fully closed and the return air dampers are 100% open. At 65 °F, the controller's output signal is 10 volts, the outside air dampers are fully open and the return air dampers are completely closed. If the mixed air temperature falls below 59 °F, the outside air dampers cannot be closed any further so the loop operates out of control. It will return back into control once the return or outside air temperature increases sufficiently, bringing the control point above 59 °F.

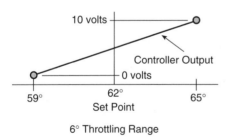

Application 6.1

6.4.2 The Proportional Band of a Process

Proportional band is another term that is also used to describe the linear operating range of a process. The proportional band indicates the *percent of the control loop's sensor span* required to stroke the final controlled device from 0% to 100% of its operating range. The units of proportional band are always expressed as a percentage (%). This differs from the units of the throttling range which has the same units as the controlled variable (°F, % rh, inches of water, etc). Proportional band is calculated using sensor's input span and the process throttling range. It is calculated by dividing the throttling range by the sensor's span (see Example 6.8).

$$Proportional\ Band = \left(\frac{Throttling\ Range}{Sensor\ Span}\right) \times 100$$

Although proportional band is often used interchangeably with throttling range, there is an important difference between their definitions. The difference is based upon the units used by the throttling range and those used for the proportional band. The units for the throttling range are the same as those of the controlled variable. The units of proportional band are always percent of sensor span. If a linear range of a processes' proportional band is labeled in units of the controlled variable, the term is being misapplied.

Example 6.8

A room temperature process uses a direct acting 45 to 90 °F sensor. The throttling range of the process is equal to the set point $+/- 1.5$ °F. Calculate the proportional band (PB) of the process.

$$PB = \frac{throttling\ range}{sensor\ span} \times 100$$

$$PB = \frac{3\ °F}{45\ °F} \times 100 = 6.7\%$$

6.4.3 The Controller's Proportional Gain

The controller's gain is calculated in the same manner as the sensor's gain. Gain is equal to Δ Output/Δ Input, where Δ Output is equal to the controllers output signal span and Δ Input is equal to the *throttling range* of the process. The throttling range is used in this calculation because it is the permissible change in the input signal that maintains a linear response. Proportional band cannot be used because it has the incorrect magnitude and units (%). If the proportional band is given in a problem, it must first be converted into the throttling range before the gain can be calculated (see Example 6.9).

Example 6.9

A hot water temperature control loop maintains the temperature of the water leaving a steam to hot water heat exchanger within a throttling range of 8 °F. A reverse acting controller modulates a NC steam valve to maintain a set point of 150 °F by generating a 3 to 13 psi output signal.

Task: Calculate the controller's gain.

Solution:

$$Proportional\ gain = \frac{\Delta\ output}{\Delta\ input}$$

$$Proportional\ gain = \frac{\Delta\ output}{throttling\ range} = \frac{15\ psi}{8\ °F} = 1.9\ \frac{psi}{°F}$$

A complementary relationship exists between the controller's proportional gain and the processes' throttling range. The throttling range defines the limits of the linear range of the process along with the change in the control point that is required to stroke the final controlled device from 0% to 100%. The proportional gain indicates the amount of change in the output signal that occurs for a unit change in the control point. Therefore, as the controller's proportional gain increases, its throttling range decreases. This occurs because the controller's output signal must position the final controlled device from 0% to 100% over a smaller change in the control point. A smaller proportional gain requires a larger change in the control point in order to generate the full range of the controller output signal. Therefore, as the proportional gain decreases, the throttling range increases. This relationship plays an important part in the stability of a control loop and will be covered in greater depth in subsequent chapters.

6.4.4 The Controller Bias

The controller's bias is the magnitude of the output signal generated by the controller when the process error is equal to zero.

Another important difference between a sensor's and a controller's transfer functions is the value used for the controller's bias. Recall, the bias of a sensor transfer function is based upon the operating range of its output signal. The bias of a controller is based upon both the controller's output signal range and the installed characteristics of the final controlled device.

The controller's bias is selected by the control technician during the calibration procedure. Initially, the technician typically sets the bias equal to 50% of the final controlled device's operating range so the final controlled device will be 50% open when the process error equals zero. For example, a loop using a damper actuator that has a range of 4 to 20 mA will initially have its controller's bias set to 12 mA, positioning the damper 50% open when the error equals zero (see Example 6.10). The following formula is used to calculate the initial bias of a controller:

$$Bias = \left(\frac{FCD\ Input\ Signal\ Span}{2}\right) + Minimum\ Value\ of\ the\ FCD\ Input\ Signal$$

or

$$Bias = \left(\frac{Controller's\ Output\ Signal\ Span}{2}\right) + Controller's\ Minimum\ Output\ Signal$$

Example 6.10

A hot water temperature control loop maintains the temperature of the hot water leaving a steam to water heat exchanger within the process throttling range of 8 °F. A reverse acting controller modulates a NC steam valve to maintain a desired set point of 150 °F by generating a 3 to 13 psi output signal.

Task: Calculate the controller's bias.

Solution:

$$Bias = \left(\frac{FCD\ Output\ Span}{2}\right) + FCD\ Minimum\ Signal$$

$$Bias = \frac{13 - 3}{2} + 3 = \frac{10}{2} + 3 = 8\ psi$$

The operational characteristics of a controller and the definition of a throttling range form the basis for the selection of an initial value for the controller bias at 50% of the actuator's operating range. Remember, a controller's function is to generate a response that minimizes the process error. At any instant in time, the control point may be greater than or less than the set point. To perform correctly, a controller must be able to modulate the final controlled device in either direction, open or closed, in relation to the current control point. As the value of the control point increases, the final controlled device is modulated in one direction. When the control point decreases, the controller modulates the final controlled device in the opposite direction. The 50% signal selection for the controller bias is based on this 50/50 split of the process throttling range. Dividing the throttling range by two and using the set point as the midpoint of the linear range of the process provides the range of control that is equal in both directions. The final controlled device can move equally in either direction as the control point moves within the throttling range.

In applications where two final controlled devices are sequenced by their actuator springs, the bias is typically set to the value between the two spring ranges, where both flow devices are closed when the process error equals zero. This creates a dead band or interval where no energy or mass transfer takes place when the control point is equal to the set point. Setting the bias to this value prevents simultaneous mixing of energy sources. Properly selecting the controller's bias, based upon the installed characteristics of the system improves the operational efficiency of the process (see Example 6.11).

Example 6.11

Task: Calculate the controllers bias for the following applications:
   a. single final controlled device . . . 3 to 18 psi
   b. single final controlled device . . . 4 to 20 mA
   c. dual final controlled devices . . . 2 to 9 psi and 11 to 18 psi

Solution:

a. $\dfrac{15-3}{2}+3=9\ psi$

b. $\dfrac{20-4}{2}+4=12\ mA$

c. $\dfrac{11-9}{2}+9=10\ psi$

6.4.5 The Controller Transfer Function

The parameters of the controller transfer function are used to develop the transfer function of a linear controller. The format of a controller transfer function is identical to that used for a sensor transfer function. In both applications, an input stimuli is multiplied by the proportional gain and added to the bias to determine the magnitude of the output signal. The only difference between sensor and controller transfer functions is the input stimulus. In a sensor transfer function, Δ Input is equal to the difference between the measured value and the value of the sensor's input range when the output signal is at its bias. In a controller transfer function, Δ Input is equal to the *process error*. Changes in the error signal provide the stimuli that causes the controller to respond.

Recall, from Chapter 2, the formula for calculating the process error is:

error = set point − control point

The negative sign in front of the control point signifies the sensor's signal is fed back to the controller to reduce error when it is subtracted from the set point. This negative sign complicates the development of the controller transfer function by requiring the insertion of a (-1) constant into the equation. The negative one inverts the sign of the error so the transfer function generates the correct output signal value. Without this constant, the controller's transfer function produces a response with the opposite phase or action. To minimize computational errors due to the omission of the (-1) constant into the transfer function, the equation for calculating the error is modified to:

error = control point − set point

This slight adjustment, multiplying the original error formula by (-1), minimizes the confusion encountered when using the transfer function in the field.

6.4.6 Step-by-Step Development of a Controller Transfer Function

Developing a controller transfer function requires information about the operating characteristics of the loop sensor, final controlled device, and process. This information is used by the technician to calculate the parameters of the controller transfer function. Step-by-step Examples 6.12 and 6.13 go through the development and use of a controller transfer function.

Example 6.12

Using the following loop and process data, develop the transfer function for a controller.

Process:

Air handling unit, supply air temperature control.

Sensor Data:

A direct acting temperature sensor is designed to measure temperatures within a range of 20 to 120 °F, generating a direct acting, linear pneumatic output signal with an output range of 3 to 15 psi.

Controller Data:

A direct acting controller maintains the process within ±3 °F of the set point. It generates a linear 3 to 15 psi output signal.

Final Controlled Device Data:

The controller modulates two valves sequenced by their actuator springs, a normally open 3 to 8 psi hot water valve, and a normally closed 9 to 15 psi chilled water valve. The valves are sequenced so both valves cannot be open at the same time.

Tasks: 1. Develop the transfer function for the controller.
2. Calculate the controller proportional band.

Solution:

Step 1 Calculate the controller's operating spans.
$$Input\ span = (2 \times 3) = 6\ °F$$
$$Output\ span = 15 - 3 = 12\ psi$$

Step 2 Calculate the controller's proportional gain.

$$Proportional\ gain = \frac{\Delta Output}{Throttling\ range}$$

$$Proportional\ gain = \frac{12\ psi}{6\ °F} = 2\ \frac{psi}{°F}$$

The proportional gain indicates the output signal changes 2 psi for every one degree change in the control point. Therefore, if the temperature increases 3°, the output will increase 6 psi. If the temperature decreases 2°, the output will decrease 4 psi.

Step 3 Determine the controller's bias.
In this application, both valves will be closed when the control point is equal to the set point. The bias is equal to the pressure that will

close the hot water valve while keeping the chilled water valve closed. Therefore, the bias can be any pressure between 8 and 9 psi.
$$Bias = 8.5 \, psi$$

Step 4 Write the transfer function using the values from steps 2 and 3.

$$Output = Proportional \, gain \times (\Delta \, Input \, °F) + bias$$

$$Output = 2\frac{psi}{°F} \times (control \, point - set \, point \, °F) + 8.5 \, psi$$

$$Output = 2\frac{psi}{°F} \times (error \, °F) + 8.5 \, psi$$

Technicians use the transfer function to determine the magnitude of the controller output signal based upon the current value of the process control point. The set point is subtracted from the control point and inserted into the Δ Input field of the transfer function equation to calculate the output signal. The calculated output is compared with the actual output signal to determine the operating status of the control loop.

Example 6.13

System Overview:

The temperature of an interior zone is maintained through the use of a reheat coil located on the variable air volume box supplying the zone. A direct acting sensor with a range of 55 to 85 °F along with a controller are mounted on the zone's wall. The direct acting controller generates a 0 to 10 volt output signal to a normally open 0 to 5 volt hot water valve actuator and the 5 to 10 volt normally closed VAV box damper actuator. The zone temperature is maintained within a throttling range of 4° of the 72 °F set point.

Tasks: 1. Develop the controller transfer function.
2. Calculate the controller output signal at a zone temperature of 73.2 °F.

Solution:

Step 1 Calculate the controller's operating spans.
$$Input \, Span = 4 \, °F$$
$$Output \, Span = 10 - 0 = 10 \, volts$$

Step 2 Calculate the controller's gain.
$$Proportional \, gain = \frac{\Delta Output}{Throttling \, range}$$

$$Proportional \, gain = \frac{10 \, volts}{4 \, °F} = 2.5 \, \frac{volts}{°F}$$

Step 3 Determine the controller bias.
$$Bias = 5\ volts$$

Step 4 Write the transfer function using the values from steps 2 and 3.
$$Output = Proportional\ gain \times (\Delta Input\ °F) + bias$$
$$Output = 2.5\ \frac{volts}{°F} \times (error\ °F) + 5\ volts$$

Step 5 Calculate the controller's output signal when the temperature is 73.2 °F.
$$Output = Proportional\ gain \times (\Delta Input\ °F) + bias$$
$$Output = 2.5\ \frac{volts}{°F}\ (73.2 - 72.0) + 5\ volts$$
$$Output = 2.5\ \frac{volts}{°F} \times (1.2) + 5\ volts = 8\ volts$$

Examples 6.12 and 6.13 show that the transfer function for the controller, although slightly more difficult to develop, is used in the same manner as the transfer function of a sensor. Both equations are used in the analysis of operating control loops to determine if the loop components or the process is responding abnormally.

6.5

TRANSFER FUNCTIONS FOR FINAL CONTROLLED DEVICES

The transfer function for final controlled devices is not as widely used in field analysis of operating systems as the sensor and controller transfer functions. There are three reasons for this reduced emphasis. The first is based in the construction of the a pneumatic actuator, which is the most common actuator presently used in commercial and industrial HVAC control systems. Pneumatic actuators have no adjustments that can be made that can alter the starting and ending limits of its operating range. The actuator's operating range is based upon the feedback spring located inside the actuator housing. The spring has a non-adjustable linear constant that only permits the activator to operate within a specified range. The actuator spring range is chosen during the design phase of the application and is seldom changed after the device has been installed.

The second reason for the reduction in the application of the final controlled device transfer function is the actuator spring tends to change its operating range or *set* as it gets older. A 3 to 15 psi spring response may slowly change into a 2 to 13 psi response due to the constant compression of the spring and warm environment. To compensate for these changes and to increase the flexibility of a fixed spring actuator, another control device is available called a *positive positioning relay*. These devices are mounted to the side of an actuator and condition the input signal in a manner that permits the technician to custom set the bias (starting point) and operating span of the actuator. These devices add additional complexity to the control loop and may introduce instability into the response that may outweigh their advantages.

The third reason for the lack of emphasis on the final controlled device transfer function lies in the response of the connected flow apparatus. Valve and dampers typically have a slightly nonlinear response to changes in their actuator's position. The closer the actuator modulates toward either end of its operating range, the more nonlinear the response of the flow control device becomes. Consequently, their responses cannot be accurately represented by a linear equation transfer function. These response characteristics produce uncertainties in the linear transfer function of final controlled devices so it only generates an approximate value for the flow, not an actual value.

6.5.1 The Parameters in the Transfer Function for a Final Controlled Device

The transfer function for a final controlled device is developed using a procedure that is very similar to that used in the development of a sensor's transfer function. The span of the actuator spring range (typically the same as the span of the controller output signal) is used to calculate the change in the input signal. The range of the flow through the flow apparatus is used to calculate the change in the output signal. These two spans are then used to solve for the operating gain of the final controlled device. The proportional gain of a final controlled device's transfer function indicates the amount of change in the flow of the control agent that occurs for each unit change in the controller's output signal. The transfer function of the final controlled device is used to indicate the approximate flow into the process at a given controller output signal. The more linear the final controlled device, the better the estimate.

6.5.2 Step-by-Step Development of a Final Controlled Device Transfer Function

The following data was listed on the tags of a steam valve and actuator. This data will be used to develop the transfer function of the final controlled device.

Valve Data:

A linear steam valve is designed to modulate its flow from 0 pounds per hour to a maximum of 50 pounds per hour at the design pressure drop.

Actuator Data:

The magnetic flux actuator varies the stroke from 0 to 100% in response to a change in input signal of 4 to 20 mA.

Tasks: 1. Develop the transfer function of the final controlled device.
2. Calculate the approximate flow through the valve when the controller's output signal is 14 mA.

Solution:

Step 1 Calculate the final controlled device's input and output spans.

$$Input\ span = 20 - 4 = 16\ mA$$

$$Output\ span = 50\frac{lbm}{hr} - 0\frac{lbm}{hr} = 50\frac{lbm}{hr}$$

Step 2 Calculate the final controlled device's proportional gain.

$$Proportional\ gain = \frac{Output\ span}{Input\ span}$$

$$Proportional\ gain = \frac{50\dfrac{lbm}{hr}}{16\ mA} = 3.125\frac{lbm/hr}{mA}$$

The gain indicates the flow through the valve changes approximately 3.125 pounds mass per hour, for each one milliamp change in the controller's output signal.

Step 3 Determine the final controlled device's bias.

The bias is equal to the minimum value of the valve's flow range:

$$Bias = 0\ lbm/hr$$

Step 4 Write the transfer function using the values from steps 2 and 3.

$$Output = Proportional\ gain \times (\Delta\ Input\ mA) + bias$$

$$Output = 3.125\frac{lbm/hr}{mA} \times (\Delta\ Input\ mA) + 0\frac{lbm}{hr}$$

Step 5. Calculate the flow through the valve when the controller output is 14 mA.

$$Output = gain \times (\Delta\ Input\ mA) + bias$$

$$Output = 3.125\frac{lbm/hr}{mA} \times (10\ mA) + 0\frac{lbm}{hr} = 31.25\ lbm/hr$$

Technicians use the final controlled device transfer function when the quantity of the control agent flowing through the final controlled device is suspect. The calculated flow rate can be multiplied by the specific energy content of the control agent to calculate the total mass or energy being transferred to the process. Comparing that value to the required value may help to pinpoint problems in the process.

6.6

SUMMARY

Most of the control components used in the HVAC industry are designed and manufactured to maintain a linear relationship between changes in their input and output signals. The linear characteristic is used by technicians in the design, calibration and analysis of HVAC process control loops. The linear response is identified by the presence of a ratio or fraction in the transfer function equation. This ratio ensures that changes in the input signal produce equivalent percentage changes in the output response.

The general form of a transfer function of a linear control device has the format:

Output = gain × (Δ *Input*) + *bias*

where the gain describes the change in output signal that occurs for each unitary change in the input signal and the bias is equal to the magnitude of the output signal whenever the change in the input signal is equal to zero.

Transfer functions are used by technicians whenever they are called upon to analyze whether the device and process is operating according to the original design. By checking the calculated output value against the actual operating values, the technician can determine which component of the control loop is malfunctioning. If that device can be calibrated, the technician will use the transfer function once again, to determine the correct output signal for the present input condition. If the malfunctioning device cannot be calibrated it must be replaced. The new component must then be calibrated using its transfer function before the system can be put back into service.

6.7

EXERCISES

Determine if the following statements are true or false. If any part of the statement is false, the entire statement is false. Explain your answers.

1. Linear control devices generate changes in their output signal that are proportional to the change that occurred in their input signal.
2. The proportional gain is a ratio equal to the change in the input span divided by the change in the output span of a control device.
3. The process throttling range increases as the gain of a controller is increased.
4. Direct acting devices have output signals that change in the same direction as a change in the input signal.
5. The bias of a reverse acting sensor is equal to the maximum value of the sensor's input range.
6. The bias of a controller is initially set equal to the midpoint of the loop sensor's output signal range.
7. The response of a linear device can be represented with a straight line.
8. The bias is equal to the value of the output signal when the change in the input signal is equal to zero.

9. The general form of the transfer function of a linear device is
   Output = Proportional gain × (Δ*Input* ᵢ) + *bias*
10. A output signal of a linear device changes in the same percentage as the change that occurred in the input signal.

Respond to the following statements and questions completely and accurately, using the material found in this chapter.

1. What is the difference between the calculation of a controller's gain and a sensor gain?
2. What is the difference between a device's output range and its output span?
3. Define throttling range.
4. What is the difference between the throttling range of a controller and its proportional band?
5. What does the transfer function element *proportional gain* indicate?
6. What does the *bias* of a controller transfer function indicate?
7. What effect does changing the magnitude of the gain have on the sensitivity of a control device?
8. What are the response characteristics of a direct acting device?
9. What are the response characteristics of a reverse acting device?
10. Why shouldn't linear devices operate outside of their linear range?
11. Why is the transfer function for a final controlled device used with less frequency than other loop component transfer functions?
12. What relationship exists between the throttling range of a process and the controller's gain?
13. What does the phrase "out of control" mean?
14. Why are the values of bias different for a controller and sensor even when their output signal ranges are the same?
15. What change is made with respect to calculating the error when developing the transfer function of a controller?
16. What is the only difference between transfer functions of a direct acting device and a reverse acting device when the values of their parameters and variables are equal?
17. Why is the format of a linear control device's transfer function similar to the equation of a straight line?
18. If the proportional band of a controller is increased, will its throttling range increase or decrease? Explain your answer.
19. Can proportional band be used to calculate the controller's gain? Explain your answer.
20. Explain why the throttling range defines the operating limits of a control loop.

Solve for the unknown values in the following problems. Show all work and check your answers using the proportional change method.

1. A linear, direct acting sensor measures temperatures within a range of −20 to 220 °F and generates an output signal of 0 to 10 volts.
   a. Calculate the sensor gain.
   b. Determine its bias.
   c. Develop its transfer function.
   d. Calculate the sensor output at 155 °F.
   e. Calculate the temperature that produces an output signal of 3.2 volts.
2. A linear, reverse acting sensor measures temperatures within a linear range of 20 to 130 °F and generates an output signal of 4 to 20 milliamps.
   a. Calculate the sensor gain.
   b. Determine its bias.
   c. Develop its transfer function.
   d. Calculate the sensor output at 62 °F.
   e. Calculate the temperature that produces an output signal of 15.2 milliamps.
3. A linear sensor has the following characteristics:
   Gain = 0.10 mA/°F   bias = 4 mA   input span = 160 °F
   a. What are the input and output ranges of the sensor if its output signal is 14.8 mA when the temperature is 68 °F?
   b. Calculate the output signal at 100 °F.
4. A direct acting controller has a throttling range of 5 °F. It generates an output signal range of 3 to 15 psi. The set point is 61 °F.
   a. Calculate the controller gain.
   b. Determine its bias if the final controlled device range is also 3 to 15 psi.
   c. Develop its transfer function.
   d. Calculate the output signal when the control point is 62.6 °F.
   e. Calculate the output signal when the control point is 59.3 °F.
   f. Calculate the control point when the output signal is 12 psi.
5. A reverse acting controller has a proportional band of 4%. It generates an output signal of 3 to 18 psi. The set point is 140 °F. The loop's sensor measures a range of -20 to 180 °F and generates an output signal of 3 to 18 psi.
   a. Calculate the controller gain.
   b. Determine the controller's bias if the final controlled device has an output signal range of 3 to 18 psi.
   c. Calculate its throttling range.
   d. Develop the controller's transfer function.
   e. Calculate the controller's output signal when the control point is 136 °F.
   f. Calculate the sensor output signal if the control point is 140 °F.
   g. Calculate the value of the control point that will produce an output signal of 8.3 psi.

PART 2

Operational Details of Control Devices

CHAPTER 7

Sensors

Sensors measure a variable of a process and generate an output signal that is used to maintain the process requirements. Chapter 7 describes the construction, operation, and installation characteristics of the more commonly used HVAC sensors.

OBJECTIVES

Upon completion of this chapter, the reader will be able to:

1. Define the operational characteristics used to select a sensor.
2. Describe the proper location requirements for HVAC sensors.
3. List the design and the operational characteristics of various HVAC sensors.

7.1

MEASURING THE CONTROL POINT

All sensors have an element that undergoes a physical change that is related to the change in the magnitude of the measured variable. A change in the length, depth, position, resistance, or capacitance of the element is converted into a usable output signal that is compatible with the other control devices in the loop. Sensing elements are manufactured in a variety of configurations using different materials and construction techniques to accommodate the large variety of conditions found in HVAC processes.

The sensing element performs the actual measurement of the control point. Therefore, it must be immersed within the controlled medium in order to respond to variations in the process. The element responds by altering one of its physical properties by an amount that is related to the measured change in the process variable. This relationship between the amount of change in the measured variable and the corresponding change in the element's attribute establishes a predictable output response for the sensor. The response remains constant for all changes in the measured variable that occur within the sensor's linear operating range. Once the measured variable exceeds the linear operating range of the sensor, the relationship between the measured change in the process variable and the sensor's output signal varies. Under these conditions, the sensor is no longer generating useful information and the process requirements cannot be maintained.

Specific materials are selected for use as sensing elements because their transfer function (Δ Output /Δ Input) remains constant throughout the life of the sensor. The selection of a particular material for a given application is based upon the conditions that can be expected to exist in the controlled medium. A sensing element must be able to withstand the expected thermal range, velocity, vibrations, turbulence, and level of cleanliness found within the process. Most importantly, the sensing element cannot experience any physical or response degradation when operating within the normal conditions found at the sensor's location. If the controlled medium were to exceed the limitations of the sensing element, the sensor's output signal would become unreliable and the sensor may be irreversibly damaged. Therefore, the range of expected process conditions must be evaluated before a sensor is selected for a particular application. When properly selected, the sensor can remain in useful service for decades.

Minor variations in a sensor's operating parameters occur naturally over time. These changes result from the thermal and physical stresses constantly being applied to its element. Consequently, the output signal of the sensor tends to drift over time. To extend their useful life, most sensors are constructed so they can be periodically calibrated by a technician to bring its response back to original design specifications.

7.2

SENSOR OPERATING CHARACTERISTICS

A sensor is selected so its operational characteristics match the environmental conditions found in the process to provide a reliable interface between the control loop and the process. The following sensor characteristics are used to evaluate the quality and applicability of different sensors for a given application.

Full Scale Range—the operating range of the sensor.

The full scale range defines the operating range of the input signal. The sensor's response remains within its design specifications throughout the full scale range. Sensors will measure values beyond the limits of their full scale range but the change in their output signal may no longer be related to the measured change in their input signal. To minimize the possibility of a sensor operating beyond its full scale range, it is selected so:

1. the process set point falls within the center portion of its full scale range. As the input signal moves toward either end of its full scale range, the sensor's response becomes slightly nonlinear, reducing the accuracy of the information contained in its output signal (see Example 7.1).
2. the expected operating limits of the process fall within the sensor's full scale range (see Example 7.1).

Example 7.1

Determine the full scale range and applicable process conditions of a sensor with the characteristics:

Input range: -20 to 140 °F
Output Range: 0 to 5 volts.

1. This sensor has a full scale range of -20 to 140 °F.
2. This sensor can be used for processes whose set point falls within the center portion (40 to 80 °F) of the full scale range.
3. This sensor can be used for processes whose control point remains within the full scale range (-20 to 140 °F).

Signal Span—the difference between the maximum and minimum values of the sensors operating range.

Span indicates the maximum operating difference between the limits of the input and output signal ranges. The input signal span is calculated by subtracting the values of the sensor's full scale range. The output signal span is calculated by subtracting the minimum value of the output signal's range from the maximum value of the output signal's range. As demonstrated in the previous chapter, the input and output signal spans are used extensively in the development of sensor transfer functions.

The input span and the sensitivity of a sensor are inversely related. As the input span increases, the sensor's sensitivity to changes in the measured variable decreases. Conversely, as the input span decreases, the sensitivity of the sensor increases. A sensitive sensor is able to maintain the control point closer to the set point because it responds to smaller changes in the measured variable. When selecting a sensor, minimize the input signal span to increase the sensitivity of the device (see Example 7.2).

Example 7.2

Determine the signal spans of a sensor with the characteristics:

Input Range: 20 to 100 °F
Output Range: 0 to 10 volts

1. The sensor has a full scale range of 20 to 100 °F. Therefore, its input span is:
    $$100 - 20 = 80°$$
2. This sensor has an output span of:
    $$10 - 0 = 10 \text{ volts}$$

Accuracy—the accuracy of a sensor indicates the maximum difference between the actual value of the measured variable and the value indicated by the sensor.

Accuracy is calculated with the formula:

$$\frac{|\, Measured\ value - Actual\ value\,|}{Sensor\ Input\ span} \times 100$$

The *measured value* is the magnitude of the input variable currently being measured by the sensor. The measured value is calculated using the sensor's transfer function and its current output signal. The output signal's value is input into the transfer function, along with the sensor's proportional gain and bias to calculate the corresponding input signal. Once calculated, this value is entered into the *measured value* field of the equation.

The *actual value* is the current value of the measured variable being measured by an instrument other than the sensor being evaluated. This instrument provides a means to compare the output of the sensor against that of the instrument. The measuring instrument must be calibrated so it generates an accurate representation of the measured condition. The instrument's sensor is placed next to the sensor's sensing element to measure the actual condition at that location. Both sensing elements must be positioned next to each other for best results. If they are placed at different locations, process conditions may create real differences between the two locations. If a real difference is present, the actual accuracy of the control loop's sensor cannot be determined. The value measured by the instrument is entered into the actual value field of the equation.

The two vertical lines enclosing the numerator of the equation indicates the *absolute* value of the subtraction is used. The absolute value of a number has no positive or negative sign associated with it. The absolute value of $|(6 - 8)|$ is $|-2| = 2$. The absolute value of $|(14 - 8)|$ is $|+6| = 6$. Finally, the sensor span is the difference between the input range of the sensor. Accuracy has units of percentage (%), therefore, the calculated value is multiplied by 100 (see Examples 7.3 and 7.4).

Example 7.3

Task:

A sensor having a full scale range of 40 to 240 °F generates an output signal indicating a measured temperature of to 73.5 °F when the actual temperature measured with an instrument of known accuracy is 75 °F. Calculate the accuracy of the sensor based upon its full scale range.

  a. Measure value = 73.5 °F
  b. Actual Value = 75 °F
  c. Span = 240 − 40 = 200°F

Solution:

$$\frac{|\,Measured\ value\ -\ Actual\ value\,|}{Sensor\ Input\ span} \times 100$$

$$\frac{|\,73.5\ -\ 75.0\,|}{200} \times 100$$

$$\frac{|-1.5\,|}{200} \times 100$$

$$0.0075 \times 100 = 0.75\%$$

This sensor has an accuracy of +/− 0.75% of its full scale range. This means the measured output signal of the sensor is guaranteed to be within 1.5 °F (0.0075 × 200) of the actual value of the measured variable. *Smaller* values of accuracy signify a more precise sensor whose output very closely matches the actual value of the controlled variable. A control loop using a sensor having a full scale accuracy of 0.5% is better equipped to maintain its set point than if the designer used a sensor with a 2% accuracy.

Accuracy is an important characteristic of a sensor. If the sensor's percent of full scale accuracy is too large, the sensor will only be able to *estimate* the control point of a process. Consequently, the control loop will not be able to maintain its set point because the sensor is only estimating the actual value of the control point.

Example 7.4

Task:

A sensor having a full scale range of 110 to 230 °F and an output signal range of 4 to 20 mA generates an output signal of 13.5 mA. The temperature measured with an instrument of known precision is 182.9 °F. Calculate the accuracy of the sensor.

  a. Measured value = 181.4 °F. This value was calculated using the transfer function of the sensor and the output signal value:

$$Proportional\ Gain = \frac{16\ mA}{120\ °F} = 0.133\ \frac{mA}{°F}$$

$$Output = Proportional\ gain \times (\Delta\ Input\ °F) + bias$$

$$13.5\ mA = 0.133\ \frac{mA}{°F} \times (x\ °F\ -\ 110\ °F) + 4\ mA$$

$$\frac{(13.5 \ mA \ - \ 4 \ mA)}{0.133 \frac{mA}{°F}} = (x \ °F - 110 \ °F)$$

$$181.4 \ °F = x$$

b. Actual Value = 182.9 °F
c. F.S. Range = 230 − 110 = 120 °F

Solution:

$$\frac{|Measured \ value \ - \ Actual \ value|}{Sensor \ Input \ span} \times 100$$

$$\frac{|181.4 \ - \ 182.9|}{120} \times 100$$

$$\frac{|1.5|}{120} \times 100$$

$$0.0125 \times 100 = 1.25\%$$

Sensitivity—indicates the amount change in the magnitude of the output signal that occurs in response to a change in the input signal.

Mathematically, the sensitivity is equivalent to the proportional gain of the sensor's transfer function. It is equal to the sensor's output signal span divided by its input signal span. The sensitivity indicates how much the output signal changes when the measured variable changes one unit. As stated in Chapter 6, the terms proportional gain and sensitivity can be used interchangeably when referencing the input/output relationship of a control device. In sensor applications, the term sensitivity is more often used than proportional gain. The equation for the sensitivity of a sensor is the same as that presented in the transfer functions of Chapter 6:

$$Sensitivity = \frac{\Delta \ Output \ Signal}{\Delta \ Input \ Signal}$$

Example 7.5

Task:

The output signal of a humidity sensor changes from 9 psi to 10.2 psi as the room's humidity changed from 42% to 50%. What is the sensitivity of the device?

a. Δ Output = 10.2 − 9 = 1.2 psi
b. Δ Input = 50 − 42 = 8%

Solution:

$$Sensitivity = \frac{\Delta Output}{\Delta\,Input}$$

$$Sensitivity = \frac{1.2}{8} = 0.15\,\frac{psi}{\%\;rh}$$

The sensor in Example 7.5 generates a change of 0.15 psi in its output signal for every one percent (1%) change in the measured relative humidity. Sensors with high sensitivities generate measurable changes in the output signal in response to small changes in the measured variable, making it possible for the controller to respond to small changes in the controlled variable. Loops incorporating sensors with high sensitivities also respond quickly to process changes, prohibiting the process error from becoming noticeable by the room occupants. Sensors with lower sensitivities allow the controlled variable to change a greater amount before the control loop responds.

Resolution—the resolution of a sensor is the smallest change in the measured variable that is required to produce a detectable change in the sensor's output signal.

Due to the dynamic nature of HVAC processes, their sensor's are always measuring tiny fluctuates in the measured variable that are caused by fluid turbulence, radiation effects, load changes, and other events. Not all of these minute changes generate measurable changes in the sensor's output signal. The sensor's resolution indicates the minimum amount of change in the measured variable that must occur before a measurable change in the output signal will occur. Therefore, higher resolution sensors can measure smaller changes in the controlled variable, allowing their loop to respond quickly thereby minimizing the side of the process error.

Changes in the sensor's output signal produced by a high resolution sensor are only useful if the other loop components can react to these small changes in their input signals. A sensor with a resolution of 0.25° is not required for a loop that uses a controller that can only respond to changes greater than 1°. The cost of the higher resolution sensor is lost when the remaining process components cannot respond to its signal. Also, keep in mind that it is unnecessary to control comfort processes within $\frac{1}{2}$° because people cannot differentiate between changes in temperature of less that 2° F. Therefore, a sensor is selected so its output resolution matches the input resolution of the other components that use its signal.

Lag Time—the time it takes for a sensor 's output signal to change 63% of the total change in response to a change in the input signal.

The lag time gives an indication of the speed of response of the sensor and consequently, the control loop, to measurable changes in the process variable. Sensors with small lag times respond quickly to changes in the process. Whenever a sensor responds quickly, the loop can minimize the process error. Sensors are selected with smaller lag times to increase process efficiency and comfort. The lag time of the sensor is also known by the term *time constant*.

7.2.1 Sensor Specifications

Sensors can be manufactured with different grades of specifications. Two common grades used by control technicians are HVAC grade and industrial grade. HVAC grade sensors have reduced accuracies and resolutions in comparison to their industrial grade counterparts. The reduction in these characteristics does not adversely impact their effectiveness in HVAC processes but does considerably reduce their cost of manufacture. Industrial grade sensors are more expensive due to their higher accuracy and resolution. They will not produce any better control because their precision characteristics are usually too extreme for HVAC comfort applications. Industrial grade sensors are only used in HVAC applications where the process requirements outweigh the added expense of the sensor.

7.3

TYPES OF HVAC SENSORS

Resistance temperature detectors (RTDs) are a family of temperature sensing elements that produce a change in electrical resistance in response to changes in the surrounding temperature. There are two types of RTDs, *metal* devices and *semiconductor* devices. The characteristics of each type are described below.

7.3.1 Metal RTDs

All metals experience changes in their electrical resistance with variations in their temperature. As the temperature increases, the resistance of a metal wire sensing element increases by a known amount. This increase results when thermal energy is added to the wire because it increases the molecular vibrations of the molecules within the metal. This response is common in all metals, making them an ideal material for temperature sensor elements. Platinum and nickel wire are widely used in electronic sensor applications. Both of these metals have linear response characteristics and produce a measurable change in their output resistance as their temperature changes within the ranges typically found in HVAC applications. Platinum has a slightly better linearity characteristic than nickel, but nickel is less expensive. These two qualities are evaluated by the sensor manufacturer before the metal is chosen for a given application. Critical control applications use the platinum elements because its response is more linear and the added cost of the sensor is outweighed by the process requirements. When resolution and accuracy requirements can be maintained with the nickel element, it will be chosen because of its reduced cost.

Metal RTD sensing elements are constructed of a very small diameter filament of platinum, nickel, or other metal wire. The filament is first coiled and then wound around a support. The fine wire has a small mass and therefore, reacts very quickly to changes in the measured temperature. This reduces the sensor's lag time and the overall response time of the sensor. Elements having a larger mass have longer response times because their entire mass must absorb or dissipate heat before the sensor generates an accurate change in its output signal. The longer lag times permit the process error to increase needlessly before the controller responds to the change in the sensor's output signal.

Excessive vibration and shock can easily break the fine wire of a metal RTDs' sensing element. To minimize breakage, the element is usually wound around a ceramic support to increase its stability and reduce its chances of breaking under normal installation and operating conditions. The element and its support are inserted inside a metal tube or shield to provide additional protection. Unfortunately, the addition of the shield around the sensing element increases the lag time of the sensor. To minimize the increase in lag time, a conductive gel can be applied between the RTD element and the shield to reduce the thermal resistance of the path between the element and the shield.

Metal RTDs have a low sensitivity and low resistance at room temperature (70 to 74 °F). Platinum devices have sensitivities of approximately 0.1 (one tenth) ohm per degree Fahrenheit (0.1 Ω/°F). Nickel elements have a sensitivity that is slightly larger. The nominal resistance of the platinum element at room temperature (72 °F) is approximately 100 ohms. Figure 7.1 depicts the response of a metal RTD. Note the small change in output resistance that occurs over the full scale range of the device.

The small nominal resistance of a metal RTD combined with their low sensitivity make it imperative that a *signal converter* be used with these sensing elements. The signal converter inputs the small changes in electrical resistance generated by the metal sensing element and converts them into a voltage or current output signal (4 to 20 mA, 0 to 10 v, etc.) that is related to the change in the input signal. If the signal converter were not used, the small changes in the sensing element's resistance would be overcome by the resistance of the wires connecting the sensor to the controller. Consequently, the controller would not know if the change in the resistance was due to the sensing element or the connecting wires and would respond incorrectly.

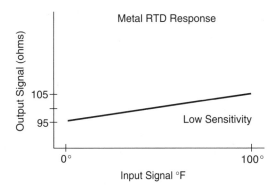

Figure 7.1   Metal RTD Response to Changes in Temperature

The current and voltage output signals produced by the signal converters are generally unaffected by the connecting wire's resistance. Therefore, the signals can be transmitted over much longer distances. The converter is typically mounted in the sensor housing so the sensing element wires can be terminated on the converter's input signal screws. Metal RTDs are very accurate, reliable and responsive and are commonly used in HVAC applications. Figure 7.2 shows a metal RTD outside air sensor, its shield and signal converter.

7.3.2 Semiconductor RTDs

RTDs are also manufactured from semiconductor materials such as silicon or germanium. The response characteristics of a semiconductor material have some of the conduction properties of metals along with the insulation properties of silicon. They are manufactured in different compositions to produce sensing elements with different ranges of operation and responses. Semiconductor RTDs are more commonly called *thermistors*. The word thermistor is derived from the phrase *"thermal affected resistor."* Thermistors belong to the family of resistance temperature detectors (RTDs) because they also experience a change in their resistance that is related to a change in their surrounding temperature.

Thermistors have much higher sensitivities than those found in metal sensing elements. They have sensitivities over 100 Ω /°F. This is 1000 times greater than the sensitivity of a metal RTD. They also have a higher nominal resistance at room temperature of approximately 10,000 Ω. The high sensitivity of the semiconductor element along with its high nominal resistance at room temperature produces a sensor that generates an output signal that is unaffected by the resistance of the wire used to connect the sensor to the other devices in the control system. The change in output resistance is so large that a signal converter is not required to amplify changes in their output resistance. They can be connected directly to the controller by a pair of wires up to 500 feet in length. This reduces the cost of a system's installation by simplifying field connection of thermistors in a control system.

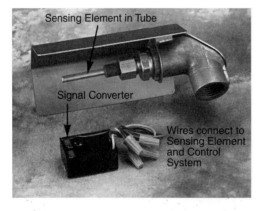

Figure 7.2   Metal RTD and Signal Converter

Unlike metal RTDs, thermistors have a *nonlinear* response to changes in the measured temperature. This means that their change in resistance per degree of temperature change (sensitivity) is not a constant value as with metal RTDs. Instead, it changes its value throughout its operating range. Consequently, the transfer function for a linear sensor developed in Chapter 6 cannot be used to calibrate a thermistor because its response is nonlinear. To determine the output signal for a given temperature, a look-up table is supplied with the sensor. The table lists the resistance of the sensor at various temperatures. If the resistance of the sensor at a given temperature does not match the value listed in the table, the sensor must be replaced. The response of a thermistor's varying sensitivity is shown in Figure 7.3.

The response of the NTC labeled thermistor shown in Figure 7.3 shows that the resistance of the sensor decreases as its temperature increases. Therefore, by the definition of the phase element of a transfer function, this thermistor is reverse acting. The output signal changes in the opposite direction with respect to the change in the input signal. Thermistors with this response are said to possess a negative temperature coefficient (NTC). Negative temperature coefficient thermistors are widely used in HVAC control systems.

Thermistor sensors can also be manufactured with a nonlinear, positive temperature coefficient (PTC). These devices experience the same change in the direction of their output signal as the change in the input signal. As the temperature decreases, the output signal also decreases. Positive temperature coefficient thermistors are not typically used in HVAC systems. Although there is no advantage of using a NTC thermistor over a PTC thermistor, they are not interchangeable. If one were used in place of the other, the controller and consequently, the entire control loop and its process would respond incorrectly. The response of a PTC thermistor is also shown in Figure 7.3.

Thermistors can be manufactured in a variety of shapes and sizes to meet almost any HVAC application. Their thermal mass and physical strength can be varied to meet the environmental conditions found in the controlled medium. Thermistor shapes range from small beads to disks the size of a quarter. They are

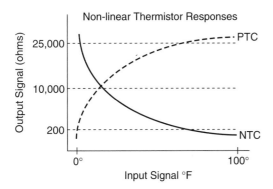

Figure 7.3 NTC & PTC Thermistor Responses

constructed so they are more resistant to vibration than metal RTDs. This ruggedness allows some thermistors, especially those used in thermostats, not to require a protective shield. This reduces the lag time of their response.

Thermistors have a tendency to drift over time but cannot be calibrated in the field because they lack a signal converter. Systems that use thermistors are designed to accept a compensation term (constant) that can be added or subtracted from the sensor's value to compensate for its drift. In similarity with metal RTDs, thermistors are selected so they measure temperatures that fall within the center (more linear) portion of its full scale range.

7.3.3 Pneumatic Temperature Sensors

Pneumatic sensors generate changes in their output signal pressure in response to changes in the measured temperature. These sensors and all the other pneumatic control components are mechanical devices constructed with internal levers, springs, bellows, and diaphragms that operate using a force-balance principle. The force-balance principle describes a response where any change in a components input signal disrupts the internal balance state that exists between its spring and pressure forces thereby initiating a change in its output signal. The output signal continues to change until the internal forces within the device return to a balanced state.

The forces inside a pneumatic device are produced by the control air pressure working against the surface area of small diaphragms. Internal passages are cast in the component's body and provide the paths between different chambers and diaphragms within the device. As the air enters a chamber, it changes the forces working on the diaphragms, causing them to be repositioned. The force on the diaphragm works against opposing forces produced by small internal and external springs. As a diaphragm is displaced, it causes levers to be repositioned or internal air flowrates to change, altering the output signal of the device. As the output signal changes, feedback passages within the device work to restore a balanced condition by varying the pressures on the diaphragms until they are returned to the center position within their chamber. Once the internal forces reach equilibrium, the output signal stabilizes.

Pneumatic control devices are connected to a 15 to 30 psi supply of instrument quality air to provide the energy they need to operate. Supply air enters the control device through a tiny orifice (0.005" - 0.007") called a *restrictor.* The restrictor limits the volume flowrate of control air that enters the device. The restrictor is sized so it passes less air into the device than the maximum flowrate through an unobstructed bleed port. This allows the device to generate an output signal of 0 psi when its bleed port is 100% open. If the restrictor were to be removed from the supply line, the output signal pressure will remain equal to the supply pressure at all times.

Figure 7.4 shows a section of a pneumatic sensor and a pressure gauge. Variations in the quantity of control air leaving the bleed port produce changes measured by the pressure gauge. The strip representing the sensing element moves

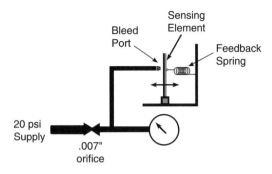

Figure 7.4   Pneumatic Restrictor/Sensor

with respect to the bleed port, varying the amount of control air exiting through the port. As the element moves toward the port, it increases the bleed port's exit resistance, reducing the amount of air leaving the port and increasing the pressure in the signal line. As the element moves away from the nozzle, it decreases the bleed port's resistance, increasing the amount of air leaving the port and decreasing the pressure in the signal line. This response is typical of all pneumatic sensors. The sensing element is configured so the output signal pressure rises as the temperature increases.

All pneumatic components operate using this principle. To ensure trouble-free operation, the control air feeding these component must be clean and dry. Their small passages and orifice plates can easily become obstructed with impurities, rendering the device inoperable. Accumulations of dust on the exterior of the device can also lead to improper operation. Due to their mechanical nature, cleaning and calibrating is recommended at least once per year to bring the unit back to its design response.

Figure 7.5 shows a schematic representation of a section of a *bimetal* pneumatic sensing element typically found in zone thermostats. The bimetal element is constructed with two metals that have different coefficients of expansion. These two strips are fused together into a single strip. Therefore, as the temperature surrounding the element changes, one side of the strip expands more than the other, causing the entire strip to bend. As it bends closer to the nozzle, it increases the resistance of the bleed port's path, reducing the flowrate of the control air leaving the pilot chamber of the sensor. The increase in pressure on its left side causes the diaphragm to move toward the right. The movement of the pilot diaphragm ultimately results in an increase in the sensor's output signal. Conversely, as the bimetal strip bends away from the bleed port, the output signal decreases.

Other sensing elements used in pneumatic temperature sensors use vapor-filled bellows, refrigerant-filled capillary tubes, or bimetal wire to position the a strip or disk that varies the amount of control air leaving through the bleed port. Pneumatic sensors are commonly called *transmitters.*

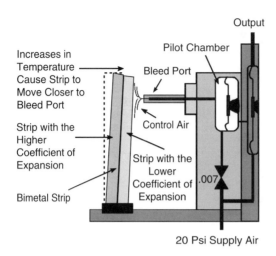

Output

Pilot Chamber

Increases in
Temperature
Cause Strip to
Move Closer to
Bleed Port

Bleed Port

Control Air

Strip with the
Higher
Coefficient of
Expansion

Strip with the
Lower
Coefficient of
Expansion

.007

Bimetal Strip

20 Psi Supply Air

Figure 7.5   Pneumatic Zone Temperature Sensor

7.3.4 Electronic Humidity Sensors

Humidity sensors are constructed using *hygroscopic* materials that expand and
contract with changes in the ambient moisture. Hygroscopic materials readily
absorb and release moisture, making them ideal sensing elements for measuring
relative humidity. As they absorb moisture, their length or depth changes in a predictable, repeatable manner. The amount of movement is related to the amount of
moisture absorbed by the material. These dimensional changes are used to generate the output signal of the sensor. Electronic humidity sensors use the dimensional changes of a hygroscopic material to produce a varying capacitance signal
in the sensing element. The change in capacitance is transformed into a linear voltage or current output signal by a signal converter.

A capacitor is an electrical device constructed of two conducting plates separated by an insulator. In sensor applications, the insulator is typically the air
between the plates. The capacitor stores an electrical charge measured in farads.
As the plates become closer together, the amount of capacitance increases. As the
distance between the plates increases, the capacitance decreases.

Capacitance sensors have one stationary conductor that acts as one plate of a
capacitor. Another conducting plate is mounted to the hygroscopic material, to act
as the other plate of a capacitor. As a hygroscopic material absorbs moisture it
expands, repositioning the moveable plate closer to the stationary one. This
response generates a change in the capacitance of the sensing element. The
amount of change is measured in pico farads (farads $\times\ 10^{-12}$) or trillionths of a
farad. These small capacitance signals are fed into a signal converter that generates a linear output signal such as 4 to 20 mA, 0 to 10 volt, etc. The converter also
linearizes the response of the sensing element, making it possible to use the transfer functions developed in Chapter 6 to analyze and calibrate the device. Elec-

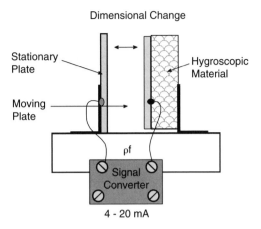

Dimensional Change

Stationary Plate

Hygroscopic Material

Moving Plate

ρf

Signal Converter

4 - 20 mA

Figure 7.6 Electronic Humidity Sensor

tronic humidity sensors are accurate and reliable and are used for humidity control and enthalpy based control strategies.

Pneumatic humidity sensors use a hygroscopic nylon tape that changes its length as the quantity of moisture absorbed increases. The tape is connected to a flapper/bleed port mechanism, similar to that used in temperature sensor applications. It varies the flow from the bleed port in response to changes in the measured humidity. As the moisture level increases, the flapper moves closer to the bleed port, increasing the output signal pressure. These devices are not as accurate or reliable as the electronic type of humidity sensor. They are also maintenance intensive, requiring a great deal of cleaning and calibration to remain accurate. Because their effectiveness in a control system is marginal, they are gradually being replaced with the electronic sensor in humidity control applications.

7.3.5 Pressure Sensors

Electronic sensors that are used to measure pressures also incorporate a capacitance sensing element to measure variances in the process. The design of a pressure sensor is similar to that of a humidity sensor. The major difference is that the moveable plate is connected to a diaphragm instead of a hygroscopic support. The diaphragm varies its position in response to changes in the measured pressure. As the pressure increases, the moveable plate on the diaphragm moves closer to the stationary plate, increasing the output capacitance of the element. As in all capacitance based sensors, a signal converter linearizes the change in capacitance and generates the related current or voltage output signal.

Static pressure sensors used in HVAC process control have two isolated chambers separated by the diaphragm. These sensors are designed to measure the *differential* pressure between two input signals or locations. Each side of the diaphragm generates a force proportional to its pressure signal. The force created by the lower signal is canceled out by the opposing force from the higher pressure

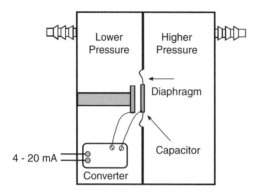

Figure 7.7   Electronic Static Pressure Sensor

that works in the opposite direction. The output signal is related to the difference between these two forces acting on the diaphragm.

The pressure signals entering the differential pressure sensor must be connected to the correct side or the sensing element will not work correctly. The HIGH pressure port receives the input signal from the higher relative pressure location and the LOW port is connected to the lower of the two input signals. This prevents the diaphragm from being displaced in the wrong direction, preventing the moveable plate from coming closer to the stationary plate of the sensing element. Figure 7.7 depicts a schematic representation of a differential pressure sensor.

7.4

LOCATING SENSORS

The location of any process sensor is critical to the proper operation of the control loop. Since the purpose of a control loop is to respond to changes in the control point, its sensor must be able to accurately detect the variations in the process as they occur. If the sensor is installed in a location or position that inhibits its ability to respond quickly and accurately to changes in the measured variable, the loop and consequently, the process, will be poorly maintained. Therefore, it is prudent to analyze the conditions of the process at the proposed sensor location before installing the device.

Temperature sensors require a constant movement of the controlled medium across their sensing element in order to accurately measure the control point. The faster the flow, the quicker the response time of the sensor. Fast moving fluids are also better mixed (reduced stratification) and provide an reliable representation of the actual process conditions. The high velocity also prevents the buildup of an insulating layer of stagnant fluid from surrounding the sensing element. These insulating layers increase the lag time of the sensor, allowing the error to grow larger before the control loop responds.

Sensors must be located in a position that is free from drafts and currents that are not related to the measured variable. Openings around mounting holes and wire entrances can allow air currents from inside the wall to flow over the sensing element. These drafts affect the sensing element, causing it to respond to a temperature that does not represent the actual control point. These conditions can cause the process to be maintained at a temperature other than the set point even when the control loop is operating correctly. This occurs because the sensor is responding to a temperature that is different than that felt by the occupants. Drafts can usually be avoided with an application of caulk or thermal insulation to the back of the mounting plate when the sensor is being installed. Avoid installing sensors above heat sources, within the air currents produced by diffusers, or in stagnant fluid locations.

Temperature sensors must also be shielded from the adverse affects of radiant energy. Whenever a sensor is located within visual sight of a high (or low) temperature source that is substantially different from the control point, an energy transfer occurs between the sensing element and the source. This transfer causes the sensing element to respond to a temperature that is different from the actual control point. When a radiant source that can adversely affect the sensor is present, a shield is used to prevent the sensor from "seeing" the source, thereby reducing its negative effects. Outside air temperature sensors are installed with a shield to reduce the radiant effects of the sun on the sensing element. Figure 7.2 shows an outside air sensor with its sun shield.

Humidity sensors should also be located in areas where the controlled medium is well mixed and constantly moving. Their sensing elements must be kept free of accumulations of dust and other impurities that may inhibit the rapid transmission of moisture between the hygroscopic material and the controlled medium. These practices result in a reliable measurement of the moisture level of the process.

Static pressure sensors have the opposite location criteria than those required for temperature and humidity sensors. Static pressure sensors must be located in an area of *laminar* or nonturbulent, layered flow. Under laminar flow conditions, the fluid flows in layers of constant velocity and pressure. These are the ideal conditions for measuring small pressure signals. Turbulent flow produces rapid fluctuations in pressure that cause the pressure sensor to generate an erratic output signal. Locating static pressure sensors at least 5 duct diameters (five times the diameter or width of the duct) downstream from any turn, transition or obstruction will reduce the possibility of experiencing turbulent flow.

When measuring static pressure, a straight tube is used to connect the high or low pressure port to the controlled medium. The tube is mounted perpendicular to the duct or room wall. If the sensor is inserted at an angle, it will measure static plus some velocity pressure which does not represent the controlled variable.

7.5

SUMMARY

Sensors measure the magnitude of the controlled variable of a process. A sensor is selected based upon its operating characteristics and the conditions found within the process. Accuracy, sensitivity, lag time, and full scale range are some of the sensor's characteristics that are evaluated during the selection process. To maximize the linear response of any sensor, the operating point of the process should fall within the center portion of the full scale range.

Sensors must be installed in the proper location to insure an accurate measurement of the controlled variable can occur. Vibrations, fluid velocity, radiant energy transfer, and corrosion potential must be considered when selecting the sensor's location.

7.6

EXERCISES

Determine if the following statements are true or false. If any part of the statement is false, the entire statement is false. Explain your answers.

1. Sensors measure the set point of a process.
2. Accuracy describes how close the sensor's output is expected to match the actual value of the measured variable.
3. As the lag time of a sensor increases, the process error increases.
4. Semiconductor RTDs have a higher sensitivity than metal RTDs.
5. Hygroscopic materials change their capacitance with changes in their moisture content.
6. Signal converters can be used to linearize sensing element signals.
7. Temperature sensors should be located in areas of turbulent flow.
8. Metal RTDs can be manufactured with a PTC or a NTC response.
9. A negative temperature coefficient sensor generates an increase in its output signal resistance as the measured temperature decreases.
10. The response of a sensor becomes more linear as the measured value approaches either end of the full scale range.

Respond to the following statements and questions completely and accurately, using the material found in this chapter.

1. What is the purpose of a sensor?
2. Describe the response of the force-balance principle of a pneumatic device.
3. What is the primary requirement of a sensing element?

4. Describe the differences between metal and semiconductor RTDs.
5. Why do metal RTDs require a signal generator?
6. How do capacitive type sensors generate changes in the output signal in response to changes in the measured variable?
7. List and describe six characteristics of a sensor that are used to evaluate sensors.
8. Summarize the installation requirements of a temperature, static pressure, and relative humidity sensors.
9. What is a hygroscopic material? Describe its application in HVAC sensors.
10. What is differential pressure?

Provide answers to the following problems. Show your work.

1. A sensor generates an output signal that represents a measured value of 32% relative humidity. An instrument of verifiable accuracy measures the same condition at 33% relative humidity. Calculate the accuracy of the sensor.
2. A static pressure sensor has the following operating characteristics:
   input range: 0 to 2 inches of water
   output range: 0 to 5 volts
   a. What is the full scale range of the sensor?
   b. Calculate the sensor's signal spans.
   c. Calculate the sensor's sensitivity.
3. A temperature sensor with a full scale range of 20 to 120 °F generates a 3 to 18 psi output signal. What is the temperature of the process if the output signal is measured to be 11.2 psi?
4. What is the accuracy of the sensor in Problem 3 if an accurate temperature meter measured a process temperature of 72.6 °F when the sensor's output signal was 11.2 psi?
5. A temperature sensor with a full scale range of 50 to 90 °F generates a 4 to 20 mA output signal. Calculate the accuracy of the sensor when the room temperature is measured to be 75 °F when the sensor's output signal is 12.3 mA.

CHAPTER 8

Controllers

Controllers evaluate the process condition and generate an output signal that creates a balance between the load and the mass or energy being transferred with the process. The controller's mode defines how the process responds to changes in its control point. The mode describes how the controller alters its output signal in response to changes in the control point. There are several controller modes used in HVAC applications. Chapter 8 describes these different configurations, their operational characteristics, and their application requirements.

OBJECTIVES

Upon completion of this chapter, the reader will be able to:

1. Define the primary characteristics of the different controller modes.
2. Describe the operational differences between both two-position modes.
3. Describe the characteristics of the different gains used in modulating controller transfer functions.
4. Associate basic HVAC applications response criteria with the proper controller mode.

124

8.1

OPERATIONAL CHARACTERISTICS OF CONTROLLERS

Controllers respond to changes that occur to the control point of the process. Whenever the load changes, its effect on the control point is transmitted to the controller by its sensor. In response to the change in the sensor's output signal, the controller calculates the related change in the process error and uses this value to produce a corresponding change in its output signal. The change in the controller's output signal repositions the final controlled device, modifying the flow of mass or energy into the process. Altering the flow of the control agent reduces the change in the control point to zero. Once the process stabilizes, its control point remains constant until another change in the load or the set point occurs.

A controller calculates the error using the information contained in the sensor's output signal (control point) and the process set point. Analog controllers have a section that performs the subtraction called the *summing junction* or *error generator*. The summing junction adds the positive set point signal to the negative (feedback) control point signal to produce a resultant signal that represents the magnitude of the process error. Digital controllers calculate the error with a section of their microprocessor called the *arithmetic logic unit* (ALU).

Once the error is calculated, a signal representing its magnitude is sent to the section within the controller that generates the change in its output signal. This section is called the *output signal generator*. The signal generator produces changes in the output signal that are a function of the size, direction and timing characteristics of the error signal. The controller's output signal is sent to the final controlled device to produce a corresponding change in the flow of mass or energy into the process.

8.1.1 A Controller's Response to Dynamic Characteristics of the Error

The controller is the only component found in a control loop that can be field calibrated (tuned) so it responds to the *dynamic* or time based characteristics of the process. By incorporating additional gain parameters into its transfer function, a controller can be configured to respond to the *rate of change* and/or the *length of time* an error exists.

Recall from Chapter 5, the proportional gain parameter in a controller's transfer function responds to changes in the *size and direction* of the process error. Integral gain can be added to the transfer function along with proportional gain to create a proportional + integral response that enables the controller to respond to the size, direction, and *length of time* an error exists. Derivative gain can also be included in the transfer function to create a proportional + integral + derivative response that reacts to changes in the size, direction, length of existence, and *rate of change* of the process error. Using the dynamic characteristics of the process improves the response and efficiency of the control loop and its process.

A controller that is only configured with proportional gain in its transfer function is called a *proportional only* controller. A controller using proportional gain along with integral gain in its transfer function is called a PI controller. A controller that is configured with all three gains is called a PID controller. These are the three most popular types of modulating controllers used in HVAC processes. Proportional gain can also be used with derivative gain to create a PD controller that responds to changes in the size, direction and *rate of change* of the error. Neither integral gain nor derivative gain can be incorporated into a transfer function unless proportional gain is also used.

8.2

CONTROLLER MODES

The controller mode describes the characteristics of a controller's output response to a change in the process error.

Controller transfer functions can be divided into seven categories based upon the characteristics of the response the controller generates when a change in the set point or control point occurs. These seven categories are called *controller modes* and are listed below in their order of complexity:

- two-position mode
- timed two-position mode
- floating or three-point mode
- proportional mode
- proportional + integral mode
- proportional + derivative mode
- proportional + integral + derivative mode

The characteristics, advantages, and disadvantages of these different transfer function modes are detailed in the following sections.

8.2.1 Two-Position Controller Mode

Two-position mode describes a controller that generates a binary change in its output signal in response to variations in the size and direction of the control point.

A two-position controller can only generate two possible output signals corresponding to the minimum and maximum values of its output signal range. Therefore, their output signal is *discontinuous* because it is impossible for the controller to produce a signal that falls between its output signal range. Figure 8.1 shows that a two-position controller's output signal snaps between its operating limits without stopping at an intermediate value. The dotted lines represent the transition of the output signal between the minimum and maximum signals.

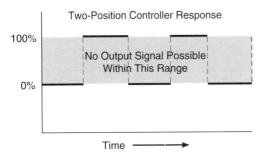

Figure 8.1   Two-position Controller Output Response

As a consequence of the two state response, the flow of energy into the process is also constrained to either 0% or 100% of the maximum flowrate of the control agent. Both the controller's output signal and the final controlled device pass through these values on the way to their final open or closed position. Two-position controllers are available for use with all signal types for use in pneumatic, analog electric, and digital electronic systems.

Controllers configured with a two-position mode have the simplest response of all seven controller modes. They are the easiest mode to calibrate and their response is the easiest to analyze. Two-position controllers can only be used for applications where the mechanical equipment can be cycled on and off to maintain the control point within acceptable range of the set point without adversely affecting occupant comfort, safety, or equipment operation.

Two-position controllers pulse energy into the process by cycling equipment on and off, or opening and closing the final controlled device in response to variations in the control point. These pulses maintain the controlled variable within an acceptable range of the set point. In these applications, the controller cycles its output signal whenever the control point exceeds a predetermined value on either side of the set point.

Figure 8.2 shows the response of a closed loop, two-position controller that cycles a direct expansion air-conditioning unit in response to changes in the room temperature. As the temperature increases above the contact closure point of the controller (76 °F), the air-conditioning unit cycles on. Heat (energy) is transferred to the evaporator and transported out of the zone, cooling the room and reducing the control point. The temperature must fall *below* the lower switching temperature (74 °F) of the controller before its output signal commands the unit off. If the control point falls below the set point without exceeding the controllers lower switching value, the output signal will not change state and the unit remains on to maintain the temperature within a few degrees of set point.

Cycling equipment on and off with a two-position controller pulses mass or energy into the process. The averaging effect of these pulses maintains the control point within an acceptable range of the set point. Since two-position control strategies cannot modulate the control agent into the process they cannot maintain the

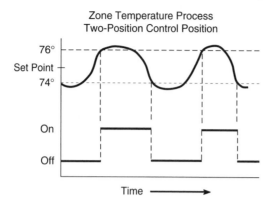

Figure 8.2   Two-Position Controller Response Maintains the Control Point Near the Set Point

control point equal to the set point, just within range of it. Two-position controller mode is most often used with electrical equipment because they are easily cycled using relays and contactors.

Two-position mode is also used to prevent the control point from exceeding a predefined safety limit. When a two-position controller is used in these applications, equipment is commanded to change state whenever the control point exceeds the design limit. Figure 8.3 shows the response of another two-position control loop used in direct expansion air-conditioning equipment. A high limit controller commands the air-conditioning equipment off whenever the discharge pressure exceeds its maximum set point. The unit is only allowed to return to normal operation after the control point has dropped below the high limit set point. In some safety applications, the system must be manually reset before it can be returned into service. This ensures someone has been made aware of a problem that caused the safety loop to respond.

8.2.2 Two-Position Controller Mode Differential

A control differential is a characteristic of the two-position transfer function that creates a delay between changes in the output signal to reduce the possibility of the output signal repeatedly cycling on and off whenever the control point is near the set point.

As depicted in Figure 8.3, a two-position controller is not designed to switch its output signal's state as soon as the control point exceeds the set point. Two-position mode requires a *delay* incorporated into its transfer function to prohibit the output from switching until the difference between the set point and the control point exceeds a predetermined value. This built-in delay is called the *control differential*.

The control differential prevents the controller from changing its output state until the process undergoes a sufficient change in its control point to prevent its output signal from rapidly cycling whenever the control point hovers around the set point. This rapid cycling is called *hunting* or *oscillating* which causes excessive

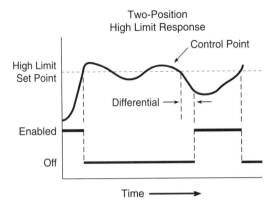

Figure 8.3   Two-Position Controller Response for a Safety Application

wear on the mechanical parts of the control loop and the process equipment. Rapid cycling also overheats relays and motors because the locked rotor current is constantly being pulsed into their coils without allowing sufficient time to dissipate the heat. Rapid cycling also produces excessive arcing of electrical contacts, causing pitting and increasing the resistance of the contact surface. The control differential is also known as a signal or process *deadband*. A deadband is an interval where no change in controller's output signal can occur.

The control differential is measured in units of the controlled variable, °F, % relative humidity, psi, or inches of water. The control range is defined as the set point + / − one half of the control differential. Therefore, range of control of a zone temperature process with a set point of 74° and a control differential of 4° is 74° +/− 2°. The control point must exceed the set point plus one half of the control differential before the controller changes the state of its output signal. Typical differentials for HVAC applications are:

1. Temperature processes: 2 to 3° (set point +/− 1 to 1.5 °F)
2. Humidity processes: 3% to 5%
3. Static pressure processes 0.1 to 0.15 inches of water.

The size of the control differential is selected by the control technician during the calibration procedure.

8.2.3 Two-Position Mode Operating Differential

An operating differential is the range of the process control point that results from the combined effects of the control differential and process time lags.

No response or change occurs instantaneously. All dynamic processes take time to respond to changes in their load or set point. The total time lag is a combination of the delays that occur while the sensor measures the change in the controlled variable plus the time it takes to send the information through the control loop and back to the process. All the while these delays take place, the control point continues to change.

As stated previously, two-position controller mode *pulses* energy into a process by cycling the final controlled device open and closed or on and off. This response produces oscillations in the process control point. When the effects of the time delays are added to the pulsations of the control agent, the resulting response shows the control point exceeding the control differential by a value related to the length of the time delays. The range of the control point is called the *operating differential* of the process. The operating differential is always greater than the control differential. Figure 8.4 depicts the effect of time delays on the control point of a dynamic process.

Figure 8.5 shows the switching points corresponding to the control differential of a two-position controller. Each dot represents a point where the control point

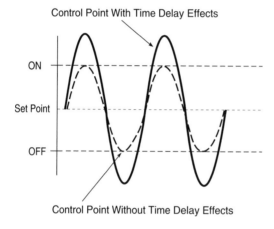

Figure 8.4   Operating Differential Caused by Time Lags Coupled with the Control Differential

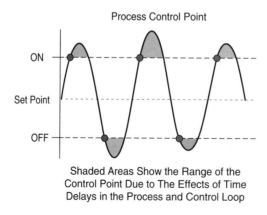

Figure 8.5   Range of the Control Point Caused by Time Lags

exceeds the set point $+/-\frac{1}{2}$ of the control differential span. The operating differential is the difference between the minimum and maximum values of the control point. The extended range of the control point is produced by the control differential coupled with the effects of time delays which allow the load to continue to change until the energy transferred by the control agent is large enough to stop the growth of the error.

8.2.4 Effects of Operating Differentials on Processes

Excessive operating differentials hinder the ability of the control loop to maintain the process close to its set point in closed loop applications. The extended range of the control point caused by the combined effects of the control and operating differentials reduces operating efficiencies and the level of occupant comfort. In a properly calibrated two-position control loop, the *control differential* is kept as small as possible while still preventing short cycling of the final controlled device. The *operating differential* can also be minimized through the proper sizing of the mechanical equipment serving the process.

The control and operating differentials do not adversely affect two-position, *open loop* applications. These applications prevent the control point from exceeding a predetermined operating limit and the delay caused by the operating differential only becomes apparent when the controller is resetting the system after the fault condition has cleared. Once the condition that initiated the controller output signal's change of state has been eliminated, reflecting the elimination of the unsafe condition, the controller resets the system back to its normal operating mode. Therefore, any additional delay caused by the operating differential only ensures the system is in fact, ready to return to normal operation.

8.3

TIMED TWO-POSITION CONTROLLER MODE

Timed two-position mode incorporates a method of reducing the operating differential in electric, two-position temperature control applications that use a thermostat.

The response of a two-position controller mode in temperature applications can be improved by adding a *heat anticipator* to the sensor's location in electric thermostats. A thermostat, similar to those used in home heating applications, has its sensor and controller circuits mounted on the same circuit board. An anticipator circuit, mounted under the sensing element, uses a resistor to add heat to the sensing element to speed up its response to changes in the room temperature. This additional heat deceives the sensor into generating a false signal indicating to the controller that the room temperature is warmer than it actually is. This false signal causes a heating system to cycle off earlier or a cooling system to cycle on earlier because the additional heat makes the room appear warmer than it is. Two-position controllers that incorporate this anticipator circuit are described as having a *timed two-position* controller mode. Because of the electrical nature of the anticipator, the timed two-position mode is only available on electrically based controllers. It is not available in pneumatic thermostats or those electric temperature control systems where the sensor and controller are two separate devices.

The response produced when a *heat anticipator* is added to a two-position controller is an improvement over the operational characteristics of a straight two-position controller in comfort temperature applications. The human body can sense changes in the ambient temperature when it drifts beyond 2° or 3° of the zone set point. As previously stated, a two-position controller also has a two or three degree control differential to prevent short cycling. Any additional overshoot caused by the process and system time delays will cause the occupants to begin to feel uncomfortable. Therefore, straight two-position controller mode should not be used in comfort temperature control applications.

Timed two-position mode reduces the operating differential of a temperature process improving occupant comfort. In heating applications, the anticipator applies a small amount of additional heat to the sensing element whenever the controller output signal is commanding the final controlled device to supply heat into the process. The anticipator's additional heat causes the controller to command the heat transfer equipment off before the actual room temperature equals the set point plus one-half of the control differential. In reality, the room is slightly cooler than this temperature but the controller does not know that because the information contained in the sensor's output signal indicates a different temperature exists. The controller cycles the final controlled device off earlier, reducing the size of the overshoot that occurs above the set point. Figure 8.6 depicts a timed two-position response for a heating process.

Cooling anticipators apply additional heat to the sensing element when the condensing unit is off, causing the controller to cycle the unit on before the zone actu-

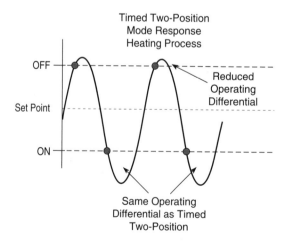

Figure 8.6   Reduced Operating Differential of a Timed Two-Position Controller Mode

ally reaches the set point temperature plus one half of the control differential. Figure 8.7 depicts a timed two-position response for a cooling process.

In any timed two-position application, the controller mode reduces the operating differential of the process by applying a small amount of heat to the sensing element, raising its temperature to a value that is slightly higher than the actual temperature (control point) of the room. The false signal from the sensor causes the controller to alter the state of its output signal earlier in time, reducing some of the negative effects of the system's time delays. This reduces the overshoot of the controlled variable which also reduces the temperature swing felt by the zone occupants. The heat anticipator does not affect the size of the control differential. The controller must still measure a change in control point equal to the set point plus or minus one-half of the control differential before it can change the state of its output signal.

8.3.1 Advantages and Disadvantages of Using a Two-Position Mode

The primary advantage of using either of the two-position modes is the simplicity of the control loop. Their control devices are easy to calibrate and analyze. When the process equipment is correctly sized, these modes will not generate an unstable or oscillating response to changes in the process load. Straight two-position mode is used for safety applications and other processes where the mass or energy can be pulsed into the process without producing excessive oscillations in the control point. Timed two-position mode should be used to maintain comfort temperature processes where the control differential and process delays could produce operating differentials that adversely affect occupant comfort. Straight two-position control should not be used for comfort temperature applications.

A disadvantage of using a two-position controller modes for closed loop processes is the pulsating characteristic of the response. These pulses produce

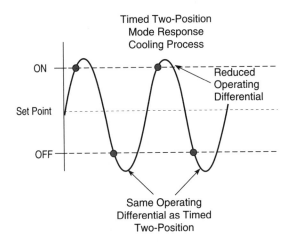

Figure 8.7   Timed Two-Position Response for a Cooling Application

oscillations in the control point that indicate a balance is not being maintained between the energy transfer and the process load. Whenever the balance is not maintained, the process efficiency is being reduced. In many processes, excessive oscillations of the control point are not permitted, further reducing the applications that can incorporate a two-position controller mode.

Two-position mode can be used with any HVAC control signal type, pneumatic, electric, or electronic. Timed two-position mode is used exclusively in electrical or electronic systems because electric current is needed to supply the anticipator circuit heat. Timed two-position mode is limited to residential and light commercial facilities where zones have their own individual heating and cooling equipment.

8.4

FLOATING CONTROLLER MODE

Floating controller mode generates a three state output signal. This signal positions a three state actuator that modulates the flow of the control agent into the process.

Floating mode controllers have operating characteristics that incorporate the advantages of the two-position mode while overcoming some of its disadvantages. Floating mode combines the simplicity of a binary controller with a *modulating* (not two-position) final controlled device. This creates a control loop that can reduce the operating differential of a two-position controller response by modulating the flow of the control agent into the process instead of transferring it in pulses. The final controlled device used with a floating mode controller is made specifically for these applications and cannot be used with any other controller mode.

A floating mode controller generates a binary output signal of a fixed magnitude, similar to that generated by a two-position controller. The difference between these two modes is that a floating mode can send its binary (on/off) signal over one of two circuits that connect the controller's output to the final controlled device. One circuit modulates the final controlled device open while the other modulates it closed. The modulating movement of the final controlled device creates a balance between the mass or energy transfer and the load. When the control point is within acceptable range of the set point, the controller stops sending a signal to the actuator and the final controlled device remains at its current position. It remains there until the controller commands it to another position by applying a signal to one of its circuits.

Due to the wiring configuration of the controller and final controlled device, floating mode is also called *three-wire float* mode. A floating final controlled device has three wire terminals usually labeled clockwise (CW), counter clockwise (CCW), and common. When a signal is applied to the common and clockwise terminals, the actuator turns in one direction. When the signal is applied to the common and counter-clockwise terminals, the actuator moves in the opposite direction. When no signal (hold) is applied to either the clockwise or counter-clockwise terminals, the flow control device remains stationary at its present position.

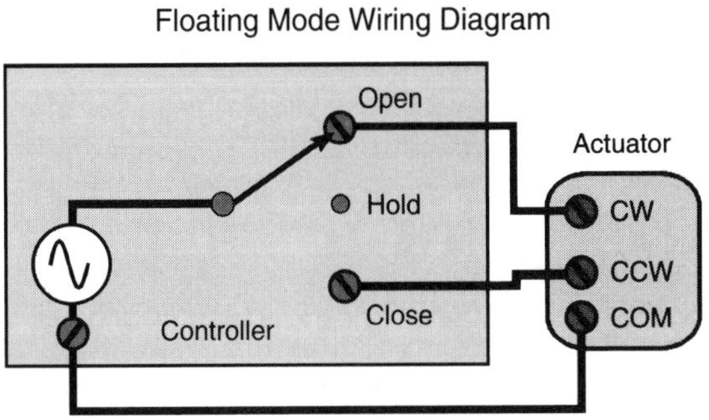

Floating Mode Wiring Diagram

Figure 8.8 Floating Controller Terminal Detail

Figure 8.8 shows that if the final controlled device is to be opened, a signal is sent to the clockwise rotation circuit over the top and common wires that connect the actuator with the controller. Upon a command to modulate in the closed direction, the bottom and the common wire complete the counter-clockwise circuit between the controller and the actuator. When the control point is within acceptable range of the set point, the controller opens both circuits and the actuator remains in its current position. The requirement of having the final controlled device remain stationary whenever the controller interrupts its output signal limits floating mode to electronic control systems because a pneumatic actuator automatically returns to its failsafe position when the controller's output signal is interrupted.

8.4.1 Operational Characteristics of a Floating Mode Final Controlled Device

The operating characteristics of the final controlled device make it possible for the control agent to be modulated into the process. The floating mode actuator converts the two-state signal from the controller into three distinct operating routines, open, closed, and hold. Recall, the hold signal is not found in either two-position modes. It is this third state that allows the final controlled device to be positioned at any point within its operating range.

The final controlled device has an electric motor that rotates at a *fixed rate of speed* (revolutions per minute) when a signal is applied to its CW or CCW terminals. Therefore, the actuator shaft also extends (or rotates) at a fixed speed producing a modulating response from the binary controller signal. When a signal appears on the CW or CCW terminal of the actuator, the motor begins to turn at a fixed rpm. It continues to rotate until the signal from the controller indicates the actuator must hold its present position. Whenever the controller issues a hold command a braking circuit in the actuator holds the current position of the flow control device. The actuator remains in that position until the controller commands it to move in either direction. The final controlled device's motor repositions the valve or damper until the control point moves back into acceptable range of the set point or until the actuator reaches the physical end of its stroke. The constant speed of rotation along with the bi-directional motor and internal braking circuit limit the floating final controlled devices to floating controller mode applications.

8.4.2 Floating Controller Mode Response

Figure 8.9 shows the floating controller's response for a cooling process. It is similar to the two-position process depicted in Figure 8.2. Note the differences between both controller mode responses. Recall, a two-position controller mode generates either an on signal or an off signal at all times. A floating controller generates one of three possible signals, CCW, CW, or HOLD. During hold intervals, the final controlled device is not being commanded open or closed but holds its last commanded position until a change in signal occurs.

The response in Figure 8.9 shows an area surrounding the set point similar to the deadband produced by the control differential of a two-position mode. The basis for the similarity is that a control differential or deadband is a characteristic of all controllers that generate binary output signals in order to minimize short cycling or oscillations. Therefore, a floating mode controller, which also generates a binary output response, has a control differential called the *neutral zone*. This deadband prevents the controller's output from oscillating between the CW and CCW terminals whenever the process control point is near the set point.

The neutral zone is the shaded area shown in Figure 8.9. During the periods when the control point is within the neutral zone, the controller generates a hold signal. It maintains the hold response until the control point exceeds either edge or limit of the neutral zone. Once the control point exceeds either limit, the controller commands the final controlled device to modulate in a direction that forces the control point back into the neutral zone. When the control point moves back within the neutral zone, the controller signals the actuator to hold its present position, maintaining the flow through the final controlled device at its current value.

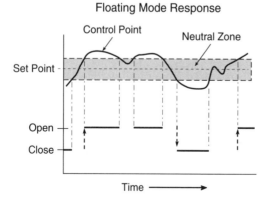

Figure 8.9   Floating Controller Response

Whenever a hold command is executed by the controller, the control point is allowed to "float" within the neutral zone. If the load of the process is constant, the controller may not generate a command to change the position of the final controlled device for hours at a time. This response is shown in Figure 8.10.

8.4.3 Advantages and Disadvantages of Floating Controller Mode

Floating controller mode is only applicable to closed loop applications. A feedback path is required to maintain the control point within the neutral zone of the process. One important advantage of using a floating mode over closed loop two-position mode lies in the modulating action of the final controlled device. Energy is not pulsed into the process as it is in two-position applications so a floating mode final controlled device has an infinite number of positions and therefore, flowrates, making it possible to balance the energy being transferred with the process load. This is a major improvement over the two-positions available with a two-position mode.

Another advantage of floating mode over two-position mode is that the control point does not have to transverse the entire neutral zone before a change in the flowrate of the control agent can be initiated. During times of increasing (or decreasing) load, the control point can oscillate along one edge of the neutral zone, slowly incrementing the flow of the control agent to offset slight variances in the load. This maintains the control point closer to the set point than is possible with two-position controllers. In two-position applications, the control point is actually forced through the entire control differential before the controller responds to changes in the control point.

The only disadvantage of floating controller mode is the presence of the neutral zone in the controller's transfer function. This response deadband creates a delay in the process that allows the control point to deviate away from the set point as part of its normal response. Although its response is better than two-position mode, the neutral zone is still a disadvantage in a controller mode if comfort and efficiency are to be optimized.

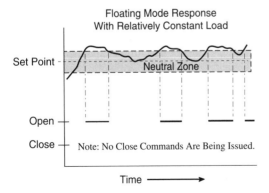

Figure 8.10   Floating Controller Response to Constant Load

8.5

PROPORTIONAL CONTROLLER MODE

Proportional controller mode is the simplest of the truly modulating controller modes. Proportional controllers generate an immediate change in the position of the final controlled device in response to changes in the process load.

Proportional mode controllers generate changes in their output signals that are *proportional* to the changes that occur in their control points. As soon as any measurable change occurs in the process, the controller generates a corresponding change in its output signal to vary the flow of the control agent. This immediate response distinguishes this mode from the two-position and floating modes. In proportional mode there is no need for a control differential or neutral zone in the transfer function because the controller generates an analog output signal, not binary signal. Consequently, there is no possibility of the final controlled device short cycling when the control point hovers near the set point.

The output signal of a proportional controller is always present. Its magnitude varies in response to changes in the control point to maintain a balance between the energy transferred into the process and its present load. The response of a proportional controller is shown in Figure 8.11. Notice the controller's output signal follows a pattern similar to the changes that occurred in the control point. The similarity between the two curves results from the response of the proportional gain in the controller's transfer function. As the control point changes, the controller's transfer function adjusts its output signal in proportion to the change in the control point. Notice that during unexpected load changes, the control point can move beyond of the limits of the throttling range. During these intervals the process operates out of control, the controller's output signal is at one of the limits of its range (0 or 100%), and the final controlled device will be commanded fully open (or closed). The process remains out of control until the load changes sufficiently to cause the control point to move back within the limits of the throttling range.

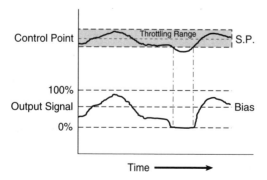

Figure 8.11   Proportional Mode Response

Proportional mode can only be used in closed loop applications. Its final controlled device must be capable of being positioned at any point between the minimum and maximum limits of its operating range by a modulating control signal. Unlike floating controller mode, if the signal to a proportional mode final controlled device is interrupted, most actuators will return the controlled device to its normal or failsafe position.

Proportional mode improves over two-position and floating mode responses by removing the control differential and neutral zone from the transfer function. The elimination of this characteristic allows the controller to change its output as soon as it receives a signal from the sensor indicating a change in the control point has taken place. This response maintains the control point closer to the set point, improving comfort and operational efficiency. Proportional controller mode can be used in pneumatic, analog, and digital electronic control systems.

8.5.1 Steady-State Offset

The purpose of proportional mode is to create and maintain a balance between the energy being transferred with the process and its present load. Once this balanced condition has been achieved, the controller's output stabilizes and the control point is at some value within the throttling range of the process. Recall, whenever the control point is within the throttling range, the loop is considered to be *in control* because the throttling range indicates the acceptable limits of the control point for a given process.

Since the proportional mode is only designed to balance the energy transfer with the load and *not* to maintain the control point equal to the set point, the control point seldom equals the set point in applications using proportional controller mode. Instead, the control point stabilizes at a value as soon as the energy transferred into the process balances with its load. With proper calibration, the control point will always be very close to the set point.

The offset between the control point and the set point that remains after the controller output signal stabilizes is called the *steady state error*. The size of the steady-state error is inversely related to the controller's proportional gain. As the proportional gain increases, the steady-state error decreases. A larger proportional gain generates a greater change in the flow of the control agent in response to small changes in the control point. Therefore, the loop responds faster, minimizing the growth of the error and consequently, the steady-state error. Figure 8.12 shows the steady-state error remaining after the controller has stabilized.

8.5.2 Advantages and Disadvantages of Proportional Mode

The advantages of using proportional mode over floating mode is the absence of the deadband in the controller's transfer function. By removing this delay from the transfer function, the total delays of the control loop are reduced. This improves comfort and process efficiency. The only delays in proportional mode control loops are due to those associated with the process. The lack of a deadband

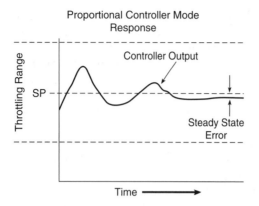

Figure 8.12   Proportional Mode Steady-State Error

and the modulating characteristics of the control loop allow the control point to remain closer to the set point over all load changes.

The only disadvantage of using proportional controller mode is the presence of the steady-state error, indicating some inefficiencies are present within the process. When a proportional mode controller is properly calibrated, the steady-state error is minimized, resulting in a tighter control response than either two-position or floating modes.

8.6

PROPORTIONAL + INTEGRAL CONTROLLER MODE (PI)

Integral mode is added to a proportional mode transfer function to modify the controller output signal in a manner that forces the steady-state error to zero.

Integral mode is used in conjunction with proportional mode in modulating, closed loop control applications to eliminate the steady-state offset that remains after the proportional mode has performed its initial response. Integral mode eliminates the only disadvantage associated with a proportional mode controller.

The integral function is made possible by adding the integral mode element to the proportional mode transfer function. Therefore, the PI controller mode has two gain parameters, proportional gain and integral gain. The function of the proportional gain is to make an initial response that will balance the energy transfer with the new load. The function of the integral gain is to modify the controller's output signal in proportion to the size and length of time the error exists. Once the proportional gain has completed its modification to the output signal, the integral mode alters the output signal in proportion to the time weighted error. This addition to the output signal disrupts the balance established by the proportional gain, providing more mass or energy transfer than needed, forcing the control point back toward the set point. The contribution from integral mode is equal to the mathematical integration of the error signal, which determines how large the error is and how long it has existed, multiplied by the integral gain parameter of the transfer function. The integral mode gain parameter is called I gain or reset.

The amount of signal added by the integral mode varies in proportion to the length of time which the error exists. As the length of time a process error is present grows longer, the contribution from the integral mode element of the transfer function increases.

This relationship also holds true with respect to the size of the error. The greater the magnitude of the process error, the greater the contribution the integral mode makes to the output signal of the controller. The integral portion of the mode continues to change the output signal until the loop error equals zero. When no error exists, the output stays at its present signal level until another load change occurs. Figure 8.13 shows the response of a proportional plus integral (PI) mode after the loop has reached steady state. Notice that there is no error remaining once the controller output signal stabilizes.

Controllers incorporating integral gain are susceptible to a phenomenon called *integral windup*. Windup describes a condition that occurs when the output signal is driven beyond the limits of its linear range due to some abnormal process condition that prohibits the control point from being reset back to the set point. Under these circumstances, the integral portion of the transfer function continues to add to the controller's output signal because the length of time the error exists continues to grow. Integral windup locks the final controlled device at one of its operating limits and the process remains out of control until it is manually reset. To minimize the possibility of windup occurring, controllers incorporating the integral gain algorithms have built-in anti-windup strategies. These strategies limit the amount of change the integral mode can add to the output signal during a response.

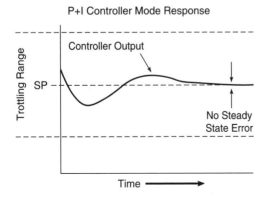

Figure 8.13   Steady-State Response of a P + I Controller Mode

8.6.1 The Advantages and Disadvantages of Proportional + Integral Mode

Integral gain restores the control point back to the value of the set point after all load changes. This response eliminates the disadvantage of the existence of a steady-state error that is associated with a proportional only controller mode. When properly calibrated, a PI control loop will improve the operating efficiency of a process beyond the response of a proportional control strategy.

The two disadvantages associated with PI mode are the possibility of the loop windup during system start up or unusual operating conditions and the increased difficulty in calibrating multiple gain control loops. A PI controller has two gain parameters that must be carefully selected during the calibration process. If either of these gains are incorrectly set, the loop has an increased possibility of becoming unstable. An unstable loop oscillates for long periods of time, decreases the operating efficiency of the process, increases wear on loop components, and generates comfort problems.

8.7

PROPORTIONAL + INTEGRAL + DERIVATIVE MODE (PID)

PID controller mode is also used in closed loop modulating applications. It is used for processes that require a quick response to all load changes with no steady-state error remaining after the loop has stabilized. Derivative mode provides an anticipatory response to the controller by altering its output in proportion to the rate of change of the process error.

The derivative mode has a gain parameter called the *D gain* or *rate.* The derivative component modifies the controller's output signal in proportion to the rate at which the error is changing. The contribution from derivative element is equal to the derivative of the error signal, which calculates how fast the error is changing, multiplied by the derivative gain parameter. This value is added to the output signal produced by the other elements present in the controller's transfer function. Therefore, derivative mode only plays an important part in the loop response when the load is changing quickly. Under these circumstances, its contribution to the controller's output signal is greatest. When the load changes slowly, the contribution of derivative element is minimal because its gain is multiplied by a small signal (derivative of the error). Under fast load changes, the derivative mode limits the amount of change in the controller's output signal. This braking action limits the growth of the error and the oscillations that may be produced by fast load changes. By anticipating the effect of the load change, the derivative mode quickly positions the valve or damper so the error does not have the opportunity to grow.

Derivative mode is typically used in conjunction with proportional or proportional plus integral mode to allow the loop to respond to the speed of change in the error. The net effect of the derivative component is to oppose changes in the magnitude and direction of the control point by altering the controllers output signal in proportion to the rate of change of the error. This restraining action keeps the control point closer to the set point during fast load changes.

8.7.1 The Advantages and Disadvantages of Proportional + Integral + Derivative Mode

PID loops produce the most advanced modulating control response available. The three gain elements of the controller's transfer function allow the controller to respond to the change in the size, direction, speed at which the control point is changing along with the magnitude of an offset and the length of time it exists. When properly calibrated, a PID control loop will improve the operating efficiency of a modulating control loop under all expected changes in process load.

The only disadvantage of a PID controller is the need to calibrate a controller with three gain parameters. Their values must be carefully selected during the calibration process to maintain the process at set point with a minimum number of oscillations during the response period. If this mode generates an unstable response, the loop can oscillate uncontrollably without stabilizing. Calibration techniques are presented in subsequent chapters that guide the technician in the tuning of these controllers.

8.8

ANALYZING THE RESPONSE OF MODULATING CONTROLLER MODES

To analyze the response of a controller, a load change is initiated and the controller's output signal is graphed over time. An actual load change would appear as a curved line similar to the one shown in Figure 8.11. The curved response is due to the dynamics of a real process change. Consequently, the response of a proportional controller would also be curved, making it difficult to identify the change in the controller's output signal that results from each of the elements in its transfer function. To simplify controller response analysis, the load change is commonly represented by a step change in the process load or set point (see Figure 8.14). This method uses a straight vertical line to represent a load change that instantly changed a finite amount. This procedure is called a step change because it has the appearance of a step or stair. Using step changes to represent load changes simplifies the explanation of the controller's response to a change. In real applications, a step change is impossible to stimulate because of time delays in dynamic processes.

8.8.1 Proportional Mode

Figure 8.14 shows a step change occurring in the process error (load) and the response of a direct acting proportional controller. Note, when the proportional

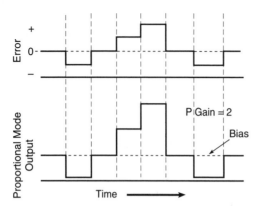

Figure 8.14 Proportional Mode Response to a Step Change in the Error

controller receives a signal from the sensor that indicates a step change has occurred in the load, a corresponding step change in the process error is calculated by the controller's summing junction. This step change representing the process error is multiplied by the proportional gain of the controller's transfer function and added to the output signal bias to generate the output signal. The proportional gain generates a response that duplicates the step changes in the error.

Whenever the error experiences a step change, the output of this direct acting controller changes in the same direction, by an amount proportional to the change in the error. Positive errors produce increasing controller output signal responses while negative errors generate a complementary controller response. Note, when the error is equal to zero, the controller's output signal is equal to its transfer function bias. The differences between the size of the step change in the controller output signal and the change that occurred in the error is a function of the proportional gain. If the proportional gain is set equal to one, the error and controller response curves will be identical. When the gain is greater than one, as depicted in Figure 8.14, the controller's response curve is larger than the error's curve.

8.8.2 Proportional + Integral Mode

Figure 8.15 shows the response of the integral mode to step changes in the process error. The graph shows that initially, when there is no error and the output of the controller is stable, the integral element produces no change in the controller's output signal. When a negative error occurs, as represented by the first step change in the negative direction, the integral mode generates a negative signal that *ramps* or inclines downward. The slope of the ramp is a function of the magnitude of the change that occurred in the error signal. The greater the error changed, the steeper the slope.

The straight line symbolizes the response that indicates the total area under the error curve is steadily increasing as time elapses. This sloped line increases in length until the error is eliminated at which point the output signal of the con-

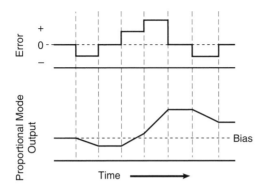

Figure 8.15   Integral Mode Response to a Step Change in the Error

troller stabilizes and remains constant until another load change occurs. This is represented by a horizontal line depicting a leveling off of the controller's output signal. Note that the controller's output signal does not return to the bias value where it originally started. This is an identifiable difference between proportional and proportional plus integral controller modes.

The next load change shown in Figure 8.15 represents a positive step change in the error. Under these conditions, the integral mode generates a positive sloped ramp. As the load continues to grow, it generates a greater error. The corresponding change in the controller's output signal is represented by an increase in the slope of the output signal's line. The increasing output signal elevates the rate of energy transfer taking place, thereby forcing the error toward zero. This pattern of response continues as the load changes. Remember, the slope of the line is related to the magnitude of the error while the length of the line is related to the length of time the error exists.

Figure 8.16 shows the combined response of both elements of a PI controller. The graph shows the integral gain's response being added to the proportional gain's response, generating the total change in the controller's output signal. In this graph, when an error occurs, the proportional mode instantly produces a step change in the output signal that is related to the size and direction of change in the error. The integral element adds its response onto the end of the proportional step change, increasing the controller's output signal beyond the level attained by the proportional step response. The controller's output signal continues to increase as long as an error is present. Ultimately, the increases in the output signal forces the offset back to zero.

When the error is finally driven to zero, the contribution produced by the proportional mode is also driven to zero because the error (0) times the proportional gain equals a change of zero units. Remember, when the error is driven to zero, the controller output *does not* equal the controller's output signal bias as it would if the controller had a proportional only mode.

The integral gain parameter is sometimes called the controller's *reset time.* Reset time is the amount of time it takes for the integral mode to change the controller's output signal by an amount equal to the initial step change produced by the

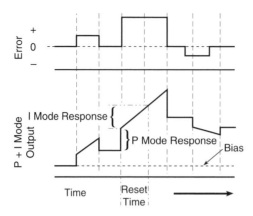

Figure 8.16   P + I Response to a Step Change in the Error

proportional element of the transfer function. For example, if the initial step change produced by the proportional element changes the output signal one unit, the reset time is equal to the amount of time (minutes) it takes for the integral element to add an additional one unit of signal to the controller's output. A long reset time is represented by a long response line having a small slope that approaches a horizontal line. When the reset time is minimized, the slope of the response line is greater (approaching vertical) and the length of the line is shorter.

The reset time is inversely related to the value of the integral gain. As the value of the integral gain is increased, the reset time decreases. Decreasing the reset time allows less time for the integral element to match the proportional response. To generate the same change in less time, the slope of the ramp must increase and is represented by a steeper line.

8.8.3 Proportional + Integral + Derivative Mode

Figure 8.17 shows the response of the derivative element to a change in the error. In this representation, step changes are not used to represent the error because a vertical rise represents an *infinite rate of change.* An infinite rate of change in the error would generate an infinite change in the controller's output signal which is not a logical response. Instead, the derivative response analysis uses sloped lines to represent changing error signals.

The derivative element's contribution to the controller's output is directly related to the rate of change of the error. As the rate of change increases, as represented by a steeper sloped line, the derivative element's contribution to the output signal increases. When the error is not changing, the derivative element's contribution is zero because derivative gain times a derivative of zero equals zero.

A PID response combines the individual responses of all three modes into one curve. The complexity of the response increases as a function of the error's dynamic characteristics. Figure 18.18 depicts a response of a PID controller.

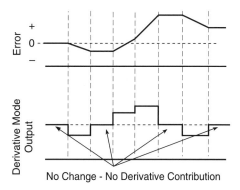

Figure 8.17   Derivative Mode Response to a Step Change in the Error

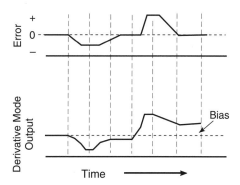

Figure 8.18   PID Response to Changes in the Process Error

8.9

SUMMARY

There are seven controller modes commonly used in HVAC applications. The simplest mode, two-position, is used in open loop applications where a safety limit needs to be maintained. It is also used in closed loop applications where the control point is maintained within an acceptable range of the set point by pulsing mass or energy into the process. Two-position mode is not used in comfort temperature applications because the operating differential will be noticed by the occupants.

Timed two-position mode adds an anticipation circuit to the control loop to reduce the operating differential, making it the choice of two-position modes for comfort applications. The anticipator circuit adds heat to the sensor's sensing element causing the controller to respond before the actual process temperature exceeds the control differential of the controller.

Two-position modes have a control differential that limits the rapid cycling of the controller's output, reducing wear on the mechanical equipment. The control differential creates oscillations in the control point that are magnified by the time delays that are present in all dynamic processes. The operating differential is the range of the process control point. Its size is a function of the size of the control differential and the amount of time delays that occur in the process. The magnitude of the operating differential must be minimized in order to maximize the efficiency of the process.

The problems associated with the two-position mode's operating differential are reduced using a floating controller mode. Floating mode uses a binary controller output signal to position a modulating final controlled device. This hybrid of two-position and modulating controls improves the process response by ramping mass or energy into the process instead of pulsing it in. Some of the same inefficiencies associated with the control differential exist with floating mode because it incorporates a neutral zone to limit oscillations in the controller's output signal whenever the control point is close to the set point. The ramping flow response reduces the negative effects of the neutral zone, minimizing its effects on occupant comfort.

Proportional controller mode is used to improve operational efficiencies and comfort. This mode generates a modulating output signal to position a modulating final controlled device. The controller's analog output signal does away with need for a the control differential or neutral zone, making it possible for the controller to alter its output signal as soon as a change in the process is measured. The response of the controller balances the mass or energy transfer with the process load. Consequently, a steady state error always remains after the controller has stabilized.

Integral mode can be added to a proportional controller's transfer function to improve the operational efficiency of the process. Integral mode responds to the size and length of an error's existence. The combined response of proportional and integral modes generates an output signal that continues to change until the control point equals the set point, maximizing comfort and efficiency.

To incorporate the advantages of proportional plus integral controller mode in fast changing load applications, derivative mode is added to the controller's transfer function. Derivative mode anticipates and minimizes the effects that fast changing loads have on the process. It produces a braking action that limits the amount of output signal change created by the other modes in the controller's transfer function. The controller's output signal will quickly change to a value that will minimize the growth of a fast changing error. The overall effect of the PID response is to minimize the range of control point during fast process changes.

8.10

EXERCISES

Determine if the following statements are true or false. If any part of the statement is false, the entire statement is false. Explain your answer.

1. Timed two-position controller mode pulses energy into a process.
2. The integral gain responds to the rate of change of the process error.
3. As the rate of change of the error decreases, the change in the PI controller's output signal increases.
4. A heat anticipator of a timed two-position mode turns off the furnace heating process earlier than a two-position mode would.
5. The control differential of a two-position mode is greater than its processes' operating differential.
6. A floating mode controller generates three possible output signals.
7. As the length of an error's existence increases, the contribution of the derivative element of a PID transfer function also increases.
8. The more gains there are in a transfer function, the greater the chances are that the loop may become unstable.
9. Integral and derivative gain elements can be used without proportional gain in a controller transfer function.
10. When the error is brought back to zero in a PID loop, the output signal from the controller equals its output signal bias.

Respond to the following statements and questions completely and accurately, using the material found in this chapter.

1. What is the purpose of a control differential?
2. What causes the operating differential of a process and why is it always larger than the control differential?
3. Why isn't two-position mode used in zone temperature control applications?
4. What advantages does floating mode have over timed two-position mode?
5. What is the purpose of the neutral zone in a floating mode transfer function?
6. Do floating mode process controllers have an operating differential? Explain your answer.
7. Describe the response of a floating mode final control device when
   a. the control point exceeds the neutral zone in an air-conditioning process.
   b. the control point returns into the neutral zone.
   c. the control point falls below the lower limit of the neutral zone.
8. What is a disadvantage of floating mode?
9. Why is proportional mode better than two-position and floating modes?
10. What is a disadvantage of proportional mode?
11. What is a step change and why is it used in controller response analysis?
12. What are some advantages of higher proportional gain settings?
13. What is the function of integral mode?
14. Does the slope of the integral mode response increase or decrease when the process error increases in magnitude?
15. Does the slope or length of the response line of the integral mode increase when the length of time an error exists increases? Explain your answer.
16. What is the advantage of PI mode over proportional only controller mode?
17. What is the function of derivative controller mode?

18. Why is derivative mode sometimes called rate mode?
19. As the rate of change of the error decreases, will the contribution of the derivative mode increase or decrease? Explain your answer.
20. What effect does derivative mode have on the steady-state error of fast changing process using a PID controller mode when compared to the same process using a proportional only mode?

CHAPTER 9

Modulating Final Controlled Devices

Valves and dampers are the most common flow control devices used in HVAC processes to vary the amount of air, steam, or water being delivered to a process. These devices are available in a variety of sizes and configurations to match the requirements of the application. Chapter 9 describes the construction and operating characteristics of the common valves and dampers used in HVAC applications.

OBJECTIVES

Upon completion of this chapter, the reader will be able to:

1. Describe the construction and operational characteristics of valves and dampers.
2. Explain how pressure drop effects the response of a flow control device.
3. Describe the adverse affects of improperly sized devices on loop operation.

9.1

VALVES AND DAMPERS

Valves are used to vary the quantity of *liquids, vapors, and gases* flowing through *piping* systems. *Dampers* are used to control the flow of *air* and *gases* within building *duct* systems. Flow control devices from both categories are used to vary the amount of the control agent that is being delivered to the process. The output signal of the loop's controller positions the actuator of the final controlled device, regulating a variable opening of the flow control device through which the control agent flows on its way to the process. As the flow area in the flow control device increases, its resistance to flow decreases, allowing more control agent to be transported to the process. Conversely, when the actuator moves to close the valve or damper, the size of the flow area decreases, increasing in the resistance of the flow path and reducing in the amount of control agent flowing to the process.

Valves come in several different configurations to meet the diverse requirements of HVAC processes. They are constructed with different materials in order to match the temperature, pressure, and corrosion characteristics of the control agent. Valves and their actuators are configured so the final controlled device fails safely, either open, closed or maintain its current position upon the loss of its control signal. They are also available in many different sizes to meet the capacity requirements of the process.

Dampers perform functions similar to those performed by valves but they are manufactured in fewer configurations. Dampers are available in different sizes to match the capacity and pressure specifications of the process. The construction and operational characteristics of valves and dampers are presented in the remaining sections of the chapter.

9.2

VALVE CONSTRUCTION

A control valve is comprised of a *valve body, bonnet*, and *internal trim components*. The valve body is typically a metal casting having one or more passages for the control agent to enter and leave the valve. The body has inlet and outlet connections to attach the control valve to the piping system. These connections are available in different styles to properly mate with the type of piping used in the system. A valve can be specified with flanged connections to bolt the valve to the flanges on the pipes in the system, flange-less connections for compression installation applications, smooth fittings for welded installations, flare or threaded connections.

Valves are also classified as having a two-way or three-way body. Two-way valves have one inlet and one outlet flow connection. Two-way valves are used in applications where the valve is placed in *series* with the heat transfer coil. In these series applications, all of the control agent entering the valve is delivered to the process. Many of the two-way valves used in HVAC applications are *globe* valves. A globe valve gets its name from the globular shape of its valve body. They are primarily used in automatic control applications to vary the flow of the control agent in response to changes in the controller's output signal. Figure 9.1 shows a two-way globe valve with flange connections.

Butterfly, gate, and ball valves are other types of two-way valve body styles used in HVAC applications. Butterfly valves are used to modulate large flowrates and are typically used in cooling tower applications. Gate and ball valves are used to manually isolate coils, strainers, control valves, and other system components so they can be removed from the piping system for repair or service.

Three-way valves have a valve body with three flow connections. They are used in applications that either *mix* two fluid streams into a single stream or *divert* one fluid stream into two separate flow streams. A three-way *mixing* valve has two *inlet* connections and one *outlet* connection. Figure 9.2 shows the flow connections of a three-way mixing valve. A three-way *diverting* valve has two *outlet* connections and one *inlet* connection. The two types of three-way valves are not interchangeable for a given application. If a mixing valve is used in a diverting application or a diverting valve used in place of a mixing valve, the process will not respond as required.

Mixing valves are used to vary the temperature of the control agent in a heat exchanger coil by modulating the quantity of water entering the coil and bypassing the remaining flow around the coil. Figure 9.3 shows a piping diagram of a mixing valve application. The control agent is supplied to the heat transfer coil through a pipe tee. One outlet of the tee is connected to the coil's inlet and the other is connected to the bypass pipe. One of the valve's inlet connections is piped to the coil's outlet connection and the other is connected to the pipe that is used to bypass the control agent around the coil. The mixing valve combines the flow from these two paths into one flow stream that is returned to the mechanical equipment that initially transferred the energy to the water. The loop's controller regulates the volume of control agent entering the coil and therefore, the energy transferred to the process. If there is too much heat being transferred by the

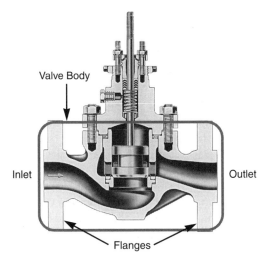

Figure 9.1   Two-Way Valve Body

3 Way Valve

Inlet A   Outlet AB

Inlet B

Figure 9.2   Flow connections of a 3 Way Mixing Valve

3 Way Mixing Valve

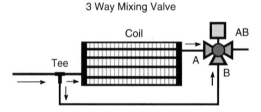

Figure 9.3   Mixing Valve Application

control agent, the controller will reduce the flow of water entering the coil, thereby increasing the volume being bypassed and reducing the energy being transferred to the process.

Three-way *diverting* valves have a configuration that mirrors the piping of a mixing valve. Diverting valves have one inlet connection and two outlet connections and are installed on the upstream side of the heat exchanger. A pipe tee is located on the outlet side of the heat exchanger allowing the bypassed and coil fluid streams to be recombined into one flow stream.

Figure 9.4 depicts a diverting valve application that maintains a constant chiller head pressure by maintaining an 85° condenser water temperature set point. The diverting valve modulates the quantity of condenser water flowing through the cooling tower. In this application, all of the control agent enters the valve, located upstream of the tower. Some of the control agent is allowed to be cooled as it flows through the heat exchanger tower while the remaining volume is diverted around the tower. The two streams are recombined and sent to the condenser at the

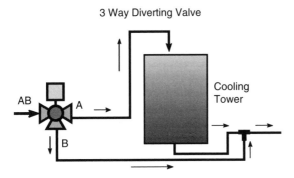

Figure 9.4   Diverting Valve Application

proper temperature. Therefore, when the condenser water temperature increases, the loop's controller diverts more water to the cooling tower to lower the temperature of the recombined fluid stream.

To simplify identification and installation of three-way valves, their valve bodies are cast or stamped with arrows or letters (A, B, & AB) that indicate the *required* direction of fluid flow through the valve body. **A** and **B** designate the two individual flow streams and **AB** indicates a combined flow stream. These designations assist the installer in determining the inlet and outlet connections of the valve. Figures 9.3 and 9.4 show the AB designations of the flow connections.

9.2.1 Valve Trim Components

The valve's *trim components* are the group of parts that work together to control the flow of fluid through the valve's port. The trim package includes the valve's *plug*, *stem*, and *seat ring*. The *plug* is the trim component that varies the area of the flow passage within the valve body. When the controller commands a change in the flowrate of the control agent, the actuator repositions the plug, altering the resistance of the flow passage. As the passage resistance changes, the flow of the control agent through the valve also changes.

The *seat ring* surrounds the edge of the valve *port*. The port is not a piece of valve trim it is the opening cast in the center of the valve body that provides the passage between the inlet and outlet connections of the valve. The seat ring provides the flat, smooth surface that the plug seals against to stop the flow of the control agent. As the plug moves away from the seat, flow through the valve is reestablished.

The valve *stem* connects the actuator to the valve plug. The stem repositions the plug in response to changes in the controller's output signal. The stem is machined with a smooth surface so it moves easily through the valve's packing. The packing prevents the control agent from leaking around the surface of the stem where it passes through the valve bonnet. Stems are made of stainless steel to minimize the formation of rust on their surface that can inhibit free movement of the stem. Figure 9.5 shows the various trim components in a valve.

Valve Trim Components

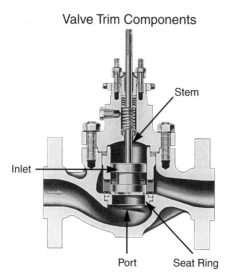

Figure 9.5   Valve Trim Components

9.2.2 Valve Ports

The *port* is the orifice located in the middle of a valve body where the fluid flow is modulated (two-way) or directed (three-way) by the valve plug. Valve bodies can be purchased in one port or two (double) port configurations. A single-port valve has one flow orifice, seat ring, and plug. Globe, butterfly, gate and ball valves are all single port valves. Double port valves have two seat rings, two flow orifices, and a special plug configuration. A double-ported valve can have a single plug that has two contoured surfaces (top and bottom) or it can have two individual plugs mounted on the same valve stem.

Both mixing and diverting valves are double-port valves. Flow is directed between their ports and connections by altering the position of their plug(s) with respect to their ports. When the valve stem is positioned at one end of its stroke, one of the ports will pass 100% of the flow and the other will be closed. When the valve is commanded to the other end of its stroke, the other port allows full flow of the control agent. The total flow is split proportionately between both ports whenever the actuator's signal falls between 0% and 100% of its operating range.

Two-way valves are also available with double ports. These valves are used in larger flowrate or pressure applications where the actuator must generate large forces to close against the flow of the control agent. The double-port configuration splits the fluid flow into two streams before they pass through the valve ports. These separate streams generate forces on the plug that are in opposite directions. One force works to close one plug while the other force works to open the other plug. These opposing dynamic forces cancel each other reducing the force requirements of the actuator. Therefore, a smaller actuator is required to overcome the unbalanced force that exists across the valve plugs.

The addition of the second port and its associated clearances prevents double-ports valves from providing tight shutoff against the flow of the control agent.

When the valve is commanded closed, one port will be seated tightly, but the other usually leaks a small percentage of fluid. Consequently, the advantages of a smaller actuator must be weighed against the inability to stop the flow of the control agent when the controller commands the valve closed.

A *balanced port valve* can be used to provide tight shut off capability in high fluid pressure applications instead of using a double-port valve. A balanced port valve has a port and plug configuration similar to other single port valves. The difference is that the plug has a small hole drilled through it's center that allows the pressures across the valve to equalize when the valve is closed. Equalizing the pressures effectively cancels the forces produced by the fluid pressure difference across the valve port. This configuration permits a tight shutoff because the actuator can firmly seat the plug against the single valve seat. It also reduces the actuator's force requirements.

Butterfly valves differ in appearance from the other valves previously described. These valves do not have a globular body. Instead, they appear as a short sleeve of pipe that has the same inside diameter as the system's piping. A disc with a slightly smaller diameter is mounted in the center of the valve body. When this disc is rotated by the actuator, it restricts the flow area and consequently, reduces the flowrate through the valve.

Butterfly valves require an actuator that generates rotational motion in order to turn the disk within its valve body. The valve has as an edge seal on the disk or a seal around the interior circumference of the valve body that permits a tight shutoff. These valves are used to control large volumes of fluids with a minimal pressure loss. Two butterfly valves can be coupled together by linkage or control signals to create a three-way valve configuration as shown in Figure 9.6.

9.2.3 Valve Bonnet Assembly

The bonnet assembly supports and guides the valve stem so it can correctly position the plug within the center of the seat ring. The valve stem extends out the top

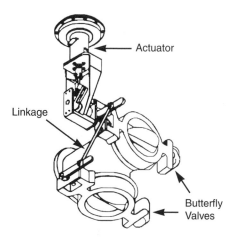

Figure 9.6 Butterfly Valves in Three-Way Configuration

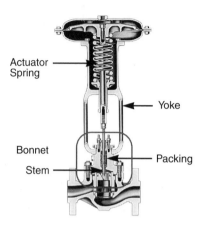

Figure 9.7   Valve Bonnet, Yoke and Packing Location

of the bonnet where it is coupled to the valve's actuator. The bonnet also contains a packing module that seals the opening at the top of the bonnet while allowing the stem to move freely in and out of the bonnet. The packing prevents leakage of the control agent as the stem moves up and down in response to changes it its control signal. The type of packing method used in an application is a function of the pressure, temperature, and chemical characteristics of the control agent.

9.2.4 The Valve Actuator

The valve actuator converts the controller's output signal into linear or rotational movement. This movement repositions the valve plug within the seat ring. The actuator's size is chosen so it can generate the force needed to close the port against the forces generated by the moving control agent. The actuator must also be strong enough to modulate against the forces created by its own internal actuator spring and the packing in the bonnet. A *yoke* connects the actuator to the valve body assembly. This rigid support permits all the movement created by the actuator to be transferred to the stem and plug assembly without any appreciable reduction in accuracy. Figure 9.7 shows the bonnet and actuator assembly.

9.3

HOW FLUIDS REACT TO CHANGES IN THE FLOW AREA OF THE VALVE PORT

Valves modulate the flow of fluids in a system by altering the open or free area through the port. Varying this area generates changes in the resistance of the flow path along with the proportions of potential and kinetic energy in the moving fluid. These changes in the properties of the fluid must be understood before a technician can properly analyze the response of a control loop that contains modulating valves.

9.3.1 Gases and Vapors

A gas is a fluid that exists at a temperature that is much higher than its saturation temperature. The saturation temperature of a gas is the temperature at which the fluid changes from its liquid state to a vapor, at a given pressure. The saturated temperature can also be stated as the temperature at which the vapor condenses back into its liquid state. Air flowing through an air handling units is considered a gas because it exists at a temperature that is over 200 ºF above its saturation temperature. At atmospheric pressure, air is unable to condense into its liquid state within HVAC processes.

Vapors are gaseous fluids that exist at temperatures close to their saturation point. Saturated steam is an example of a vapor. In HVAC processes, it usually exists at a temperature that is relatively close to its saturation temperature. As steam gives up its latent heat within a heat transfer coil, it changes its state from a vapor back into a liquid.

The energy absorbed when a liquid is converted into a gas is used to break the chemical bonds between the liquid molecules. Once separated from their companion molecules, each vapor molecule is free to move within its container. Therefore, gases and vapors always expand to fill the volume of their storage vessel. Space fills the void between the gas molecules within their container. Consequently, gases and vapors are *compressible*. Compression forces work to move the molecules closer together, thereby increasing the pressure and density of the gas. As soon as the compressive force is removed, the gas reexpands to fill the container's volume.

9.3.2 Liquids

Liquids are fluids that take the shape of the container that they are stored in. The liquid will form a free surface at the boundary between itself and the atmosphere. Liquids do not diffuse to fill the surrounding volume because of the strong cohesive forces that exist between its molecules. Therefore, a liquid can be confined in an open container whereas a gas or vapor cannot. The molecules at the surface remain joined together until enough heat is added to the liquid to permit the surface molecules to break their bonds, changing their state into a vapor. Unlike gases and vapors, liquids are *incompressible*. Their molecules are so closely packed together that any force applied to its surface generates an immediate displacement of the liquid.

9.3.3 Liquid Flashing

Liquids can undergo a change in state in response to an increase in their heat content or by a reduction in their surface pressure. This reaction is based upon the thermodynamic relationship that exists between a liquid's saturation temperature and its saturation pressure. Lowering the pressure on a liquid's surface reduces its boiling point. Raising its pressure increases its boiling point. Consequently, if a liquid exists at a temperature close to its saturation temperature and its pressure is reduced, enough heat already exists in the fluid to cause part of the liquid to rapidly change state into a vapor. This rapid change of state is called *flashing*.

In HVAC applications, liquid control agents are maintained at a relatively constant temperatures as they are transported from the central equipment to the heat transfer coils. As the liquid travels through the piping system, its total pressure decreases in order to overcome the piping resistance. If the liquid's pressure drops below the saturation pressure corresponding to its present temperature, the liquid flashes into a vapor. Flashing is a real concern in valve applications. It can destroy the seat ring, plugs and body. Once the trim's shape is altered, the controller can no longer maintain the correct response of the process to changes in its load.

Whenever fluid flows through a valve seat, its velocity increases in response to the reduction in the fluid stream's diameter. The smallest diameter experienced by the fluid stream is called the *vena contracta*. The diameter of the fluid stream at the vena contracta is smaller than the diameter of the valve port. The vena contracta forms a small distance downstream of the valve port. The fluid reaches its highest velocity as it passes through the vena contracta. This increase in fluid velocity causes a proportional reduction in the fluid's static pressure. Therefore, this is the point in the system where flashing is most likely to occur. Figure 9.8 depicts the vena contracta's location in a schematic of a valve port.

If the fluid's static pressure in the vena contracta is below the saturation pressure corresponding to its present temperature, some of the control agent flashes. Consequently, there is no longer a solid stream of liquid exiting the valve. Instead, a mixture of liquid and vapor now flows through the outlet port and adjoining piping. The vapor bubbles occupy a greater volume of the flow path than the liquid component of the flow. This leads to a compounding of the flashing response as the vapor further reduces the flow stream's diameter, increasing the fluid's velocity and further reducing its static pressure, causing additional flashing. If sufficiently large quantities of vapor are generated downstream of the valve port, the flow of liquid through the valve becomes *choked*. Choked flow limits the quantity of control agent that can pass through the valve prohibiting the control loop from maintaining the process set point.

In addition to limiting flow through a valve, flashing also destroys its trim components. The excessively high fluid velocity cuts though the soft metal of the seat ring and the valve plug destroying the flow characteristics of the valve. Valve

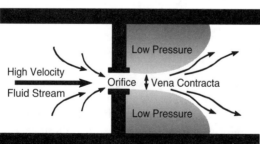

Figure 9.8   The Vena Contracta - Location of the Highest Fluid Velocity and Lowest Static Pressure

plugs that show signs of flashing appear as if a knife sliced channels into its metal. A damaged seat ring also prohibits tight shutoff of the control agent which can lead to uncontrollable operation and poor equipment operation.

9.3.4 Flow Cavitation

After the fluid passes the vena contracta, its velocity decreases as the stream fills the larger volume at the outlet of the port and the connecting pipe. As the fluid's velocity decreases, its static pressure increases back toward the value the fluid had before it entered the valve. This response is called static head *regain*.

If liquid flashed in the vena contracta and the static regain increases the fluid's pressure above its saturation pressure, the vapor bubbles in the fluid stream violently implode back into their liquid state. These implosions convert their energy into shock waves that chip off particles of metal from nearby trim and piping. This erosion creates a gritty appearance on the surfaces of the affected components. The mechanical integrity of the piping system is reduced by cavitation, increasing the possibility of bursting or leaking pipes. In similarity with damage by flashing, valves damaged by cavitation cannot correctly modulate flow or provide a tight shut off. Both cavitation and flashing are minimized by properly sizing and selecting the control valve.

9.4

VALVE-FLOW CHARACTERISTICS

Changes in the controller's output signal are received by a final controlled device's actuator, which repositions the valve's stem and plug in response to changes occurring in the process. Moving the valve plug alters the flowrate of the control agent in a predictable manner. The amount of change in the flowrate that occurs for each incremental change in the valve stem's position defines a response called the valve's *flow characteristic*. This response is a function of the shape of the valve plug and seat. The flow characteristic remain the same throughout the life of the valve. *Quick opening, linear,* and *equal percentage* are the flow characteristics used in HVAC processes. These responses are described in the following sections.

9.4.1 Quick Opening Flow Characteristic

A quick opening valve is used in applications that incorporate a two-position controller. These valves are constructed with plugs that cannot modulate the flow of control agent through their port. Their plugs are flat, disk-shaped devices that are made to seat tightly against the valve's seat ring to completely stop flow when required. As the two-position controller's output signal changes state, the plug moves off the seat. After a relatively small change in the position of the stem, the entire flow area of the port is opened, allowing full flow through the valve.

Valves with a quick opening flow characteristic are used in two-position control loops where the controller's output signal positions the valve either fully open or fully closed whenever the control point exceeds the set point $+/-\frac{1}{2}$ of its control differential. Quick opening valves have a high gain response. Ninety percent of their maximum flowrate is achieved when the stem is only moved approximately twenty percent of its entire stroke. This limits their application to processes that can be maintained by a two-position response. Valves using the quick opening flow characteristic pulse mass or energy into a process using a binary flow response (0% flow or 100% flow). Figure 9.9 shows the change in flowrate through a quick opening valve as a function of changes in the stem's position.

The proper installation of a valve is critical to maintaining its flow characteristics. Fluid must flow through the valve in the proper direction to maintain the proper response. An arrow is typically stamped or cast into the side of the valve body indicating the proper direction of flow. The reason for this requirement is based upon the forces created by the momentum of the moving fluid. These forces must *oppose* the closing force generated by the actuator if the valve's response is to remain stable as it approaches its closed position.

If the flow was inadvertently applied in the direction opposite of that indicated by the arrow on the valve body, the force created by the moving fluid would cancel some of the opposing actuator spring force that is needed to keep the plug off the seat as it approaches its closed position. At some point, the actuator's spring will no longer be able to hold the plug off the seat. Consequently, the plug rapidly and destructively slams closed against the seat, damaging the valve. Once the plug slams into the seat, flow stops and the forces created by the moving fluid quickly diminish to zero. Once this force is eliminated, the actuator's spring lifts the plug off the seat, reestablishing flow. When the flow returns, the forces in the valve revert back to their previous levels and once again, the plug slams closed. This rapid opening and closing of a valve plug is called *chatter*. Chatter destroys the valve's seat and plug. Each time the plug abruptly closes the port, it generates a damaging shock wave that travels through the incompressible liquid, generating a loud noise (water hammer) and weakening the piping system.

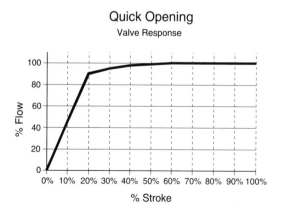

Figure 9.9 Two-Position (Quick Opening) Valve Response

9.4.2 Linear Valve Flow Characteristics

A valve with a linear flow characteristic varies its flow in *proportion* to the change that occurred in the controller's output signal. For example, if the controller's output signal changed 12% of its span, then the actuator moves the stem 12% of its stroke and the flowrate through the valve also changes 12%. The proportional gain of the valve, (Δ flowrate/Δ stroke) is established at the time of the valve's manufacture and remains constant throughout its life. Figure 9.10 shows the flow characteristic of a linear valve. Note, the response becomes slightly non-linear at either end of its stroke.

Valves having a linear flow characteristic are used in processes that have a linear response characteristic. These processes experience a change in their control point that is proportional to the change that occurred in the flowrate of the control agent. Steam heating and chilled water applications are linear processes. In steam applications, the vapor condenses and transfers its latent heat of vaporization. Since the latent heat is a constant (970 to 1000 btu/lb), the temperature of the controlled medium changes as soon as the quantity of steam entering the process changes. In chilled water applications, the temperature difference between the air and the water is relatively small. Consequently, if the flowrate of chilled water is varied, a proportional change in the temperature of the controlled medium occurs. In both processes, the change in temperature is proportional to the change in the flowrate of the control agent passing through the valve.

9.4.3 Equal Percentage Valve Flow Characteristics

Equal percentage valves have a nonlinear flow characteristic. The flowrate through these valves is *not* proportional to the change in their stem's stroke. Instead, the change in the valve's flowrate is based upon *equal percentage* changes in the stem's stroke. Figure 9.11 shows the response of an equal percentage valve

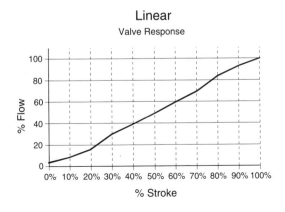

Figure 9.10   Linear Valve Response

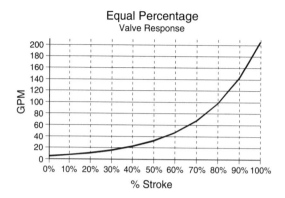

Figure 9.11   Equal Percentage Valve Response

that changes its flowrate by 45% of its *present* value for each 10% change in its stroke. Therefore, if the flow through the valve is currently 32 gpm and the stem moves the plug an additional 10% open, the flow through the valve will increases 45% of 32 gpm or 14.4 gpm. The new flowrate through the valve is $32 \times 1.45 = 46.4$ gpm. If the stem is moved an addition 10% open, the flow through the valve increases 20.8 gpm to 67.3 gpm.

This valve's response shows that changes in the stem's position at higher flowrates makes a larger change in the flow through the valve. At reduced flowrates, the change in the stem's position makes smaller changes in the flow through the valve. This is a nonlinear response. Remember, to have a linear response, each change in the valve stem's position must produce the same amount of change in the flow through the valve, no matter what its present flowrate is.

Valves with equal percentage flow characteristics are used to linearize the response of nonlinear processes. By linearizing these processes, the entire system operates with the relationship that a change in the load will produce a proportional change in the flowrate of the control agent. This response makes systems easier to calibrate and analyze.

Hot water applications are nonlinear processes whenever the water is at a temperature that is over 50 °F higher than the process set point. Under these circumstances, a reduction in the water's flowrate does not produce a proportional reduction in the process temperature. There is still considerable energy left in the hot water remaining in the coil to continue to raise the air's temperature after the valve is closed. This is a characteristic of a nonlinear process. When a nonlinear equal percentage valve having a complimentary response is used to control the coil, the total process response becomes nearly linear as depicted in Figure 9.12.

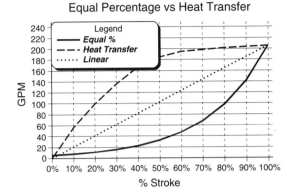

Figure 9.12 Equal Percentage Valve Response Installed in a Nonlinear Process

9.5

VALVE FLOW TERMINOLOGY

The following valve and process characteristics are used by the control technician to correctly select a new or replacement control valve:

- rangeability
- turndown ratio
- flow coefficient
- pressure drop at shutoff
- maximum temperature
- maximum pressure
- inlet and outlet connection size and type
- critical pressure drop

9.5.1 Rangeability

The rangeability of a valve is a ratio equal to the valve's maximum controllable flowrate divided by its minimum controllable flowrate. The maximum controllable flowrate is the quantity of control agent that flows through the control valve when it is wide open. The minimum controllable flowrate is the smallest modulated flowrate that the valve can maintain (see Example 9.1).

The minimum controllable flowrate is a function of the manufacturing tolerances that exist between the valve's plug and seat. Every valve has a gap between the sides of its plug and the inside surfaces of its seat ring to prevent the plug from binding with the seat when the valve is closed. This would inhibit the valve from opening. As a consequence of this gap, a minimum flowrate is established as soon as the plug moves off the seat. This is an uncontrollable flow equal to a small percentage of the maximum flowrate of the valve. The valve does not actually begin to control the flow until the flowrate exceeds this initial *cracked* flow condition.

The following formula is used to calculate the rangeability of a valve:

$$Rangeability = \frac{Maximum\ Controllable\ Flow}{Minimum\ Controllable\ Flow}$$

Example 9.1

Calculate the rangeability of a control valve having a minimum controllable flow of 1.2 gpm and a maximum flowrate of 20 gpm.

$$Rangeability = \frac{Maximum\ Controllable\ Flow}{Minimum\ Controllable\ Flow}$$

$$Rangeability = \frac{40}{1.2} = 33.3:1$$

Most control valves used in HVAC processes have a rangeability that falls within the range of 30:1 to 50:1. Linear valves have rangeabilities of approximately 30:1. Equal percentage valves have a rangeability that falls within a range of 40:1 to 50:1. The larger values of rangeability have better control during periods of light loads when the system is operating with reduced flowrates.

When selecting a control valve, the minimum controllable flowrate must be smaller than the lowest flowrate required during normal light load operation. If the minimum controllable flowrate of the valve is greater than the light load process requirements, the valve will be unable to maintain a balance between the energy transfer and the load. This occurs because too much control agent will be passing through the valve once it is cracked open. If this occurs, the control loop will become unstable, cycling the valve open and closed as it tries to obtain a balanced condition.

9.5.2 Turndown Ratio

The valve's turndown ratio is very similar to its rangeability. The turndown ratio is the ratio of maximum *required* flowrate to its minimum *controllable* flowrate. The maximum required flowrate is the quantity of control agent that is needed to balance the load during *design load* operation. When the design load flowrate is less than the maximum controllable flowrate of the valve, the turndown ratio is used to give a better indication of the range of control of the valve. The formula for the turndown ratio is:

$$Turndown\ Ratio = \frac{Maximum\ Required\ Flow}{Minimum\ Controllable\ Flow}$$

Example 9.2

Calculate the turndown ratio of a valve having a minimum controllable flow of 0.5 gpm and a maximum flowrate of 16 gpm.

$$Turndown\ Ratio = \frac{Maximum\ Required\ Flow}{Minimum\ Controllable\ Flow}$$

$$Turndown\ Ratio = \frac{16}{0.5}$$

$$Turndown\ Ratio = 32{:}1$$

The turndown ratio defines the controllable range of the control agent and the resolution possible with that particular valve and process. The turndown ratio can be equal to but is typically less than the rangeability of the valve.

9.5.3 Tight Shut-off

The maximum pressure difference that occurs across a closed valve port must be known in order to select a valve that can maintain a tight shut off when commanded closed by the controller. Single-seat valves can provide a tight shut off as long as the pressure drop across the valve does not exceed the valve's design pressure rating or the maximum force the actuator can generate to keep the valve closed. Tight shutoff allows the control loop to maintain better control over the process because it can stop the flow of the control agent completely in order to maintain a balance with the load. If some of the control agent leaks by the plug and seat after the controller commanded the valve closed, the process will drift out of control. It will remain out of control until a load change occurs that requires the valve to open beyond the flowrate that is leaking by the closed valve.

Some valves are designed so there is no leakage between the plug and the seat when the valve is closed. Others, like double-port and three-way valves are incapable of providing a leak-free barrier. These valves typically have 1% to 3% leakage due to the reduced tolerances between both of the valve plugs and their seats.

9.5.4 Valve Capacity Rating

The valve capacity rating is used to size the valve for a given application. It indicates the flowrate of 60 °F water that passes through a wide open valve port when a 1 psi pressure drop is maintained across the port. The valve capacity rating is also called the valve flow coefficient, and is noted by the variable C_v. The formula for calculating C_v of a liquid is:

$$C_v = \frac{gpm \times \sqrt{Specific\ Gravity}}{\sqrt{Pressure\ Drop}} \qquad Liquid\ Flow$$

For steam, the formula to calculate the valve's flow coefficient is:

$$C_v = \frac{\dfrac{lb.}{hour}}{2.1 \times \sqrt{\Delta p} \times (P_{inlet_{abs}} + P_{outlet_{abs}})} \quad \text{Steam Flow}$$

Both of these formulas and their applications are described in the next chapter.

9.5.5 Maximum Pressure and Temperature

The maximum pressure and temperature characteristics of the control agent are used to select the metal used to cast the valve body along with selecting the trim and packing materials so they can withstand the operating properties of the control agent. Increasing fluid temperatures reduce the maximum allowable pressure that can be applied to a valve body.

9.6

DAMPER CHARACTERISTICS

Dampers perform the same function in modulating air flow applications as valves do in liquid and gas flow applications. They are used in HVAC systems to control the flowrate of air in air handling units, ductwork, and zones. Like valves, dampers come in different configurations to perform two-position and modulating control. They also have response characteristics comparable to those of valves.

The most commonly used dampers are called parallel blade and opposed blade dampers. Both configurations generate a nonlinear response to changes in their actuator's stroke. By correctly sizing the dampers wide open resistance to flow, they can be made to respond in a more linear manner.

9.6.1 Parallel Blade Dampers

Parallel blade dampers are generally used in two-position applications where the required flow is either 0% or 100% of the total flowrate. This type of damper can be identified by the way the adjacent blades move with respect to each other when the shaft is moved by its actuator. Parallel blade dampers move their adjacent blades in the same direction when their actuator is stroked. This response tends to generate a directional change in the airstream rather than modulating its flow. Changing the direction of higher velocity air streams increases their turbulence as they pass through the damper blades. This can be advantageous in reducing stratification of air streams that are being mixed in the air handling unit. Figure 9.13 shows a rendering of a rack of parallel blade dampers.

In similarity with quick opening valves, parallel dampers provide little restriction to the flow of air over most of the actuator's stroke. They do not actually begin to restrict the airstream until they are about 80% closed. Figure 9.14 shows

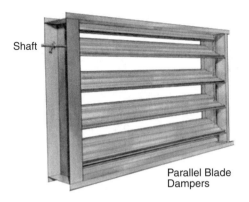

Figure 9.13 Parallel Blade Dampers

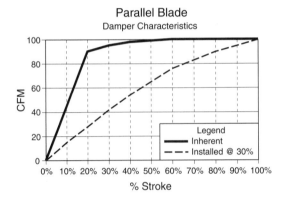

Figure 9.14 Parallel Blade Damper Response

the response of parallel blade dampers. Note the similarities between the flow characteristics of parallel blade dampers and a two-position valve. Parallel blade dampers allow nearly 100% flow when positioned approximately 20% of the actuator stroke. The only modulating portion of the damper's response occurs within first 20% range of the actuator's stroke.

When the pressure drop across the wide open parallel blade dampers is significantly increased, the flow characteristic of the dampers begins to take on a somewhat linear response as shown by the dashed line in Figure 9.14. To create this response, the wide open pressure drop must exceed 30% of the damper's flow path resistance. Unfortunately, this requires a considerable increase in the system's fan horsepower. This additional energy is dissipated as heat and noise as the air moves through the dampers. The larger fan horsepower requirements also increase the amount of noise that is carried through the ducts into the zone. Therefore, it is better to use the parallel blade dampers in two-position air flow applications and use a damper with a more linear flow characteristic for modulation applications.

9.6.2 Opposed Blade Dampers

Opposed blade dampers are used in modulating air flow applications where the flow must be varied between 0% and 100% in response to changes in the process load. The adjacent blades of an opposed blade damper assembly move in opposing (opposite) directions. This produces a throttling effect as soon as the actuator begins to reposition the dampers. The opposing movement generates an immediate change in the flow area of the dampers as the actuator moves. Figure 9.15 shows a rendering of opposed blade outside air dampers on an air handling unit's mixing plenum.

The response of opposed blade dampers resembles the nonlinear flow characteristics of an equal percentage valve. By carefully selecting the pressure drop across the dampers, their flow characteristics can be made more linear. When opposed blade dampers are designed to drop about 10% of the flow path's resistance, their flow characteristics approach the desired linear response. The lower pressure drop requirement of opposed blade dampers makes them more efficient than parallel blade dampers in modulating applications. Figure 9.16 shows the response of opposed blade dampers.

9.6.3 Damper Selection

Dampers do not provide a tight shut off capability unless special seals are applied to the edges of the blades. A typical damper may leak 10 to 25 cfm per square foot of surface area when one inch of water pressure is applied across the closed damper assembly. This infiltration of unnecessary air requires the expenditure of additional energy to condition it before it is sent to a zone. It also increases the possibility that coils will freeze whenever the unit is off and the dampers are com-

Opposed Blade
Dampers

Figure 9.15   Opposed Blade Dampers Mounted in an AHU Mixing Plenum

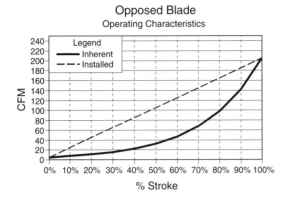

Figure 9.16 Opposed Blade Damper Response

manded closed. Low leakage dampers have seals on the blades that reduce the amount of leakage that occurs when the blades are closed. These dampers require a larger actuator to overcome the additional friction created by the seals. Damper sizing is covered in Chapter 10.

9.7

SUMMARY

The modulation of a fluid is accomplished by varying the amount of resistance a valve or damper presents to the flow of the fluid. Increasing the resistance of the flow path reduces the flowrate through the device. Conversely, decreasing the resistance of the flow path by opening the device increases the flowrate into the process. Valves are commonly used to control the flow of liquids and steam in HVAC applications. Dampers regulate the flow of air within HVAC equipment, ducts, and zones. Both of these final controlled devices are available in different configurations to meet the response requirements of the application.

When a liquid flows through a valve, its velocity increases as it flows through the reduced area of the port. The increase in its velocity causes a decrease in its static pressure. The reduction in static pressure can result in uncontrolled vaporization or flashing of the liquid. Flashing only occurs when the static pressure of the liquid drops below its saturation pressure. Flashing cuts the surface of valve plugs and seats, altering the flow characteristics of the valve.

Cavitation occurs when the static pressure of a vapor increases above its saturation pressure. Cavitation describes the violent implosion of flashed vapor back into its liquid state. Cavitation occurs downstream of the valve plug, after the velocity decreases, converting its energy back into static pressure. Cavitation chips away at the surface of the plug and downstream piping, changing the flow characteristics of the valve and weakening the valve body and piping.

9.8

EXERCISES

Determine if the following statements are true or false. If any portion of the statement is false, the entire statement is false. Explain your answers.

1. Saturated steam is a gas.
2. Three-way valves have a single port and seat ring.
3. A quick opening response is available in three-way mixing valves.
4. A diverting valve has one inlet and two outlet connections.
5. The rangeability of a valve can never be greater than its turndown ratio.
6. Opposed blade dampers are used to modulate the flow of air in a duct.
7. A nonlinear process uses an equal percentage valve to linearize the entire process response.
8. Parallel blade dampers require a lower pressure drop to produce an installed linear response than opposed blade dampers.
9. Cavitation produces knife cuts in the valve plug.
10. Static regain increases the possibility of liquids flashing in a valve.

Respond to the following statements and questions completely and accurately, using the material found in this chapter.

1. What is a valve?
2. What is a damper?
3. List the different types of two-way valves and one identifying characteristic of each.
4. Why is the arrow stamped or cast on the side of a two-way valve body?
5 What would happen to the response of a two-way valve if it were installed with the flow in the direction opposite of the arrow stamped on the valve body?
6. Name the components classified as valve trim and explain their function in a valve.
7. What is a valve bonnet and what is its function in a valve?
8. What is the valve plug?
9. Describe liquid flashing and how it occurs.
10. How can flashing be prevented in a system?
11. Describe how cavitation occurs.
12. Can cavitation occur if the pressure in the system never drops below the saturation pressure of the fluid? Explain your answer.
13. Are steam applications likely to experience cavitation or flashing? Explain your answer.
14. Sketch the response characteristics of a two-way, linear, and equal percentage valve. Briefly describe the flow characteristics of each of their responses.
15. Why are parallel blade dampers used for two-position control instead of in modulating control applications?

CHAPTER 10

Sizing and Selecting Control Valves and Dampers

Chapter 9 described the operational characteristics of the valves and dampers commonly used in HVAC applications. This chapter introduces the reader to some of the formulas and relationships used to size and select these devices. Many examples are given to illustrate the application of these formulas.

OBJECTIVES

Upon completion of this chapter, the reader will be able to:

1. Correctly size a control valve.
2. Correctly size a damper.
3. Select the proper valve for a given application.
4. Select the proper damper for a given application.

10.1

MODULATING FLOW CHARACTERISTICS OF VALVES

Specific procedures have been developed by valve manufacturers for measuring the flowrate and classifying the flow characteristics of a valve. These procedures insure that a valve manufactured by Company A responds the same as a valve of the same size and configuration manufactured by Company B.

When testing a valve's response to changes in its stem's stroke, the pressure drop across its port is set to one pound per square inch. The testing equipment maintains the pressure drop as the valve stem is slowly stroked from 0% to 100% in small increments. At each step, the flowrate through the valve is measured and recorded. A graph of the flowrate verses stem position is plotted similar to those shown in the figures in Chapter 9. This plot depicts the *inherent* flow characteristics of the valve. The pressure drop across the port must be maintained at 1 psi in order to determine a valve's inherent flow response.

Once a valve is installed in the field, its flow response changes. Whenever a valve is installed in a piping system, the pressure drop across its port changes as its stem is repositioned. The pressure drop increases as the valve modulates from its wide open (full flow) position toward its closed position. Consequently, the *actual* flow characteristics of an installed valve differ from its inherent flow characteristics which were based upon a constant pressure drop. This response is called the valve's *installed* flow characteristics.

When a valve is selected from a catalog, the technician has the responsibility of determining the wide open pressure drop that will appear across the valve after it has been installed in the system. This pressure drop is the resistance the valve adds to the piping system. In a typical heat transfer application consisting of a coil, balancing valves, connecting piping, fittings and a control valve, the ratio of the valve's wide open pressure drop to the sum of the pressure drops from the remaining piping system components establishes the installed flow characteristics of the control valve.

To minimize the differences between a valve's installed and inherent flow characteristics, the pressure drop across the wide open valve must be carefully selected. If a large difference exists between its installed and inherent flow characteristics, the valve will not respond as intended. An equal percentage valve can respond like a linear valve or a linear valve can respond with a quick opening characteristic. By correctly selecting the valve's pressure drop, a linear process response can be maintained.

10.1.1 Pressure Drop

The flowrate through a pipe is a function of the available pressure of the source and the resistance of the circuit. A differential pressure between the inlet and the outlet of a piping system will permit flow if the resistance of the path is not too large. The resistance in the piping circuit is a function of the size of the pipe, the material its made from, the length of the piping, number and type of fittings and the velocity of the fluid flowing in the pipe. The source of pressure in an HVAC piping application can be supplied by a pump or by the differential pressure that exists across the supply and return mains of a distribution network. The source pressure across a piping system is usually constant. As fluid flows through the piping, valve, fittings and coil, their individual resistances dissipate the source pressure in order to maintain flow through the system. The amount of pressure dissipated by a component is called its pressure drop. The maximum flowrate is based upon the amount of pressure drops in the circuit and the available source pressure. Therefore, as the resistance of a piping circuit increases, the flowrate through the circuit decreases. Conversely, as the piping circuit's resistance decreases, the flowrate through the circuit increases.

Selecting the correct valve pressure drop is based upon the other pressure drops in the process piping. Figure 10.1 shows a schematic representation of a simple heat transfer piping circuit. The total pressure drop (resistance) that must be over-come by the source of pressure is equal to the combined full load (maximum flowrate) pressure drops of the piping, fittings, coil and the control valve, whose port will be wide open. Once P1, P3 and P4 have been calculated under full load flow conditions, the valve's wide open pressure drop can be selected. To keep the installed flow characteristics of a valve close to its inherent flow characteristics, the valve's pressure drop must be equal to or greater than the combined total of the other pressure drops in the system (P1, P3 & P4).

When the valve's wide open pressure drop is equal to or greater than the remaining system resistances, it can modulate flow of the control agent through-out its entire stroke. If another component or section of the piping system has more resistance than the valve's wide open value, that component will establish the maximum flowrate through the circuit, not the control valve. The higher resis-tance component will maintain the controlling influence over the flowrate until the pressure drop across the valve port becomes the greatest pressure drop in the piping circuit. For example, if a valve were to be selected with a wide open (full load) pressure drop that was less than the full load pressure drop of the heat exchanger coil, the coil would determine the minimum flowrate of the process. The valve would have no control over the flowrate until its resistance increased

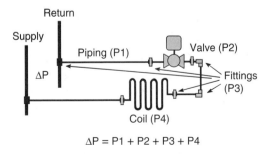

$$\Delta P = P1 + P2 + P3 + P4$$

Figure 10.1   Heat Exchanger Circuit Pressure Drops

above the resistance of the coil. The valve's pressure drop increases as it begins to modulate toward its closed position. Once its resistance became greater than the coil's, the valve would begin to modulate flow and would remain the controlling influence of the circuit throughout the remainder of its stroke.

Whenever another component in the piping circuit has a greater pressure drop than the control valve, the valve is sized too large for that application. It will have a larger port diameter and consequently, smaller resistance to flow than required. As an oversized valve begins to close, its resistance increases as its flow area becomes restricted by the plug. Once the resistance of the control valve exceeds the resistance of the remaining circuit, it begins to modulate the flowrate of the control agent. Therefore, oversized control valves do not modulate the process flowrate over their entire stroke. Consequently, their installed characteristics take on a response similar to that of a two-position valve, resulting in the introduction of a nonlinear response into the process and control problems will result.

In addition to reducing the controllability of the process, oversized valves also cause other problems. Since an oversized valve only begins to modulate after its plug is positioned close to the seat, the velocity of the control agent is higher than necessary during most of the valve's operation. This increases the potential for damage by liquid flashing and cavitation. High velocity fluids can also deform or draw the edges of the seat ring, producing a sharp, fragile lip that rises above the surface of the ring. This phenomenon is called *wire drawing*. The displaced metal interferes with the tight shut off capabilities and alters the flow characteristics of the valve.

When a valve is selected with a wide open pressure drop that is too high, its port is too small to meet the process requirements. Consequently, the flow through the control valve will be restricted during full load operation even though its port is wide open. Under these operating conditions, the loop will operate out of control. Undersized valves also require larger pumps to overcome their unnecessarily high pressure drop, and imposing additional energy requirements on the operation of the system. This additional pump energy does not increase fluid flow. It is dissipated as heat and noise at the valve, reducing the operating efficiency of the process.

10.1.2 Selecting the Correct Pressure Drop for a Valve

Engineering analysis by valve and system designers has determined that a modulating control valve operates effectively throughout the entire range of its stroke when its wide open pressure drop is between 1 and 1.5 times the sum of the remaining pressure drops in its circuit. In other words, if the valve's wide open pressure drop is between 50% and 70% of the total circuit's pressure drop, the control valve will modulate efficiently. Selecting a modulating valve's wide open pressure drop outside of this range results in installed characteristics that deviates too much from its inherent characteristics, reducing process controllability, capacity and efficiency.

Two-position valves do not modulate flow and therefore do not require a large pressure drop across their ports during a wide open operation. Typically, the pressure drop is set to 20% of the available system pressure drop. This small pressure drop minimizes the system pump size by reducing the pressure drop required for a system. The following sections detail the formulas used to properly size valves for liquid and gaseous flows.

10.2

FLOW COEFFICIENT SIZING FORMULAS

The flow coefficient or capacity coefficient (C_v) is used to determine the *size* of a valve for a specific application. The flow coefficient is a parameter that is also used to standardize the size of valves made by different valve manufacturers. This allows the technician to select valves with the same C_v rating and inherent flow characteristics from any manufacturer. C_v is a dimension-less number, meaning it has no units. A valve's flow coefficient is calculated using the valve's full load (design) pressure drop and the flowrate of the control agent required during full load operation.

The flow coefficient is a measurement of the quantity of 60 °F water, measured in U.S. gallons per minute, that pass through a wide open valve port that has 1 psi of differential pressure across the port. For example, if a valve passes 20 gpm of 60 °F water when its port is wide open and a ΔP of 1 psi is maintained across its port, its flow coefficient (C_v) equals 20. To calculate C_v for hot or chilled water, use the following formula:

$$C_v = gpm \times \frac{1}{\sqrt{\Delta P \ (psi)}}$$

Example 10.1

On a winter design day, a hot water heating process requires 35 gpm to balance its load. The pressure drop across the wide open control valve is measured to be 6 psi. Calculate the flow coefficient of the valve.

$$C_v = gpm \times \frac{1}{\sqrt{\Delta P}}$$

$$C_v = 35 \ gpm \times \frac{1}{\sqrt{6 \ psi}}$$

$$C_v = 35 \ gpm \times \frac{1}{2.45 \ psi} = 35 \ gpm \times 0.41 = 14.3$$

When the pressure drop across the control valve is given in units of feet of head, it must be converted into units of psi before calculating the flow coefficient. To convert a measurement from units of feet of head into psi, divide the head by 2.31 ft/psi.

$$psi = feet\ of\ pressure\ head \div 2.31 \frac{ft.\ head}{psi}$$

$$psi = feet\ of\ pressure\ head \times 0.433 \frac{psi}{ft.\ head}$$

Example 10.2

A chilled water coil requires 120 gpm during full load operation. The pressure drop of the system, excluding the control valve, is 24.6 feet of head. Calculate the required valve flow coefficient. The first step is to calculate the control valve's pressure drop based upon the pressure drop of the remaining system. Use the design factor of 1.5 times the remaining system pressure drop to determine the valve's pressure drop.

$$Valve\ \Delta P = 1.5 \times 24.6\ ft$$

$$Valve\ \Delta P = 36.9\ ft$$

$$Valve\ \Delta P\ in\ units\ of\ psi = 36.9\ ft \div 2.41 \frac{ft}{psi} = 15.8\ psi$$

$$C_v = gpm \times \frac{1}{\sqrt{\Delta P}}$$

$$C_v = 120\ gpm \times \frac{1}{\sqrt{15.8\ psi}} = 120\ gpm \times \frac{1}{3.98}$$

$$C_v = 120\ gpm \times 0.25 = 30.1$$

10.2.1 Specific Gravity

The specific gravity is the ratio of the *density* of the control agent divided by the *density* of water. Pure water has a specific gravity of 1. Whenever the control agent is not 100% water, then the appropriate value of specific gravity must be input into the numerator of the flow coefficient formula to calculate the correct value for the control valve's flow coefficient. The specific gravity parameter adjusts the port size to account for the change in the flow resistance caused by fluids having a density different from that of 60 °F water. As the specific gravity of a fluid increases above one, the fluid becomes denser with respect to pure water. Therefore, it requires a greater force to push its mass through the port of a valve. Consequently, the port's diameter must be increased so it can pass the required flowrate of the denser liquid. Since the valve sizing formula is based on pure water, the specific gravity parameter adjusts the port size so it will have the same resistance for the denser fluid as it would for water at the same flowrate. The formula for calculating the specific gravity of a liquid is:

$$g = \frac{density\ of\ the\ control\ agent}{density\ of\ water}$$

$$g = \frac{x\ lb/ft^3}{62.4\ lb/ft^3}$$

The following equation incorporates the specific gravity parameter into the flow coefficient formula for a liquid other than water:

$$C_v = gpm \times \frac{\sqrt{g}}{\sqrt{\Delta P}}$$

where: g = specific gravity of the fluid.

Example 10.3

75 gpm of an aqueous solution of ethylene glycol is used in a chilled water system. This 40% solution has a specific gravity of 1.055. The pressure drop across the supply and return mains serving the process is 12 psi. Calculate the control valve's coefficient for a modulating application.

In applications where the pressure drop of the across the water mains that serve the process is given, the modulating valve's pressure drop is selected to equal a value between 50% and 70% of the pressure drop across the mains that supply the coil. Therefore, the pressure drop for this valve can be any number between 6 (0.50 × 12) to 8.4 (0.70 × 12) psi. In this example, the larger drop (8.4) will be used to cause the valve's installed response closer to its inherent flow characteristics.

$$C_v = gpm \times \frac{\sqrt{g}}{\sqrt{\Delta P}}$$

$$C_v = 75\ gpm \times \frac{\sqrt{1.055}}{\sqrt{8.4\ psi}} = 75\ gpm \times \frac{1.03}{2.9}$$

$$C_v = 75\ gpm \times 0.35 = 26.6$$

10.3

CALCULATING REQUIRED FLOWRATE FOR A PROCESS

Typically, a mechanical design engineer calculates the loads expected within each zone to determine the size of the equipment needed to balance the load during design load operation. The design loads are converted into a flowrate requirement for the process. The design flowrate requirements are used to calculate the valve's flow coefficient that is used to select the valve that best fits the application. The following sections outline the procedures used for calculating the design flowrate for common HVAC processes based upon the units given for the design load.

10.3.1 Given: zone load in btu/hr

When the design load requirements are given in btu's/hr, the required flowrate in gallons per minute can be calculated using the formula:

$$gpm = \frac{\dfrac{btu}{hr}}{500 \times \Delta t}$$

btu/hr = load of the process
500   = multiplication constant equal to 8.33 lb/gal $\times$ 60 min/hr $\times$ btu/lb °F (specific heat of water) = 500 btu min/gal-hr-°F
Δt = temperature difference between the water entering the coil and the water leaving the coil

Example 10.4

A zone has a design load of 16 tons of refrigeration effect per hour. The chilled water enters the coil at 41 °F and leaves at 52 °F. Calculate the design flowrate of the process and the control valve's flow coefficient if the available pressure drop for the valve is 10 psi.

$$\frac{btu's}{hr} = 16\frac{tons}{hr} \times 12000\frac{btu's}{ton} = 192{,}000\frac{btu's}{hr}$$

$$\Delta t = 52 - 41 = 11 \ °F$$

$$gpm = \frac{\dfrac{btu}{hr}}{500 \times \Delta t}$$

$$gpm = \frac{192{,}000}{500 \times 11} = 34.9$$

$$C_v = 34.9 \ gpm \times \frac{1}{\sqrt{(0.70 \times 10 \ psi)}} = 13.2$$

10.3.2 Given: cfm + Δt_{Water} + Δt_{Air} (sensible load)

When the sensible heat load requirements of a zone are given in units of volume flowrate of air across the coil (cfm) along with the design temperature difference for water and air, the flowrate through the valve can be calculated using the following formula:

$$gpm = \frac{cfm \times 1.08 \times \Delta t_{air}}{500 \times \Delta t_{water}}$$

where:

cfm   = volume flow of air across the coil (ft^3/minute)

1.08   = constant equal to 0.24 btu/lb-°F (specific heat of air) × .075 lb/ft^3 (density of air) × 60 min/hr = 1.08 (btu-min)/(°F - ft^3 -hr)

Δt_{air} = temperature difference of the air across the coil

500   = constant equal to 8.33 lb/gal × 60 min/hr × 1 btu/lb -°F (specific heat of water) = 500 btu-min/gal-hr-°F

Δt_{water} = temperature difference of the water through the coil

This formula only considers the dry bulb temperature difference of the air, therefore it can only be used for heating or sensible cooling processes.

Example 10.5

A 20,000 cfm air handling unit raises the temperature of the air flowing across the coil from 65 to 130 °F. Hot water enters the coil at 170 °F and leaves at 140 °F. The pressure drop across the process piping is 18.5 feet of head. Calculate the flow coefficient of the valve.

$$\Delta t_{air} = 130 - 65 = 65°$$

$$\Delta t_{water} = 170 - 140 = 30°$$

$$gpm = \frac{cfm \times 1.08 \times \Delta t_{air}}{500 \times \Delta t_{water}}$$

$$gpm = \frac{20,000 \times 1.08 \times 65}{500 \times 30} = \frac{1,404,000}{15,000} = 93.6$$

$$psi = feet\ of\ head \div 2.31\frac{ft}{psi} = 18.5\ feet \div 2.31\frac{ft}{psi} = 8\ psi$$

$$C_v = 93.6\ gpm \times \frac{1}{\sqrt{(0.70 \times 8\ psi)}} = 39.6$$

10.3.3 Given: cfm across the coil + btu/hr load (sensible + latent)

When the total heat load (sensible + latent) requirements of a zone is given along with the volume flowrate of air passing across the coil (cfm), the flowrate of the control agent (gpm) can be calculated using the following formula. This formula is primarily used for chilled water applications where sensible and latent cooling take place simultaneously.

$$gpm = \frac{cfm \times \frac{btu's}{lb}}{112.5 \times \Delta t_{water}}$$

where:

cfm   = volume flow of air across the coil (ft3/minute)
btu/lb   = total heat given up by the air as it passes across the coil
112.5   = constant 12.5 ft^3/lb (specific volume of the air) $\times$ 1 btu/lb-°F
             (specific heat of water) $\times$ 8.33 lb/gal = 112.5 btu-ft^3/lb - gal -°F
Δt_{water} = temperature difference of the water passing through the coil

Example 10.6

> *A 12,000 cfm air handling unit lowers the temperature of the air flowing across the coil from 78 to 55 °F. The chilled water enters the coil at 41 °F and leaves at 50 °F, removing 11 btu's of heat (sensible + latent) from every pound of air that passes across the coil. The total water pressure drop of the coil, piping and fittings in the circuit is 12 psi. Calculate the flow coefficient of the control valve*

$$\Delta t_{water} = 50 - 41 = 9°$$

$$gpm = \frac{cfm \times \dfrac{btu's}{lb_{air}}}{112.5 \times \Delta t_{water}} = \frac{12,000 \times 11}{112.5 \times 9} = \frac{132,000}{1012.5} = 130.4$$

$$C_v = 130.4 \; gpm \times \frac{1}{\sqrt{(1.5 \times 12 \; psi)}} = 30.7$$

10.4

SIZING VALVES FOR STEAM FLOW

In steam systems, the flowrate is quantified in units of pounds of steam per hour (lb/hr). The force required to produce flow in a steam system is created by the heat transferred from combustion of fuel into the water inside a boiler. The boiler adds heat to the water, changing its state to a vapor and increasing its pressure. The pressure provides the force needed to transport the steam through the system piping.

The pressure drop parameter used in steam control valve flow coefficient calculations is measured as the pressure difference between the inlet connection of the valve and the outlet of the *coil*. This is the pressure drop that regulates the flow of steam through the heat transfer piping system.

10.4.1 Critical Flow in Steam Systems

Steam valve sizing formulas are based upon the *critical* pressure drop of the control valve. The critical pressure drop is the maximum *allowable* pressure drop that can occur across the valve. If the valve's pressure drop exceeds the maximum allowable pressure drop, a choked flow condition occurs which limits the flowrate through the valve. Choked flow develops when the velocity of the vapor flowing through the vena contracta increases beyond the speed of sound. Once this velocity is reached, the maximum flowrate through the valve has been reached and any additional pressure drop does not increase the flowrate. Instead, its energy is dissipated as a loud high frequency noise. When the flowrate becomes choked, the heat transfer coil will not be able to operate at its rated capacity. The critical pressure is equal to 50% of the absolute pressure available at the inlet of the valve. The formula for calculating the critical pressure drop is:

$$\Delta P_{critical} = 0.50 \times P_{inlet_{absolute}}$$

This text uses the value 0.50 to calculate the critical pressure drop. Different valve sizing reference tables may use a different values within a range of 0.45 to 0.58.

Example 10.7

Calculate the critical pressure drop for a valve with an inlet pressure of 15 psig. (The gauge pressure must first be converted into absolute terms.)

$$\Delta P_{critical} = 0.50 \times P_{inlet_{absolute}}$$

$$\Delta P_{critical} = 0.50 \times (15 + 14.7) = 14.85 \; psi$$

The pressure drop across the valve is equal to the difference between the inlet of the valve and the saturation pressure of the condensate in the system. This is the same pressure as the condensate return system. If the saturation pressure of the condensate is greater than the critical pressure drop of the control valve then choked flow is not likely to occur. Conversely, if the saturation pressure of the condensate is less than the critical pressure drop choked flow is likely to occur.

If the pressure on the outlet side of the valve is measured in a vacuum, that reading must be converted into units of psia before it can be used in the steam valve sizing formulas. The conversion factor for inches of mercury to psi is 2.04 inches of hg/psi. The formula used to perform the conversion is:

$$10 \; inches \; hg = 10 \; inches \; hg \times \frac{1 \; psi}{2.04 \; inches \; hg}$$

$$= 4.9 \; psi \; vacuum = 14.7 + (-4.9) = 9.8 \; psia$$

10.4.2 Steam Valve Sizing Formulas

After the critical pressure drop for a steam application has been calculated, use the following formula for non-critical flow applications:

$$C_v = \frac{W}{2.1 \times \sqrt{\Delta p \times (P_{inlet_{abs}} + P_{outlet_{abs}})}}$$

W	=	required steam flowrate in lb/hr
$P_{inlet} + P_{outlet}$	=	sum of the absolute inlet pressure and the absolute saturation pressure of the condensate (return pressure of the system)
Δp	=	difference between the inlet pressure and the saturation pressure of the condensate (return pressure of the system)

Use this formula for applications where choked flow is likely to occur:

$$C_v = \frac{W}{1.6 \times P_{inlet_{abs}}}$$

where:

W = required steam flowrate in lb/hr
P_{inlet} = absolute pressure at valve inlet (psia)

Example 10.8

A preheat coil requires 2000 lb/hr of steam to temper the air temperature during winter design operation. Steam enters the valve at 100 psig and leaves the valve at 0 psig. Calculate the flow coefficient of the control valve

$$\text{Critical pressure drop} = 0.50 \times P_{inlet_{abs}}$$

$$\text{Critical pressure drop} = 0.50 \times (100 + 14.7) = 57.35 \ psi$$

$$\Delta p = 100 \ psi > 57.35 \ ... \ Choked \ Flow \ Will \ Occur$$

$$C_v = \frac{W}{1.61 \times P_{inlet_{abs}}}$$

$$C_v = \frac{2000}{1.61 \times (100 + 14.7)}$$

$$C_v = \frac{2000}{184.7} = 10.8$$

Example 10.9

A reheat coil requires 1000 lb/hr of saturated steam during full load operation. Steam enters the valve at 8 psig. The outlet pressure on the valve is 1 inch hg vacuum. Calculate the flow coefficient required for the control valve.

$$\text{Critical pressure drop} = 0.50 \times P_{inlet_{abs}}$$

$$\text{Critical pressure drop} = 0.50 \times (8 + 14.7) = 11.35 \ psi$$

$$P_{outlet} = \left(1 \ inch \ hg \times 2.04 \ \frac{psi}{in \ hg}\right) = -2.04 \ psi$$

$$\Delta p = 8 - (-2.04) = 10.04 < 11.35 \ ... \ Choked \ Flow \ Will \ Not \ Occur$$

$$C_v = \frac{W}{2.1 \times \sqrt{\Delta p \times (P_{inlet_{abs}} + P_{outlet_{abs}})}}$$

$$C_v = \frac{1000}{2.1 \times \sqrt{10 \times (22.7 + 12.6)}}$$

$$C_v = \frac{1000}{39.45}$$

$$C_v = 25.3$$

10.5

CALCULATING THE PROCESS STEAM LOAD

When steam is used as a control agent, the energy it transfers comes from the latent heat absorbed by the water as it changed state from a liquid to a vapor. When the steam condenses in the heat exchanger, the vapor transfers approximately one thousand btu's for each pound of condensate produced. The hot condensate leaves the exchanger through a steam trap at a temperature that remains close to its saturation temperature. Relatively little sensible heat transfer occurs with the condensate in steam heat processes. The following sections outline the equations required to calculate the process steam load. Each of the formulas determines the mass flowrate in pounds per hour of saturated steam required to balance the load during design load operation.

10.5.1 Given: btu/hr transferred by the Heat Exchanger

When the maximum load requirements of a process are given in units of btu's per hour, the steam load is calculated using the latent heat of steam.

$$W = \frac{btu/hr}{1000\ btu/lb}$$

where:

W   = required steam flowrate in lb/hr
1000 = latent heat of condensation for each pound of steam

10.5.2 Given: cfm across the coil and Δt_{air}

When the volume flowrate of air moving across the coil is known along with the rise in its temperature, the steam load is calculated using the formula:

$$W = \frac{cfm \times \Delta t_{air} \times 1.08}{1000\ btu/lb}$$

where:

1000   = latent heat for each pound mass of steam
cfm   = volume flow of air across the coil (ft^3/minute)
1.08   = constant equal to 0.24 btu/lb-of (specific heat of air) $\times$ .075 lb/ft^3
          density of air) $\times$ 60 min/hr. = 1.08 (btu-min)/($^\circ$F-ft^3 -hr)
Δt_{air} = temperature difference of the air across the coil

10.5.3 Given: flowrate and water temperature increase of a steam converter

When the volume flowrate of water through a steam to hot water converter is known, along with the rise in temperature of the water as it passes through the coils, the steam load is calculated using the following formula:

$$W = gpm \times \Delta t_{water} \times 0.49$$

where:

gpm	=	volume flow of water through the converter
0.49	=	constant 1 btu/lb -°F (specific heat of water) $\times$ 8.33 lb/gal $\times$ 60 min/hr $\div$ 1000 btu/lb of steam
Δt_{water}	=	temperature difference between the water entering and leaving the converter.

10.6

OTHER VALVE SELECTION CONSIDERATIONS

After the size of the valve's flow coefficient has been calculated, the structural characteristics of the valve are analyzed in order to select the correct valve body. Each valve body assembly has definite operating limits of pressure and temperature. Before selecting a valve by size, check the manufacturers specification sheet to be sure the valve body, trim and packing can withstand the:

- maximum inlet pressure of the process
- maximum temperature of the process
- corrosive nature of the control agent
- pressure drop across the valve at shutoff
- normal (failsafe) position requirements

10.6.1 Failsafe Position

All valves must be chosen so they respond safely during periods when the process losses power or when the control signal to the actuator is interrupted. To accommodate this requirement, valves are manufactured with either a normally closed (NC) or a normally open (NO) failsafe response. The term *normal* defines the position of the valve plug when there is no signal being sent to its actuator. Therefore, a *normally open* valve is wide open when there is no signal on its actuator and a *normally closed* valve has its plug firmly seated when there is no signal present on its actuator. In both configurations, the valve plug is returned to its normal position by the spring located in the actuator.

The choice of which normal position to use in a particular application is based upon the *failsafe* requirements of that process. The failsafe position of any final controlled device determines whether a process receive 0% or 100% of the flow of its control agent during times of system malfunction. The failsafe position also gives the control system the ability to provide some measure of protection to the mechanical systems and building occupants in the event of a control device failure, interruption of instrument air, a signal line is severed or some other abnormal event. It also protects the equipment during normal shutdown periods. Before a failsafe configuration can be selected, the process must be analyzed to determine which normal position will maintain the integrity of the system and protect the occupants during system shutdown or faults. The following paragraphs list some typical HVAC processes and the normal position of final controlled devices typically used in these applications.

10.6.2 Hot Water Process Failsafe Position

Hot water heating applications usually incorporate normally open (NO) valves in their system design. Whenever the control signal to a NO hot water valve is interrupted, the valve opens 100% allowing a full flow of hot water to pass through the heat exchanger. This response provides a measure of freeze protection to coils and their surroundings when the control system or building power is interrupted.

10.6.3 Chilled Water Process Failsafe Position

Chilled water valves are typically normally closed (NC) devices. When the signal to a NC chilled water valve is interrupted, the valve closes, stopping the flow of control agent through the coil. There are generally no safety concerns in these applications. If the malfunction is due to a power failure, chilled water will not be available so it is of minimal concern whether the valve fails open or closed. A normally closed valve prohibits a cold coil from lowering the temperature of the duct below its dewpoint temperature, minimizing condensation from forming on adjoining surfaces.

10.6.4 Steam Valve Failsafe Position

The failsafe position for steam valves is a function of the application requirements. Some processes require a normally open valve to protect systems from freezing while others use a normally closed valve to prevent the process equipment from bursting when the steam valve is fails open under low load operations. A steam valve used for a heating coil is usually chosen to fail open (NO) to prevent the condensate in the bottom of the coil or trap along with the coils located downstream of the steam coil, from freezing. This is similar to the hot water valve applications.

Normally closed steam valves are found in steam to hot water converter applications. When a control system malfunction occurs, this valve closes, interrupting the generation of hot water for the systems it supplies. This response prohibits the steam flowing through the converter from causing the water in the tubes to boil if the circulation pumps are also off. Other converter applications put a greater emphasis on supplying hot water to their systems for freeze protection and incorporate NO steam valves.

Steam valves used in humidity applications always fail closed to prevent uncontrolled flow of steam into a duct of an air handling unit that is inoperative. This prevents the large quantities of steam from condensing in the ducts and saturating the duct insulation, rendering it useless. Damp insulation also promotes the growth of molds and mildew that can adversely affect indoor air quality.

10.7

SELECTING A VALVE USING A MANUFACTURER'S CATALOG TABLE

After the information describing the requirements of the process and the characteristics of the fluid have been determined, the valve can be selected from a manufacturer's catalog. The process piping layout determines whether a two-way or three- way valve is required to meet the specification. The piping layout also determines the type of connection that is needed to install the valve to the piping. The operating characteristics of the control agent are used to determine the conditions that the valve body and trim components must be able to withstand without deteriorating or failing. Finally, the full load flowrate of the process must be known to determine the C_v rating required for the valve.

Tables 10.1 and 10.2 depict a sample of a cross-reference table that some manufacturers include in their catalogs to determine the C_v rating based upon the dif-

Table 10.1 Flow Coefficients for Water (GPM)

C_v	Differential Presure (PSI)						
	3	5	7	10	12	15	20
0.2	0.3	0.4	0.5	0.6	0.7	0.8	0.9
0.4	0.7	0.9	1.1	1.3	1.4	1.5	1.8
0.6	1.0	1.3	1.6	1.9	2.1	2.3	2.7
0.8	1.4	1.8	2.1	2.5	2.8	3.1	3.6
1.0	1.7	2.2	2.6	3.2	3.5	3.9	4.5
1.5	2.6	3.4	4.0	4.7	5.2	5.8	6.7
2.0	3.5	4.5	5.3	6.3	6.9	7.7	8.9
3.0	5.2	6.7	7.9	9.5	10.4	11.6	13.4
5.0	8.7	11.2	13.2	15.8	17.3	19.4	22.4
10.0	17.3	22.4	26.5	31.6	34.6	38.7	44.7
15.0	26.0	33.5	39.7	47.4	52.0	58.1	67.1
20.0	34.6	44.7	52.9	63.2	69.3	77.5	89.4
25.0	43.3	55.9	66.1	79.1	86.6	96.8	111.8
30.0	52.0	67.1	79.4	94.9	103.9	116.2	134.2
35.0	60.6	78.3	92.6	110.7	121.2	135.6	156.5
40.0	69.3	89.4	105.8	126.5	138.6	154.9	178.9

Table 10.2 Flow Coefficients for Steam With Return Pressure of 0 PSIG (lb/hr)

C_v	Inlet Steam Presure (PSI)						
	2	5	10	15	20	25	30
0.2	3.3	5.5	8.3	10.8	13.2	15.5	17.7
0.4	6.7	11.0	16.7	21.7	26.4	31.0	35.5
0.6	10.0	16.5	25.0	32.5	39.6	46.5	53.2
0.8	13.3	22.0	33.3	43.4	52.8	62.0	70.9
1.0	16.6	27.5	41.7	54.2	66.0	77.4	88.6
1.2	20.0	33.0	50.0	65.0	79.2	92.9	106.4
1.4	23.3	38.6	58.4	75.9	92.4	108.4	124.1
1.6	26.6	44.1	66.7	86.7	105.6	123.9	141.8
1.8	30.0	49.6	75.0	97.6	118.8	139.4	159.6
2.0	33.3	55.1	83.4	108.4	132.0	154.9	177.3
3.0	49.9	82.6	125.1	162.6	198.0	232.3	265.9
4.0	66.6	110.2	166.7	216.8	264.0	309.8	354.6
5.0	83.2	137.7	208.4	271.0	330.0	387.2	443.2
6.0	99.9	165.2	250.1	325.2	396.0	464.7	531.9
7.0	116.5	192.8	291.8	379.4	462.1	542.1	620.5
8.0	133.1	220.3	333.5	433.6	528.1	619.6	709.2
9.0	149.8	247.9	375.2	487.8	594.1	697.0	797.8
10.0	166.4	275.4	416.8	541.9	660.1	774.4	886.5

ferential pressure across the valve and the required process flowrate. Table 10.1 lists data for water valves and Table 10.2 lists data for steam valves. These tables can be used in place of calculating the flow coefficient or to verify the accuracy of a valve sizing calculation.

After the flow coefficient, body style and trim materials have been selected, the manufacturer's catalog is used to select a valve. The first step in selecting a valve is to find the page in the catalog that has the required body style, normal position and construction for the application. Next, look for the row that has a C_v value close to that calculated for the process. If the calculated C_v falls between two values in the table, selecting the smaller C_v will yield a valve with better modulating characteristics but may be undersized for full load operation. Selecting the valve with the larger C_v will yield an oversized valve with sufficient capacity for full load operation. The choice is a matter of evaluating the advantages of one over the other for that particular application. The row that lists all the proper requirements of the valve will also list the diameter of the pipe connections and the manufacturer's part number. Table 10.3 shows a portion of a table of valve size information typically found in a catalog. If the calculated C_v of a valve is 50, Table 10.3 indicates a $2\frac{1}{2}$ inch valve would meet the application requirements. Keep in

Table 10.3

Size (inches)	C_v
$\frac{1}{2}$	0.7
$\frac{1}{2}$	4.2
$\frac{1}{2}$	3.0
$\frac{1}{2}$	4.7
$\frac{3}{4}$	7.0
1	12
$1\frac{1}{4}$	19
$1\frac{1}{2}$	29
2	26
2	47
$2\frac{1}{2}$	51
3	83
4	150
5	240
6	350

mind, the actual flow through the valve is still a function of the pressure drop across the valve after it is installed and the system is operating normally.

10.8

DAMPER SIZING PROCEDURES

Damper sizing procedures are very similar to those presented for control valves. The design air flowrate must pass through the wide open damper assembly and generate a sufficient pressure drop to produce a linear installed flow characteristic. Recall from Chapter 9, both parallel and opposed blade dampers have a non-linear inherent flow characteristic. Parallel blade dampers have an inherent response that is similar to that of a two-position valve and opposed blade dampers have an inherent response that is similar to that of an equal percentage valve. Through the proper determination of the wide open damper pressure drop, both of these types of dampers can be made to respond in a nearly linear manner.

Calculating the pressure drop for a set of dampers differs from the procedures used to size valves. Although both procedures relate the wide open pressure drop of the device to the total system resistance, the damper calculation only considers the resistances measured between duct branches, not the entire duct system.

The system drop used in damper sizing is equal to the sum of all the resistances found in the duct flow path in which the damper is to be installed, including duct resistance, losses from flow baffles, bird screens, filters, heat transfer coils, elbows, turning vanes and any other devices found in that branch. Figure 10.2 shows three duct sections having installed dampers. R1 through R4 represent the resistances or pressure drops located in the outside air damper's path. R1 represents the pressure drop produced by the rain guard on the exterior of the building; R2 represents the pressure drop due to the bird screen; R3 is the resistance of the wide open outside air damper and R4 is the duct's pressure drop that occurs when the maximum volume of air is flowing through the duct. In the second branch, R5 represents the resistance of the duct and R6 is the return air damper's wide open resistance to air flow. The third branch is similar to the outside air damper's path. These are the pressure drops used to calculate the required pressure drop across the wide open dampers located in their respective path.

10.8.1 Sizing Dampers for a Linear Response

To correctly size opposed blade dampers for linear response, the damper's wide open pressure drop should consume about 10% of the total of all the pressure drops in the path, including the dampers. This equates to about 11% of the other pressure drops in the path, not including the dampers. Using the outside air damper circuit in Figure 10.2, the damper resistance should equal 10% of the total resistance $R_1 + R_2 + R_3 + R_4$ or $R_3 = (R_1 + R_2 + R_4) \times 0.11$:

Opposed blade drop $= 0.11 \times$ *remaining full load pressure drop*$_{branch}$

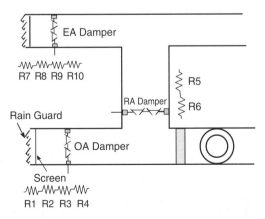

Figure 10.2   Damper Circuit Pressure Drops

Example 10.10

> *The pressure drops through the rain guard, screen and duct of a 4000 cfm outdoor air damper branch at full load, equal 0.09 inches of water. Calculate the pressure drop for opposed blade dampers that would generate a linear installed response characteristic.*
>
> $$Opposed\ blade\ drop = 0.11 \times remaining\ full\ load\ pressure\ drop_{branch}$$
> $$Opposed\ blade\ drop = 0.11 \times 0.09\ inches\ of\ water = 0.01\ inches\ of\ water$$
> $$Total\ branch\ drop = 0.01 + 0.09 = 0.10\ inches\ of\ water$$

Parallel blade dampers are not commonly used for modulating applications because they require a substantially higher pressure drop to produce a linear installed flow characteristic. These dampers require about 25% more fan energy to produce a linear modulating response. Therefore, parallel blades are used for two-position applications and opposed blade are used for modulating applications where there are other pressure drops present in the branch circuit.

If parallel blades must be used to provide a high degree of mixing and to modulate flow, the damper's wide open pressure drop must equal 30% of the total branch pressure drop ($R_1 + R_2 + R_3 + R_4$) or 40% of the sum of the other branch pressure drops. In other words $R_3 = (R_1 + R_2 + R_4) \times .40$:

$$Parallel\ blade\ drop = 0.40 \times remaining\ full\ load\ pressure\ drop_{branch}$$

Example 10.11

> *The pressure drops through the rain guard, screen and duct of a 4000 cfm outdoor air damper branch at full load, equal 0.09 inches of water. Calculate the pressure drop for parallel blade dampers that would create a linear installed response characteristic.*
>
> $$Parallel\ blade\ drop = 0.40 \times remaining\ full\ load\ pressure\ drop_{branch}$$
> $$Parallel\ blade\ drop = 0.40 \times 0.09\ inches\ of\ water = 0.037\ inches\ of\ water$$
> $$Total\ branch\ drop = 0.037 + 0.009 = 0.127\ inches\ of\ water$$

10.8.2 Damper Selection Considerations

As with valves, dampers must be chosen to fail in a safe position. Unlike valves, a damper assembly does not have a manufactured normal position. Their normal position is established when the actuator is connected to the damper shaft in a manner that the actuator's spring force holds the dampers to the required normal position upon loss of the control signal. Outside air dampers are always configured as normally closed to limit infiltration into the building when the air handling unit's fan is off. In emergency situations, the outside air dampers must fail closed to minimize the possibility of coils freezing.

Exhaust and relief air dampers are also configured as normally closed. This prohibits infiltration when the dampers are closed and also allows them to modulate in unison with the outside air damper. This strategy prevents excess air pressure from developing within the building when ventilation air is brought in to meet the ventilation requirements of the zones.

Return air dampers are configured as normally open. This allows a single controller to modulate the outdoor and return air dampers in a complimentary manner so the mixed air temperature can be controlled. As the outdoor dampers open 10% in response to an increase in the controller's output signal, the return air dampers will stroke 10% closed in response to the same signal.

10.9

ACTUATOR SELECTION

After a valve or damper is selected for an application, the actuator must be properly sized to provide the force needed to correctly position the plug in the valve's seat ring or position a set of dampers against the forces created by the fan, wind and convection forces that effect the system's static pressure. Pneumatic actuators are commonly used because of their ability to generate high forces in a small package, inherent simplicity, reliability and low cost. Electric motor driven actuators can also be used in applications where no instrument air is available or where desired for other reasons other than first cost. Electric actuators are more costly than pneumatic devices and are available with or without a return spring. If a return spring is not included in the motor actuator, the final controlled device stays in its last position or floats when the control signal is interrupted. This may be a problem in failsafe strategies where the actuator must be driven closed in an emergency situation and power may not be available.

In valve applications, the force produced by the actuator must be large enough to overcome the force created by the valve's spring, packing friction, internal friction and the force created by the momentum of the moving fluid that flows against the closing valve plug. Additional force must also be available to insure a tight shutoff when the valve plug closes against the seat ring.

Damper actuators are selected to overcome friction from the blade shafts and linkage, friction from blade seals and forces generated by the air flow impinging on the damper blade's surface. Addition force must also be incorporated into the design to keep the dampers closed against the forces created by the wind. Unlike valve applications, larger force requirements for damper racks are usually met by installing multiple actuators on the damper assembly. The same control signal is sent to all the actuators so they operate in unison to modulate the dampers.

10.9.1 Pneumatic Positioning Relays

A positioning relay is a device that can be attached to a pneumatic actuator to alter its operating range or to provide sufficient air pressure to overcome the inertia of a final controlled device.

A positioning relay or positioner is a control device that can be used to alter or customize the operating range of a pneumatic actuator. It accomplishes this function by converting the output signal from the controller into a different operating range that strokes the final controlled device from 0% to 100% of its stroke. These devices are commonly used to sequence the operation of actuators that have the same spring range. For example, two actuators with 9 to 13 psi springs can be configured to operate with 3 to 8 and 9 to 15 psi responses, effectively sequencing the valves with one controller output signal.

Positioning relays are also used to supply main air pressure directly to the actuator until it correctly positions the flow control apparatus at its correct location. They can produce up to seven times the starting force on an actuator. It performs this task by making the full 25 psi (or greater) supply air pressure available to the actuator piston instead of using the initial 3 psi or 4 psi signal from the controller. This smaller controller signal only generates a 50 pound (pi $\times$ r^2 $\times$ pressure) starting force on a 4 inch diameter actuator. A 25 psi main air signal produces a 315 pound force to initially start the movement of the same 4 inch actuator. This higher force is sometimes required to overcome the starting resistance of a damper shaft or valve stem as it begins to modulate open from its closed position.

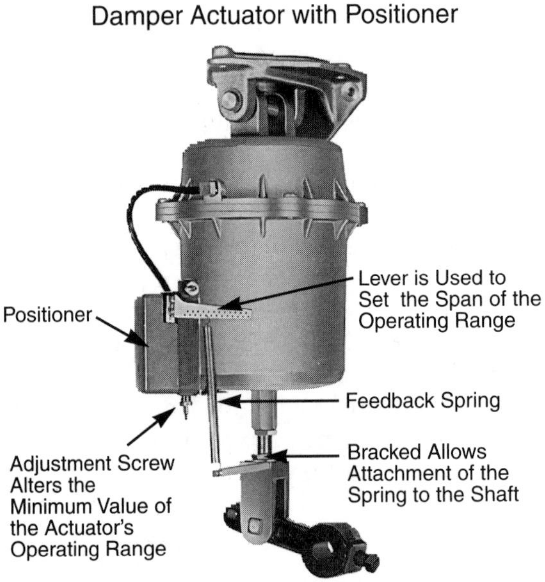

Damper Actuator with Positioner

Figure 10.3 Positioning Relay and Damper Actuator

A feedback spring connects the positioner input lever with a bracket mounted directly on the shaft of the final controlled device. As the actuator shaft moves its position is conveyed to the positioner via the spring. The farther the shaft extends, the greater the force exerted by the stretched feedback spring. When the feedback spring's force balances with the internal forces in the positioner, the shaft stops moving. Figure 10.3 shows a positioner mounted on a damper actuator.

These devices have an adjusting screw that is used to set the start point of the actuator's movement. The start or 0% point is equal to the minimum value of the desired operating range of the actuator. For a 9 to 15 psi operating range, a 9 psi signal would be applied to the input port of the positioner and the zero set screw tuned until the actuator shaft begins to move. The span of the actuator/positioner is a function of the location of the feedback spring on the positioner lever. The lever has a number of holes drilled along its bottom edge to accept the end of the feedback spring. The hole that the spring's end is inserted into determines the span of the actuator. As the spring is places farther out on the lever, the operating span is decreased. For a 9 to 15 psi actuator range, the span is equal to 6 psi (15 − 9 = 6). The spring is inserted in the hole that generates a 6 psi span. After the spring is inserted into the lever, the input signal to the positioner is varied between 9 and 15 to verify that the actuator strokes between 0% and 100%. If it does not, the spring is moved to adjoining holes in the lever until the desired response is generated.

10.10

EXERCISES

Using the formulas and procedures outlined in the chapter, solve the following sizing problems. Show all work. The answers in brackets at the end of the question.

1. A hot water coil has a design load of 18 gallons per minute. The pressure drop across the wide open valve is 2.5 psi. What is the valve flow coefficient for this application? [C_v = 11.4]
2. Determine the maximum flowrate through a valve that has a Cv of 50 and a pressure drop of 30 psi when the valve is wide open. [273.8 gpm]
3. Select chilled water valve for a cooling application that requires 75 gpm at full load and has a valve design pressure drop of 23.1 feet of head. [C_v = 23.7]
4. An ethylene glycol/water solution has a specific gravity of 1.04. Calculate the flow coefficient of the valve needed to supply 88 gpm with a wide open pressure drop of 30 feet of head. [C_v = 25]
5. Water enters an air handling unit heating coil at 155 °F and leaves the coil at 130 °F. 250,000 btu/hr of energy are transferred from the coil to the air passing through the coil fins. Determine the hot water flowrate required to offset this load. [20 gpm]

6. If the pressure drop through the coil and fittings in Problem 5 is 15 feet of head, determine the minimum desired flow coefficient of the valve. [C_v = 7.9]

7. How many btu/minute can pass through a wide open chilled water valve that has a Cv of 45 and a pressure drop of 20 psi? The water enters the coil at 43 °F and leaves at 52 °F. [201 gpm; 15,094 btu/minute]

8. If ethylene glycol was added to the system in Problem 7, the flowrate through the coil would have to be increased to 240 gpm to offset the inefficiencies introduced into the process by the mixture. If the specific gravity of the mixture is 1.028 and the pressure drop across the valve remained the same, how much would the flow coefficient of the valve change? [9.41 units increase]

9. A 25,000 cfm air handling unit warms air from 46 to 130 °F. Water enters the coil at 155 °F and leaves at 139 °F. How many gallons of water per minute must pass through the coil to transfer the required heat? [283 gpm]

10. What would the flow coefficient be for the valve serving the coil in Problem 9 if the pressure drop across the wide open valve was 10 feet? [C_v = 136.5]

11. How many tons per hour of refrigerating effect could a chilled water valve pass if the air across the coil drops 12 °F, the water temperature rises 10 °F and the flowrate through the valve is 60 gpm? [25 tons]

12. A chilled water system removes 20 btu's for each pound of air passing across the coil. The air flow rate is 5,000 CFM and the temperature difference of the water passing through the coil is 11 °F. Calculate the required water flowrate through the coil's valve. [81 gpm]

13. Calculate the valve flow coefficient for the system outlined in Problem 12. Use a pressure drop of 8 feet to size the valve. [C_v = 43.1]

14. What is the critical pressure of a steam valve if the inlet pressure of valve is 10 psig? [12.35 psi]

15. A converter requires 175 lb/hr of steam to produce hot water for the building heating system. The inlet pressure to the valve is 8 psi and the coil pressure is 0 psig. Determine the critical pressure and the flow coefficient for the valve. [$P_{critical}$ 11.35 psi, C_v = 4.8]

16. If a process requires 50,000 btu/hr of heat from a steam source, determine the steam flowrate of the process. [50 lb/hr]

17. A steam preheat coil tempers 30,000 cfm of 0 °F air to a temperature of 50 °F. How many pounds of steam per hour are required for this process? [1620 lb/hr]

18. If the steam pressure entering the valve controlling flow to the coil in Problem 17 is 20 psig, and the pressure in the coil is 0 psig, calculate the flow coefficient for the valve. [C_v = 29]

19. The pressure drop through the outside air branch excluding the damper is 0.12 inches of water at full load. Calculate the pressure drop needed for opposed blade dampers to modulate in a linear fashion. [0.0132 inches of water]

20. What would be the pressure drop requirement for parallel blade dampers if they were used to modulate flow in the system in Problem 19? [0.048 inches of water]

CHAPTER 11

Selecting a Controller Mode Based Upon the Process Characteristics

This chapter describes the attributes of HVAC processes. These characteristics govern how a process responds to changes in its load or set point. The effects that the process capacitance, resistance and time lags have on the response of its control loop and how they are evaluated to determine the correct controller mode are also presented.

OBJECTIVES

Upon completion of this chapter, the reader will be able to:

1. Define temperature, heat, thermal capacitance and resistance.
2. Describe time lags and their effect on the response of a process.
3. Select the proper controller mode based upon its process characteristics.

11.1

THERMAL PROCESS CHARACTERISTICS

Temperature, capacitance and resistance are characteristics used to evaluate the response of thermal processes. The interaction between the thermal resistance (insulation and conductance), thermal capacitance, the rate of heat transfer and the temperature differential of a process have a significant effect on the type of response generated by its control loop. The relationships that exist between these characteristics are used by a control technician to select, calibrate and analyze the operation of control loops.

Control technicians are familiar with the attributes of voltage, current, resistance, capacitance, insulation and conductance from their experiences in servicing HVAC/R equipment. Thermal systems have analogous (comparable) elements that are used to describe their processes. Both electric and thermal systems have elements of force, flow, resistance and capacitance. The relationships that govern variations in these characteristics are also similar in electrical and thermal circuits. Therefore, the response that results from a change in the temperature of a thermal process will be similar to the response of that results from a comparable change in the voltage of an electrical circuit. When reading the following sections, keep in mind that by comparing the response of a thermal circuit to an analogous response in an electric circuit will help to develop an understanding of the thermal relationships that are important in analyzing HVAC/R processes.

11.1.1 Temperature

Temperature is the force within a thermal system that causes energy to flow from warmer objects to cooler ones.

Temperature is a measure of the *thermal intensity* of a substance. Whenever a substance absorbs energy, the vibrations of its molecules increase in intensity. The energy stored in these molecular vibrations can be measured by the increase in the object's temperature. A difference in temperature must exist between two points to cause heat to flow through a thermal circuit. Once a temperature difference is established, energy flows from the warmer "heat source" to the cooler "heat sink". This is analogous to current flowing in an electric circuit whenever a potential difference is applied across a load. Temperature is measured in degrees using the Fahrenheit (°F), Celsius (°C), Rankine (R) or Kelvin (K) scales.

Energy flow that occurs as a consequence of a difference in temperature between two points is called *heat*. The quantity of heat that flows from a source to a sink is proportional to the temperature difference that exists between the two objects. For example, as the temperature difference between a heat transfer coil's surface and the air flowing across its fins increases, the rate of heat (energy) transferred to the air also increases. The change in the energy content of the air can be measured with a thermometer. Conversely, as the temperature difference between the coil's surface and the air decreases, the amount of heat transferred to the air also decreases. When the temperature difference between the coil's surface and the air stream decreases to zero, heat transfer stops. The following equation summarizes the relationships between temperature difference, heat and resistance:

$$Force = Flow \times Opposition$$

$$\Delta\ Temperature = Energy\ Flow \times Resistance\ of\ the\ Path$$

$$\Delta\ °F = \frac{Btu's}{Minute} \times Thermal\ Resistance$$

11.1.2 Thermal Energy Flow

Heat is the energy transferred between two or more objects whenever a temperature difference exists between them.

Heat is analogous to electric current. Both terms describe the movement of energy that results from a difference in force applied across two points of a circuit. Heat naturally flows from a higher temperature source to a lower temperature sink by conduction, convection or radiation. *Conduction* describes heat transfer between two objects that are in physical contact with each other. In this mode of heat transfer, the vibrations of the higher intensity molecules transfer some of their kinetic energy to the cooler molecules they are in contact with. This causes the warmer object to cool as the cooler object warms up. Heat continues to transfer until both objects reach the same temperature.

Convection is the exchange of heat between objects that have different temperatures but are not in physical contact. Convection uses an intermediating fluid as a conductive path between the objects. Heat from the source warms the fluid by conduction. The energy absorbed in the warmer fluid (air or water) is transported to cooler objects. These objects receive energy as the fluid's molecules give up energy to the cooler object's molecules by conduction, increasing the temperature of the object as the fluid's temperature decreases.

Heat transfer by *radiation* occurs between objects at different temperatures through the emission and absorption of electromagnetic waves of energy. These waves require no intermediate fluid to transfer energy between objects. Radiation heat transfer occurs naturally between all objects that are at different temperatures.

Heat is measured in units of btu per unit time (btu/sec, btu/hr). The following equation describes the relationship between heat, temperature difference and thermal resistance. Note its similarity to Ohm's Law:

$$Flow = \frac{Force}{Resistance}$$

$$btu/hr = \frac{°F}{\left(\dfrac{°F - hr}{btu}\right)}$$

11.1.3 Thermal Resistance

Thermal resistance is a measure of the ability of a material to oppose the flow of heat.

Thermal resistance opposes the transfer of heat in thermal circuits similar to the way electrical resistance opposes the flow of current in electric circuits. Figure 11.1 shows a thermal circuit through an insulated section of a wall. A temperature difference of ten degrees is applied across the surfaces of the wall, providing the thermal force needed to establish heat transfer. The quantity of heat migrating through the wall is *directly proportional* to the temperature difference and *inversely proportional* to the resistance of the insulation, wall materials and film of air that forms on the wall's surfaces. If the resistance of this path were increased while the temperature difference remained at 10°, the heat transferred would decrease in proportion to the increase in the resistance of the wall.

Thermal resistance is measured in units of °F/btu/hr. These units are derived using the resistance relationships described by Ohm's Law which states that the resistance of a circuit is equivalent to the applied force divided by the flow through the circuit. Using the analogy that exists between electrical and thermal circuits, thermal resistance is equal to the temperature difference (°F) divided by the flow (btu/hour) or °F/btu/hr. This is commonly written as °F-hr/btu. The following equation shows the relationship between thermal resistance, temperature difference and heat:

$$Opposition = \frac{Force}{Flow}$$

$$\frac{°F}{btu/hr} = \frac{°F}{btu/hr}$$

$$where \frac{°F \ hr}{btu} = \frac{°F}{\frac{btu}{hr}}$$

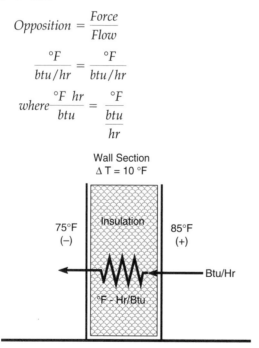

Wall Section
Δ T = 10 °F

75°F (−)   Insulation   85°F (+)

Btu/Hr

°F - Hr/Btu

Thermal Resistance Opposes Heat Flow

Figure 11.1   Schematic Drawing of a Thermal Circuit

This equation shows that larger quantities of thermal resistance reduce the amount of heat that can be transferred through a thermal circuit. This is the principle behind adding insulation to the walls and ceilings of buildings to reduce heat gains and losses to reduce the cost of operating the building.

11.1.4 Thermal Conductors

Thermal conductors provide a low resistance path between surfaces, increasing the rate of heat transfer by conduction.

Conductors are materials that have a molecular structure that transfers thermal energy (heat) efficiently. These materials have molecules that are easily stimulated by the change in their thermal energy. They efficiently transfer their vibrating energy to adjoining molecules, effectively transporting heat through the substance. Heat exchangers are constructed with materials that are excellent conductors of thermal energy. This allows heat to be exchanged between the control agent and the controlled medium though an aluminum, steel, or copper heat exchanger without allowing the two fluids to mix.

11.1.5 Thermal Capacitance

The thermal capacitance indicates the amount of heat that must be transferred with a process to produce a one degree change in its temperature.

In similarity with the other thermal circuit elements, thermal capacitance is analogous to electrical capacitance. Recall, an electrical capacitor is constructed of two conducting plates separated by a dielectric insulator. Current (electrons) is stored in its plates, raising the potential difference (volts) measured across its terminals. In a thermal capacitor, heat flows (btu) into a process (capacitor) and is stored in the materials located within the process. Heat stored in a process produces an increase in the process temperature. Both types of capacitors act as storage devices for energy. Electrostatic capacitors store energy as an electric charge that can be released later as current flow in the circuit while thermal capacitors store heat that can be used to warm cooler objects located in the process.

Capacitance is an important process characteristic. It is used to estimate the amount of time it will take for the process to complete a change in its control point. The relative relationship between the process capacitance and the rate of mass or energy transferred by the control agent is used to select the controller mode for the application. Processes with small relative capacitances change their control point quickly in response to changes in the quantity of control agent passing through the final controlled device. Conversely, processes with large relative capacitances require larger transfers of energy to produce a change in temperature and therefore, take a longer time to change their control point. The following equation describes the relationship between thermal capacitance, thermal energy and temperature:

$$Capacitance = \frac{Unit\ of\ Flow\ (charge)}{Force}$$

$$Thermal\ Capacitance = \frac{btu}{{}^\circ F}$$

Specific heat is an indicator of the amount of heat required to change the temperature of one pound of the material one degree Fahrenheit (btu/lb-°F). The quantity of energy stored in a thermal capacitor is a function of the mass and specific heat of each object found within the boundaries of the process. When the specific heat of a material is multiplied by its mass, the result equals the thermal capacitance of that object (btu/°F). This relationship is shown in the following formula:

$$Specific\ Heat \times Mass = Thermal\ Capacitance$$

$$\frac{btu}{lb\ {}^\circ F} \times lb = \frac{btu}{{}^\circ F}$$

11.2

PROCESS TIMING CHARACTERISTICS

The timing characteristics of a process are also used to properly select its controller's mode. Timing characteristics quantify the time delays incurred in a process. Some of the most common delays that occur naturally in HVAC/R processes and affect their response to changes in load are:

1. The time elapsed as mass or energy is being transported to the process location from its source.
2. The rate which the control agent flows through the final controlled device.
3. The amount of time it takes for a change to occur in the control point.
4. The time it takes for the loop's sensor to measure the change in the control point and generate its corresponding signal.
5. The time it takes for the controller to process the information being sent by the sensor and generate a charge in its output signal.
6. The time it takes for the controlled medium to mix the mass or energy being transferred by the control agent.
7. The time it takes to transmit changes in control signals between devices in the control loop.

After the timing characteristics of the process are combined with the effects of its thermal capacitance, the response of the process can be classified as either being fast or slow. Fast processes require modulating control loops to maintain their set point while slower processes can use less expensive two-position control systems without compromising their process requirements.

11.2.1 Time Lags

A time lag is an period that occurs between the initiation of an action and the appearance of its desired outcome.

Time lags are natural occurrences in all dynamic or real time processes. No process change occurs instantaneously. Consequently, all responses take some amount of time to complete. Transporting mass or energy from one location to another, mixing fluids, changing control signals, positioning actuators, and measuring changes occurring within a process all require some amount of time to elapse before the intended outcome becomes measurable. The period of time, starting at the instant a change occurs in a process and ending the moment the appropriate response is completed, is called a process *dead time lag* or simply, *lag*. All the individual lags combine together into a larger delay that adversely affects the response of a process to changes in its load. Longer lags cause the controller to respond to conditions that no longer exist due to the dynamics of the process. The difference between the actual state of the process and the condition that the controller is currently responding to is called a *phase shift*. When the phase shift becomes too large, indicating the controller is responding to a control point that is significantly different than the current value of the controlled variable, the loop's response will become unstable, resulting in a loss of control.

11.2.2 Categorizing Process Lags

HVAC process time lags typically fall into one of four categories: measuring lags, transmitting lags, transportation lags, and mixing lags. *Measuring* lags quantify the time it takes for a sensor to react to a change in the measured variable; *transmission* lags are delays that take place as control signals are transmitted between components in a control loop; *transportation* lags are associated with the movement of mass and energy from one location to the process; *mixing* lags are delays that occur as the controlled medium is being mixed within the process to offset the load.

A measuring lag is the amount of time it takes for a sensor to measure a change that occurred in the control point and generate the associated change in its output signal. The length of this delay is affected by the amount of thermal resistance that exists within the proximity of the sensing element. Recall from Chapter 7, a sensing element is typically surrounded by a metal sheath to protect it from physical harm. This adds additional thermal resistance between the controlled variable and the sensing element. Sensing elements can also be placed inside a thermal well, which is a sealed metal tube that is threaded into the side of a pipe to isolate the sensing element from the fluid. This protects the element from the turbulence of the flow stream and other harmful characteristics of the fluid. Thermal wells also make it possible to remove the sensing element for replacement without having to open the system. These and other similar types of protective components add thermal resistance to the path between the sensing element and the measured variable. Consequently, heat must pass through multiple resistances before the element can respond to changes in the control point. The interval of time that elapses while the heat is flowing through these resistances generates the measuring time lag.

The mass of the sensor also contributes to the size of the measuring lag. Some sensors are constructed in larger sizes using with more material to increase their ruggedness. This allows them to withstand the riggers of harsher environments. As the structure of a sensor increases in mass, its thermal capacitance also increases. Consequently, the time it takes for the sensing element to respond to changes in temperature also increases. This increases the measuring lag of the process. Optimum sensor design incorporates good conductors, lightweight sensing elements, and shields to produce a strong, fast responding device thereby minimizing the measuring lag of the process.

11.2.3 Transmission Lags

Transmission lags are delays that take place as signals are being sent between devices in a control loop. Transmission lags begin when the output signal of a device begins to change and end when the connected control device receives the entire change in its input signal. Electric signals travel at the speed of light and therefore, do not introduce a measurable transmission lag in HVAC control loops. Conversely, pneumatic control systems are required to change the quantity of instrument air in long lengths of signal lines whenever the control signal changes. This produces longer transmission lags that adversely affect the response of the control loop. This is one reason why pneumatic signal tubes are installed with the shortest possible lengths.

11.2.4 Transportation Lags

Transportation lags are delays that occur whenever energy or mass is transported to a process. This type of lag is related to the distance that exists between the point where energy is originally transferred to the control agent (boiler, chiller, convertor, etc.) and where it is transferred in the process. For example, in a zone heating process, the energy is added to the controlled medium at the heat exchanger's location in the air handling unit. From there, it must be transported through the ducts to get to the individual zones. The greater the distance between the coil's location and the process, the greater the transportation lag of the process.

11.2.5 Mixing Lags

Mixing lags occur when the energy contained in the controlled medium must mix with the air or water in the process in order to produce a change in the control point. In an air handling unit application, the conditioned air exits the registers into the zone where it must mix with the room air to produce a change it the zone's temperature.

11.3

TIME CONSTANT

A time constant is the amount of time it takes for a process to complete 63% of a desired change.

Time constants are used in many technical fields to indicate how long it takes to complete a process response. The size of the time constant is a function of the resistive and capacitive characteristics of the process. The resistive element of the process determines the rate of heat transfer and the capacitance element establishes how long it takes for measured variable to change one unit. As either of these process characteristics increases in magnitude, the amount of time it takes for the control point to change 63% of its total response also increases. The equation for calculating the time constant of a process is:

$$T = R \times C$$

where:

    T = time constant (seconds)
    R = thermal resistance of the process
    C = thermal capacitance of the process

The value of 63% is derived from a characteristic of all dynamic processes whose response changes in proportion to itself. Consider a heat transfer process. The rate of energy transfer between two objects is proportional to the temperature difference that exists between them. Also, the temperature of the cooler object will increase in proportion to the amount of heat transferred from the warmer object. As the heat absorbed by the cooler object increases its temperature, the rate of heat transfer decreases because the temperature difference between the two objects decreases. Summarizing, the temperature of the cooler object changes in proportion to its current temperature, therefore, one time constant is required to complete 63% of the total change in temperature.

When the response of the process is based upon its present state, it is mathematically described by Euler's Number, which is the lower case e or ln found on scientific calculators. When Euler's Number is raised to the -1 power, (e^{-1}) the result is 0.367. The-1 exponent indicates one time constant has expired. When e^{-1} is subtracted from one $(1 - e^{-1})$ the result is 0.633 or 63.3%. This number indicates that whenever a process changes in proportion to itself, 63% of its entire response is achieved in one time constant and 36.7% of the response must still take place before the process stabilizes. For example, if the hot water set point of a steam converter is increased 10 °F, the time constant is equal to the amount of time it takes for the hot water control point to increase 6.3 °F (63.3 % of 10°). The actual time can be measured with a wrist watch or clock. After one time constant has elapsed, a process will have changed 63.3% of its total response. After two time constants the process will have changed 86% $(1 - e^{-2})$; after three time constants the process will have changed 95% $(1 - e^{-3})$; after four time constants the process will have

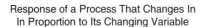

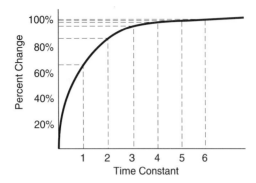

Figure 11.2   Graph of a Time Constant Response Curve

changed 98% $(1 - e^{-4})$ and after 5 time constants have elapsed, the total response or change is considered to be complete, based on the following calculation: $(1 - e^{-5}) = 0.993$ or 99.3%. Therefore, if it took 1 minute for the hot water to increase 6.3°, the entire process response, increasing the temperature 10°, will take 5×1 minute or 5 minutes. Each successive time constant produces a smaller increase in temperature over the same period of time because the differential temperature between the steam and the water is getting smaller as the water temperature increases. As the temperature difference decreases, the rate of heat transfer also decreases. The plot of a process response that changes in proportion to itself if shown in Figure 11.2. The curve shows that the response time of the process is nonlinear.

11.4

DETERMINING A PROCESS CONTROLLER MODE BASED ON THE PROCESS CHARACTERISTICS

The process characteristics described in this chapter are used to select the correct controller mode for a given application. Before choosing the controller mode for a process, each of the following factors must be evaluated to determine which of the seven controller modes will most efficiently maintain the process at its required set point:

1. The process capacitance.
2. The size of a normal load change.
3. The speed of the load change.
4. The length of the process time lags.
5. The permissible operating range of the control point.

6   The maximum cycling rate of the equipment.
7. The permissible size of the initial error and steady state offset.
8. The acceptability of a steady state error.

The following sections outline the requirements that must exist before a particular controller mode can be used to efficiently maintain the process set point.

11.4.1 Timed Two-Position Controller Mode

Timed two-position controller mode should only be used in processes that:

1. *Have a large capacitance.*
2. *Have small time lags .*
3. *Only experience small load changes.*

Large capacitance

A two-position controller pulses energy into a process by cycling its final controlled device open and closed (on/off). This pulsating response produces an operating differential in the process that appears as oscillations in the control point. The size of these oscillations is a function of the controller's control differential, the capacitance of the process and the measuring, transmission, transportation, and mixing lags in the system.

A process with a large capacitance can absorb the pulsations of energy while minimizing the size of its operating differential. Transportation lags will have a smaller effect on the control point because it takes a large quantity of energy to make a unit change in the control point in processes have a large capacitance. In these applications, when the remaining energy being transported to the process finally arrives, its size is small in relation to the process capacitance so it produces a minimal change in the control point. Processes that have a smaller capacitance will magnify the effects of the transportation lag, producing noticeable changes in the control point. Therefore, two-position controller mode should not be used in applications where the relative process capacitance is small.

The process capacitance must be sufficiently large to limit cycling of the equipment to a level below the manufacturer's design limits. The small capacitance experiences fast changes in its load because it only takes a small amount of mass or energy transfer to change the control point. When the capacitance is too small, cycling the final controlled device will produce a large changes in the control point in a relatively short amount of time, causing the controller to rapidly cycle the equipment. This rapid cycling or hunting can destroy control devices and mechanical equipment.

Small time lags

A two-position controller mode already has a time delay in its transfer function in the form of its control differential. The control differential delays changes in the controller's output signal until the control point exceeds a predetermined span.

The total system lag is determined when the remaining process lags are added to the controller's transfer function lag. When the length of the total system lag becomes too large, the phase shift between the actual value of the control point and the response of the controller becomes prohibitively large. The controller begins to alter the flow of the control agent based upon the state of the controlled variable that no longer exists. Consequently, its corrective action is no longer applicable to the actual state of the process. The increase in the phase shift of the response permits the control point to move farther away from the set point, further increasing the size of the operating differential of the process. Therefore, systems with large time delays cannot use a two-position mode because it add additional delays to the process.

Processes with a larger capacitance can offset some of the negative effects that longer time lags have on the operating differential of the process incorporating a two-position controller mode. When the capacitance is sufficiently large, the change in the control point is relatively small in response to pulses of energy being transferred into the process. This reduces the measurable effects of the system lags and the operating differential of the process. Minimizing the length of the measuring, transmitting, transportation and mixing lags through proper design and installation practices also reduces the length of the system lag.

Size of the load change

Two-position mode can only be used in applications where the process experiences relatively small load changes. A small load change is one that has a typical transfer rate that is less than the output capacity of the two-position final controlled device. During a normal load change, opening of the final controlled device will allow sufficient mass or energy to flow into the process, limiting the growth of the operating differential. Larger changes in the load produce greater operating differentials, much like the effects produced by larger time lags. If the load change is larger than the capacity of the final controlled device, the control point will exceed the acceptable limits of the process and the loop will operate out of control. A loop operating out of control produces comfort or other process problems. Therefore, the maximum anticipated load change must be smaller than the maximum flow through the final controlled device.

The size of the load change is the limiting factor in two-position controller applications, not the speed of the load change. The load may change quickly or slowly without adversely affecting the response of the process as long as rapid cycling does not occur. These capacitive, lag and size requirements apply to both two-position and timed two-position controller modes.

11.4.2 Floating Controller Mode

Floating controller mode should only be used for processes that:

1. *Have medium capacitances.*
2. *Have small time lags.*
3. *Experience slow load changes.*

Small to medium capacitances

The performance of a floating controller mode is an improvement over the response of two-position mode. The modulating action of its final controlled device reduces the range of the control point, improving process comfort and efficiency. Floating mode can be used for processes with capacitances that are too small to permit the use of a two-position controller mode or where large changes can occur in the load.

In floating mode applications, the relative process capacitance only needs to be large enough to prevent the control point from traveling beyond the limits of the neutral zone during normal load changes. Since the control agent is modulated into the process, its capacitance does not have to be large enough to absorb pulses of mass or energy. Therefore, the process capacitance does not have to be as large as that required for two-position mode.

The modulating action of the final controlled device also reduces the total length of the system's lag by removing the need for the control point to transverse the entire control differential before a change in the controller's output signal is initiated. Also, the transportation lag of a floating mode application contains less energy that has to be absorbed by the process capacitance because the flow through the final controlled device is modulated, not pulsed. The reductions in lag time coupled with the modulating response of the final controlled device reduces the capacitive requirements of floating mode processes.

Small time lags

The floating mode also has a delay inherent in its transfer function. The neutral zone postpones the response of the controller until the control point exceeds a limit of the neutral zone. As in two-position applications, when the length of the controller transfer function and process lags become too large, the controller's response shifts out of phase with the actual process conditions. Therefore, floating controller mode cannot be used in processes that have inherently long system lags.

Slower load changes

The final controlled device in a floating mode application is specially designed for use with floating mode controllers. It repositions the shaft of the flow control device at a constant rate of speed. This characteristic of the mode limits the speed of the load change to value that is slower than the speed of the actuator. When load changes occur at a speeds lower than the rate of change of the controlled device, the controller can quickly overcome the load change, maintaining the control point within the neutral zone. When a load change occurs at a rate that is faster than the final controlled device can transfer energy, the control loop is driven out of control. Under these circumstances, the final controlled device is driven to one of its output limits and stays there until a change in the load reverses the direction of the actuator. Contrary to a two-position mode of control, the speed of the load change, not its size, is the limiting factor on whether floating mode can be used.

Large load changes

Unlike two-position mode, floating mode can be used for processes that experience larger load changes. This is possible because of the modulating final controlled device can be sized larger than what would be possible in a two-position application. Since the floating mode final controlled device modulates flow, it can maintain control under part load conditions without creating large swings in the control point.

11.4.3 Proportional Controller Mode

Proportional controller mode is used in applications that:

1. *Have a small capacitance.*
2. *Have large or small system lags.*
3. *Experience large or small load changes.*
4. *Experience slow or fast load changes.*

A proportional mode controller generates a change in its output signal that is proportional to the change that occurred in the control point. Proportional mode is the simplest, fully modulating controller mode. The controller's output signal and its final controlled device are 100% modulating devices. Unlike the two-position and floating modes, the output of a proportional controller is analog in nature, capable of taking on any value within its operating range. It is a universal mode that can be used for any process, regardless of its capacitance and length of time lags or the size and speed of its load changes. Once a proportional loop has stabilized, a small offset will remain between the set point and the control point. This is the nature of proportional mode.

Due to the additional complexity and cost of these modulating devices, proportional control mode is typically used for commercial and industrial processes that have small capacitances or that experience large load changes that occur at speeds that prohibit the use of a two-position or floating controller mode.

Small capacitance

The proportional mode transfer function has an adjustable gain parameter that is tuned so that the loop's response compliments the reaction of its process. The size of the proportional gain affects the speed of the controller's response to changes in the control point. Its value is selected to reposition the final controlled device as quickly as possible without causing loop instability. Since the gain of the proportional controller can be set to any value, the controller can be calibrated to maintain any process, regardless of the characteristics of its load changes.

Proportional mode has no inherent time delay in its transfer function. There is no control differential or neutral zone that must be exceeded before the controller can initiate its response. In two-position and floating applications, the process capacitance had to be sufficiently large to overcome the effects of the control differential and the process lags. In proportional applications, the controller's output signal begins to change as soon as a measurable change occurs in the control point. Therefore, the process capacitance can be small because it does not need to absorb pulses of energy or prevent the occurrences of large phase shifts between the process condition and the controller's response.

Large or small system lags

All dynamic processes have some amount of naturally occurring time lag. The proportional gain of the controller can be selected to minimize the detrimental effects of these system lags. Processes with *long* lags are tuned with a larger proportional gain to permit the controller to make larger changes in the position of the final controlled device in response to small changes in the control point. This limits the effect that the lags have on the phase shift of the process. Small values of proportional gain are used for processes that have shorter lags because the process already changes quickly. If a larger gain were used, the loop response would become unstable and the controller would begin to respond more like a two-position device than a modulating one.

Experience large or small load changes

A proportional controller adjusts the rate of energy transfer occurring with the process as soon as a change in the control point is measured. Therefore, the loop can create a balance between the energy transferred and the process load regardless of the size of the load change. The only requirement is that the final controlled device must be properly sized, using the procedures presented in Chapter 10.

Experience slow or fast load changes

The proportional gain can be tuned to a value that reduces the detrimental effects of time lags and the speed of the load changes. When fast load changes are normally found in a process, the proportional gain will be set to a larger value. This allows the controller to make greater changes in the flow of the control agent, thereby minimizing the change in the control point. When slower load changes are the norm, the gain can be reduced to better match the response of the load.

11.4.4 Integral Controller Mode

Integral mode is used along with proportional mode to reduce any steady state offset to zero.

The integral controller mode can be used in conjunction with proportional mode for any process where the requirements impose that there cannot be a steady state error remaining after a load change. In PI mode applications, the proportional mode responds to the initial change in control point, balancing the energy transferred into the process with its load. The integral mode will continue to make changes in the controller's output signal until the steady state offset is driven to zero. When the controller's output signal stabilizes, the control point will equal the set point.

The integral mode acts as a fine tuning control, responding slowly to reduce any steady state error to zero. The integral gain is selected so the control point is driven to the set point within a reasonable amount of time after the load change occurred. Larger integral gain values will eliminate any offset quickly, but may cause the controller's response to become unstable. A smaller integral gain will reduce the error to zero without increasing the possibility of loop instability. The integral mode improves process operating efficiency and comfort by maintaining the controlled variable at its desired set point.

11.4.5 Derivative Mode

The derivative mode anticipates the maximum growth of the error based upon its present rate of change and quickly moves the final controlled device to a position that minimizes the change in the control point caused by fast load changes.

Derivative mode is used with proportional mode (PD) or proportional plus integral (PID) mode to provide an anticipating action that prevents excessive swings in the control point in response to fast load changes. Derivative mode can be used to add stability to loops that require a large proportional gain. The derivative mode's response acts to reduce the adverse effects of a high gain proportional mode during quick changing events. It accomplishes this by positioning the final controlled device quickly, thereby reducing the offset experienced by the process and consequently, limiting the amount of change in the controller's output signal that would be produced by the large proportional gain. By reducing the effect of the proportional response during fast load changes, the loop can be set up with a higher initial proportional gain for normal load changes without the fear of instability during quick load changes. Derivative mode is not used as often as other modes. The relative slowness of HVAC processes reduces the need for derivative gain's braking action.

11.5

SUMMARY

HVAC processes have force, flow, capacitance and resistive elements that affect the response of the system to load changes. The force element generates flow of mass or energy in the process. The quantity of flow is inversely proportional to the system's resistance. The flow element transports mass or energy from a source to the process. The capacitive element stores mass or energy to reduce the effects of load changes on the control point. These four process characteristics exist in electrical, thermal and liquid systems.

Lags are natural occurrences in all dynamic systems. A lag is a time delay that occurs between the initiation of an action and the appearance of the desired result. Lags produce phase shifts between the response of the controller and the actual condition of the process. Large phase shifts result in a poor loop response, an inability to maintain the control point close to the set point and loop instability. HVAC process lags can be classified as either a measuring, transmitting, transporting or mixing lag.

Time constants are a standard means of measuring a dynamic process that changes its response in proportion to its present condition. A time constant indicates the length of time it takes for a process to complete 63% of its required change. Five time constants are required for a process to complete its entire change.

The elements of a process along with the size and speed of the process load change is used to select the controller's mode for a given application. Two-position mode is used for processes having a large capacitance, small load changes and short system lags. Floating controller mode is used for processes having a medium capacitance where the speed of the load change is slower than the speed of the actuator of the final controlled device. Floating mode can also control large changes in load that could not be maintained with two-position controller mode. Proportional mode can be used for any process regardless of its capacitance, length of time lags and the speed or size of the expected load changes. This mode generates a modulating signal whose gain can be adjusted to meet the operational requirements of the process. Proportional mode can be coupled with integral mode to produce a response that leaves no steady state error. It can also be coupled with derivative mode to produce a response that is sensitive to fast load changes without causing instability.

11.6

EXERCISES

Determine if the following statements are true or false. If any portion of the statement is false, the entire statement is false. Explain your answers.

1. Heat is a measure of an object's thermal intensity.
2. The process capacitance defines how much mass or energy is required to change the control point one unit.
3. As the thermal capacitance of a process increases, the amount of time it takes to change the temperature of the process increases.
4. Dirt and other debris caught between the fins of a heat transfer coil will increase the thermal conductivity of the path between the control agent and the controlled medium.
5. Two-position controller mode can be used in processes having small process lags.
6. Floating mode and proportional mode controllers can be used for processes that experience large load changes.
7. The transportation lag of a system increases as the length of the signal lines connecting the control devices increase.

8. Proportional mode is used for fast changing processes that can have a steady state error.
9. Derivative mode is used more often than integral mode in HVAC processes.
10. A proportional mode should be used in place of a floating mode controller for a process that experiences fast load changes.

Respond to the following statements and questions completely and accurately, using the material found in this chapter.

1. List the four elements of a process and describe their relationship to each of the other process elements.
2. As the force element of a process increases, will the transfer rate of the mass or energy increase or decrease? Explain your answer.
3. What is the function of the capacitive element in a thermal process?
4. What is the difference between thermal capacitance and thermal capacity?
5. What are time lags and what causes them?
6. List four different types of HVAC time lags and describe a characteristic of each.
7. What effect does a time lag have on the phase shift and stability of a control loop response?
8. What is a time constant?
9. Where does the value 63% come from when calculating a time constant?
10. What happens to the length of the time constant of a classroom during the cold weather, when it fills up with students entering for the first class of the day?
11. A technician repositions a pneumatic thermostat in a cafeteria from the west wall to the east wall. Forty feet of pneumatic tube is needed to reconnect the thermostat. What affect will this have on the response of the process?
12. The same technician installs a tamper-proof shield over a thermostat located in a cafeteria. What affect will this have on the process response?
13. What effect will closing the dampers in a wall register have on the time constant of a process.
14. A set back thermostat changes the set point of a zone from 62° to 72° at 7:00 am. The time constant of the process is 6 minutes. How long will it take for the room to reach set point?
15. Describe the process requirements for using each of the the seven controller modes.

The Calibration and Response of Modulating Control Loops

The calibration procedure begins by reviewing the objectives that are to be met by control loop after it is properly tuned. Next, control loop is analyzed while it is operating to determine how it is presently responding to changes in its set point. Any necessary adjustments are made to the loop components to make them perform as designed or required. Finally, the response of the newly calibrated control loop is evaluated to determine if it meets one of the criteria of a properly tuned control loop. Chapter 12 explains how to calibrate a control loop so it maintains the process at set point as efficiently as possible.

OBJECTIVES

Upon completion of this chapter, the reader will be able to:

1. List and explain the objectives of a modulating control loop.
2. List and explain the characteristics of each of the five possible closed loop responses to a change in load or set point.
3. Explain the effects that changes in the controller's gain parameters have on closed loop response.
4. Explain the criteria used in the evaluation of a closed loop response.
5. Explain why the objectives and responses listed in this chapter do not apply open loop configurations

12.1

CALIBRATING AN OPEN LOOP CONTROLLER

The first step performed in any controller calibration procedure is to check the accuracy of the loop sensor's output signal. In this procedure, an instrument of proven accuracy is placed next to the sensing element of the process to measure its control point. The measured value of the process variable is compared to the output signal of the sensor. If both devices indicate the same magnitude, the sensor is calibrated and the technician can move on to the controller calibration. If the measured and sensor signals do not correlate, the sensor must be calibrated or replaced before the controller can be calibrated.

Open loop controllers are calibrated using a *static* evaluation procedure. To calibrate a two-position, open loop controller, its set point adjustment is slowly turned until the output signal of the controller changes its state. The position of the set point adjustment where the controller's output signal changes state correlates with a set point equal to the present value of the measured variable. Therefore, once the output signal changes state, the indicator of the set point dial can be aligned with the mark corresponding with the value of the control point. For example, if the measured variable was 72°, the controller's output signal will change state when the control point exceeds 72°. Therefore, at the point where the controller's output changes state, the mark corresponding to 72° on the set point dial will be aligned the dial's pointer.

To complete the calibration procedure, the control differential is adjusted to meet the requirements of the process and the set point dial is set to the desired process set point. The control point should be varied above and below the set point to verify that the controller operates as intended. Note, the entire calibration procedure is performed without the need to evaluate the control loop's response under actual operating conditions. In fact, open loop controllers can be calibrated on a bench before being installed in the field without affecting its operation. Closed loop controllers cannot be calibrated using these procedures.

12.1.1 Closed Loop Dynamic Analysis

Dynamic analysis monitors the oscillations of a controller's output signal to determine if its response meets the objectives of a properly tuned control loop.

In Chapters 6, the transfer function of a control device was used to calculate the magnitude of an output signal based upon a known input condition. To use this analysis tool, a technician enters the current value of an input signal into a transfer function and calculates the corresponding output signal. This technique is useful when checking the operation of a control device under *static* or unchanging conditions. Unfortunately, it does not indicate how the loop will dynamically respond to load changes. Consequently, it cannot be used to *tune* a modulating controller.

To effectively calibrate any closed loop, an evaluation of the *dynamic* response of the controller's output signal must be performed. Dynamic analysis examines the path or curve generated by the controller's output signal response during the period following a set point or load change. The response is compared with the control industry's standard modulating responses to determine if the loop is properly calibrated. Figure 12.1 depicts the path generated by a modulating controller's output signal in response to a change in the process load.

In dynamic response analysis, the path generated by the controller's output signal is evaluated over a interval of time called the *transition period*. A loop enters a transition period as soon as its controller begins to respond to a change in its load or set point. The transition period ends once a balance is established between the energy transfer and the load. At that point, the controller's output signal and the process control point stabilize.

Once the transition period ends, the loop enters a period of *steady state* operation. Steady state operation describes a loop condition that occurs when all the variables of a control loop stabilize, or become steady. A control loop remains in a steady state condition until another change occurs within the process. Figure 12.2

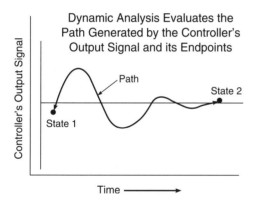

Figure 12.1   The Response Path Generated by a Modulating Controller

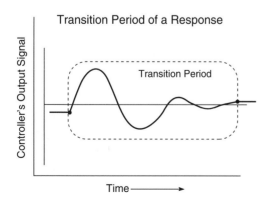

Figure 12.2   Transition Period of a Closed Loop Modulating Response

shows the transition period of a modulating loop response. The steady state periods occur on both sides of the transition period, depicted as the areas outside of the dashed lines.

12.1.2 Transition Period Characteristics

The focus of closed loop dynamic response analysis is on the path generated by the controller's output signal during the transition period. By monitoring the response of the loop during this period, the technician can determine whether the loop is calibrated correctly. The number and severity of the oscillations in the transition period along with the size of the remaining steady state error indicate the current state of calibration of the controller. These response characteristics indicate what changes need to be made to the parameters of the controller's transfer function to achieve optimal calibration.

The oscillations in a closed loop controller's response path are the result of the self-correcting action provided by the presence of the feedback path. Each oscillation represents a corrective action generated by the controller as it monitors the effect its last output signal change had on the condition of the controlled variable. Application 12.1 describes a typical closed loop response, illustrating the generation of the oscillations in the controller's output signal.

APPLICATION 12.1

Process: Classroom Temperature Control.

Objective: Maintain the classroom at a temperature of 72 °F.

Controller Mode: Proportional Only

Response:

a. At present, the room is in steady state operation at a temperature of 71.5 ° and the controller's output signal is stable as depicted at State 1 in Figure 12.1.

b. A large number of students enter the classroom, causing the control point (temperature) to increase in response to the change in the process load. The sensor generates a new output signal corresponding to the new control point. The signal is sent to the controller which calculates the new process error and generates a corresponding change in its output signal. The controller increases its output signal to close the NO heating valve. This is depicted as the first rise in the output signal following State 1 in Figure 12.1.

c. Process lags generate a phase shift between the controller's response and the actual condition of the process. Consequently, the controller's previous change in its output signal cannot balance the existing load. The sensor, via the feedback path, informs the controller that its initial response closed the valve too much, causing the control point to drop. The controller responds by modifying its output signal based on the new error. This sequence of events continues until a balance is established between the flow of the control agent and the process load.

Calibrating a modulating control loop is achieved by properly selecting values for each of the controller's gain parameters and in some applications, adjusting the output signal's bias. A proportional controller has one gain parameter; proportional plus integral (PI) and proportional plus derivative (PD) controllers have two gain parameters and PID controllers have three gain parameters that must be correctly chosen during the calibration procedure. Variations in these parameters determine the size and number of oscillations that appear in the transition period of the controller's response. By properly selecting values for these gain parameters, the loop response can be optimized to efficiently maintain the process requirements.

12.2

OPERATIONAL OBJECTIVES FOR A CLOSED LOOP RESPONSE

Operational guidelines have been established for the calibration of closed loop modulating control applications that are used to help a technician determine a correct value for each of the controller's gain parameters. These guidelines consist of three objectives that define the overall characteristics of a properly tuned, modulating closed loop response. Calibrating a loop to meet these objectives insures that the technician has completed the calibration process correctly. The three operational objectives of a control loop are:

1. Minimize the size of the initial change in the control point.
2. Minimize the length of time of the transition period.
3. Minimize the size of the remaining steady state error.

Each one of these objectives is affected by the values used for the controller's gain parameters. Proper selection of the proportional, integral and derivative gains by the careful adjustment of the throttling range, reset time and rate of the controller's transfer function will compensate for the process lags, providing a loop that meets the purpose and objectives of a properly tuned modulating control loop.

12.2.1 Minimize the Initial Change in the Control Point

The first operational objective of a closed loop response is to *minimize the initial change in the control point*. Changes in the control point result from variations that occur in the process load or set point. Whenever either of these variables changes, the loop departs from its steady state operating mode and enters a transition response period. Typically, the control point experiences its greatest variance from its previous steady state value at the beginning of the transition period.

What is the initial change in the control point?

The first oscillation in a transition period is called the *initial change* or *maximum error* of the response. It is the first change experienced by the control point after a period of steady state operation. Figure 12.3 depicts the initial change in the control point. Typically, the first oscillation will be the largest in the transition period because the controller's response is delayed by process lags.

At the start of the transition period, the control point initially moves away from its steady state value. The amount of change experienced by the control point is related to the size of the load or set point change. The control point will continue to move away from its initial steady state value until the amount of energy transferred by the control agent exceeds the rate required to balance the new load. Process lags cause the flowrate of the control agent to exceed the quantity needed to balance the new load. Consequently, the control point reverses its direction and begins to move back toward its initial value. Other oscillations will occur as the controller continues to attempt to find the balance point of the load. The number of oscillations are a function of the magnitude of the process time lags and the values chosen for the controller's proportional, integral and/or derivative gains.

What affects the size of the initial change in the control point?

The size of the process time lags have the greatest effect on the size of the initial change in the control point. Lags allow the control point to begin changing before the controller becomes aware of a variation occurring in the load. As the size of the process time lags increase, the size of the initial change in the control point also increases. Conversely, processes with small lag times experience small delays and

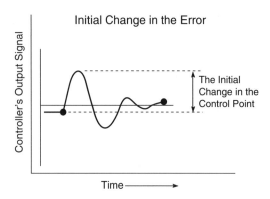

Figure 12.3   Initial Change in the Control Point

initial changes in the control point before the controller responds, minimizing the initial change in the control point.

The controller's proportional gain is used to reduce the detrimental effects of the process time lags. When the controller's proportional gain is increased, it generates larger changes in the flow of the control agent in response to small changes in the control point. Therefore, once the controller receives an indication the control point is changing, it responds quickly to dampen the growth of the initial change in the control point. When the proportional gain is set to a small value, the control point can travel outside the limits of the throttling range before the controller will make the necessary adjustments in the flow of the control agent.

Why must the initial change in the control point be minimized?

If the initial change in the control point is allowed to grow too large, the loop moves outside of the limits of its throttling range, adversely affecting the comfort, efficiency and safe operation of the process. These excessive responses are noticed by the occupants and generate service related complaints. Large changes in the control point also result in large steady state offsets after the loop has stabilized.

How is the initial change in the control point minimized?

The size of the initial change in the control point is primarily controlled by the value used for the controller's *proportional* gain, and to a lesser extent, the other gains that may be used in the transfer function. A technician alters the value of the controller's proportional gain to regulate the size of the initial change in the control point. The controller's proportional gain defines the amount of change in the output signal that is generated for each incremental change in the control point (error). The controller's response becomes quicker as its proportional gain is increased. The faster a controller can alter the flowrate of energy through the final controlled device, the smaller the initial change in the control point will become. Application 12.2 describes the effect of changing the proportional gain's value to reduce the size of the initial change in the control point.

APPLICATION 12.2

Process: Cold Deck Temperature Control Loop.

Objective: Maintain the temperature of the air leaving the cooling coil at 55ºF.

Components: A direct acting sensor and a direct acting controller that generate 4 to 20 milliamp output signals. A normally closed, 4 to 20 milliamp linear chilled water valve.

Response: As the temperature of the air downstream of the cooling coil increases due to an increase in the process load, the controller's output signal also increases, modulating the chilled water valve open.

Proportional Gain: With an initial gain of 2 mA/°F, the throttling range is 8 degrees which corresponds to the set point $+/- 4$ °F. With this proportional gain, the valve moves 12.5% (100% ÷ 8) for each one degree change in the control point. Unfortunately, the process lags produce a larger than necessary initial change in the control point.

   To improve the response of the loop, the calibrating technician increases the gain to 4 mA/°F. This equates to a 4 °F throttling range. The valve now opens 25% for each one degree change in the control point. The loop will now produce the same change in the position of the final controlled device after only one half of the previous required change in the control point. The loop is now twice as responsive with its larger gain. The reduction in response time will minimize the size of the initial change in the control point.

How is the value of the gain selected?

A technician calibrating a modulating control loop selects the controller's proportional gain *while the system is operating under normal load conditions*. The gain is set as high as possible to minimize the initial change in the control point without producing a large number of oscillations within the transition period. This strategy creates a control loop that is sensitive to small changes in the controlled variable. High gain controllers quickly position the final controlled device to allow sufficient energy transfer to inhibit any further growth in the initial change in the control point. This sensitivity prevents large initial changes in the control point from occurring whenever the load changes.

12.2.2 Minimize the Transition Period of the Response

The second operational objective of a closed loop response is to *minimize the length of the transition period of the response.*

What is the transition period of the response?

The transition period is a interval of time when the controller's output signal is in a state of change. The transition period is bounded by periods of steady state operation. Whenever a change in process conditions occurs, the optimal response of a control loop would be to immediately drive the final controlled device to the position that produces the required balance between the energy transferred and the new load. Unfortunately, time lags and the controller's proportional gain combine their effects, producing oscillations in the controller's output signal and the control point. In most control loops, there are always a few oscillations found within the transition period.

What affects the length of the transition period of the response?

The length of the transition period is affected by the amount of gain in the controller's transfer function. The proportional and integral gains have the greater effect in processes that change at a normal rate. Proportional, integral and derivative gains work together in processes that experience fast load changes. Whenever a change in the control point causes the controller to respond, all of the gains in its transfer function begin to adjust the output signal in proportion to the size of the error. The larger the overall gain of the controller, the longer the transition period of the response.

Why must the transition period of the response be minimized?

The longer the transition period exists, the greater the length of time the process operates in an unbalanced state. Consequently, the control system is supplying the incorrect amount of energy to the process. Minimizing the transition period of operation reduces the amount of time the control point is changing. It also minimizes disturbances in occupant comfort, process efficiency and mechanical wear of system components.

How is the transition period of the response minimized?

To reduce the number of oscillations within the transition period, the controller's overall gain must be set to values that limit the number of oscillations. Smaller values of proportional, integral and derivative gain slow down the response of the controller, limiting oscillations. If the gain parameters are set too low, the transition period will begin to become longer even though the controller's output signal will not oscillate. Although the oscillations in the response have been removed by the lower gain controller, it takes too long for the final controlled device to change position and the loop will go out of control before the controller's output stabilizes. Figure 12.4 highlights the differences in responses for a high gain and lower gain controller to the same change in load. Note, the lower gain response has a longer transition time.

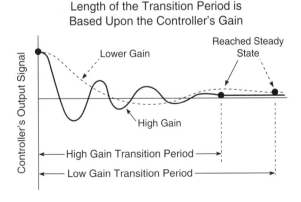

Figure 12.4   Comparison of High and Low Gain Controller Responses

How is the value of the proportional gain selected?

Higher values of proportional gain are needed to minimize the size of the initial change in the control point. Unfortunately, these higher gain values also generate more oscillations within the transition period of a response. As the controller's gain becomes larger, the number of oscillations and the transition period increase. Lowering the value of the proportional, integral and derivative gains shortens the transition period but increases the size of the initial error, which is contradictory to the first operational objective.

The calibrating technician must balance the advantages of a smaller initial change in the control point with the disadvantages of a longer transition period. A properly calibrated modulating control loop reacts quickly to minimize the growth of the initial change in the control point but also permits the loop to oscillate a minimum number of times without significantly increasing the transition time.

12.2.3 Minimize the Steady State Error

The third and final operational objective of a closed loop is to *minimize the steady state error.*

What is the steady state error?

The steady state error is the mathematical difference between the control point and the set point of a process once the loop has reached a steady state condition after a response. It is caused by the inherent response of a proportional mode controller which balances the load and energy transfer at some point within the throttling range of the process. Once the controller achieves the balance point its output stabilizes and a period of steady state operation begins. The difference between the control point and the set point when a loop is in its steady state operating mode is called the steady state error.

In a properly calibrated loop, the steady state error falls within throttling range of the process during normal load operation. The point of this objective to finish a

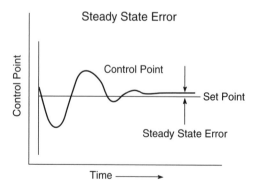

Figure 12.5   Steady State Error

response to a change in load or set point with the control point as close to the set point as possible. Figure 12.5 shows the steady state error that is present after a proportional mode controller has reached steady state operation.

What affects the size of the steady state error?

In similarity with the first two operational objectives, the proportional gain of the controller is also responsible for minimizing the size of the steady state error. When a loop is calibrated to respond quickly to changes in its process, the steady state error is minimized. Slower responding loops produce larger steady state errors.

Why must the steady state error be minimized?

The steady state error is minimized to improve the comfort and efficiency of the process. Whenever the control point is allowed to stabilize at a value that differs from the set point, the process is operating in a less than optimal condition. Minimizing the difference between the set point and control point minimizes the inefficiencies of the process.

How is the steady state error minimized?

The steady state error is minimized by increasing the proportional gain of the controller. A high gain response reduces the initial change in the control point, meeting the requirements of operational objective one. By restraining the initial growth of the control point, it remains closer to the set point during the transition period. Consequently, by the time the loop achieves a steady state condition, the control point is closer to the set point than it would be if the controller's gain was set at a lower value.

How is the value of the gain selected?

The value of gain is selected to meet the first objective, which automatically minimizes the size of the steady state error. Increasing the gain minimizes the initial change in the control point and the size of the steady state error.

12.2.4 Selecting the Proper Gain to Meet All Three of the Loop Operational Objectives

There is no single value for a controller's gain that satisfies all three operational objectives simultaneously. The selection of a gain that meets the first and third objective will not optimize the requirements of the second objective. Conversely, when the controller's gain is initially set too low, the first and third objectives are not be met in order to meet the second objective. The art of effectively calibrating modulating control loops lies in the ability of the technician to select a value for the controller's gain(s) that consider the advantages and disadvantages of all three operational objectives with respect to the requirements of a particular process. To assist the technician in selecting this value, the loop's response to a change in its load or set point can be compared with five standard modulating responses. Once the loop is calibrated so its response matches the standard response for that process, the control loop is considered to be properly tuned. The following section details the five possible responses of closed loop controls.

12.3

THE FIVE POSSIBLE LOOP RESPONSES

A modulating control loop's response to a change in is load or set point always falls into one of five standard response categories. As the controller's gain is varied in an attempt to meet the operational objectives of a control loop, the resulting response will move out of one category and into another. The final value of the gain must place the loop's response in the category that meets all three operational objectives and the process requirements. The five response categories are:

1. Overdamped
2. Critically Damped
3. Underdamped
4. Unstable with Constant Amplitude
5. Unstable with Increasing Amplitude

12.3.1 Damping

Damping describes any effect that reduces or impedes a reaction.

Damping can be considered a source of resistance that is applied to a system to reduce any change in the magnitude or direction of a response. Damping lessens or smooths out the severity of the oscillations in the transition period. Conversely, reducing the damping force applied to a system allows the rate of change of the output signal to increase, increasing the severity of the oscillations. As shown above, the term damping is used to classify the overall characteristics of a controller's response.

Applying a damping force to a controller slows the rate of change of its output signal. Controllers incorporate circuitry that make it possible for the technician to regulate the amount of damping applied to the output signals, thereby manipulating a controller's response. Controller's tuned with a large amount of damping generate responses that do not oscillate. Reducing the amount of a controller's damping increases the number of oscillations in its response. The proportional gain adjustment of the controller varies the damping factor of its response. As a technician increases the controller's proportional gain, its damping factor *decreases.* Conversely, decreasing the gain increases the damping factor.

12.3.2 Overdamped Response

An *overdamped* response is characterized by the slow rate of change in a controller's output signal. The slowness results from a small proportional gain value which creates a large amount of damping into the controller's output response. Consequently, the controller's output changes slowly in response to load changes. An overdamped loop responds too slowly to changes in the load and consequently, cannot meet operational objectives one and three. The slow response allows the control point to stabilize beyond acceptable limits of the process and produces larger steady state errors. Figure 12.6 depicts the controller's overdamped response to a change in load.

Overdamped control loops *do not* oscillate. If the slightest amount of oscillation occurs in the output signal, the loop is not overdamped and its response falls into one of the four remaining response categories.

12.3.3 Critically Damped Response

A *critically* damped controller's response has a larger gain and therefore, less damping than an overdamped response. Consequently, it will take less time for the controller to complete its response and return to a steady state condition. A critically damped controller has just enough damping to permit a fast response without allowing any oscillations to occur in its output signal. If the critically damped controller's gain is increased slightly, a small oscillation will occur in the output signal. Once an oscillation occurs, the response is removed from the critically damped category. Figure 12.7 shows a critically damped response to a change in load superimposed over the overdamped response that was shown in Figure 12.6.

The gain of the controller that generated the overdamped response of Figure 12.6 was increased to change its response to critically damped. The resulting critically damped response allows the process to reach steady state operation much quicker than the overdamped response, without allowing the output signal to

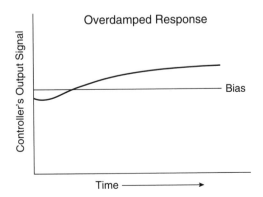

Figure 12.6   An Overdamped Response

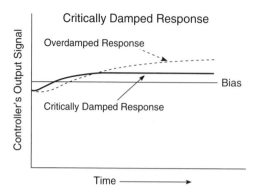

Figure 12.7   Critically Damped Response

oscillate. Note, the steady state error is smaller for a critically damped response than for an overdamped response, reflecting the increase in the controller's gain. Critically damped responses are used for processes that cannot tolerate oscillations in their control point. A critically damped response meets all three operational objectives because the loop responds as quickly as possible within the design requirements of the process.

12.3.4 Underdamped Response

When the gain of a critically damped controller is increased slightly, reducing its damping, the resulting response changes to *underdamped*. An underdamped response exhibits a reduced degree of resistance to changes in its output signal which allows the controller's output signal to initially overshoot its balance point thereby requiring a corrective action. Consequently, underdamped responses always generate some degree of oscillations. As the controller's gain is increased, the number of oscillations in the transition period also increase before the output signal stabilizes after a load change.

The advantages of an underdamped response are it ability to reduce the size of the initial and steady state errors. These reductions in the change of the process error are the result of the faster initial response of the controller. A disadvantage of an underdamped response is a longer transition period. Figure 12.8 depicts an underdamped loop response.

The operational objectives of a control loop require the controller's proportional, integral and derivative gains to be set high enough to minimize the size of the initial change in the control point, length of the transition period and the size of the steady state error. To meet these contradictory requirements simultaneously, the overall gain of the controller is increased beyond its critical damped response, permitting the generation of a few oscillations. Sufficient damping remains in the response to permit the loop to stabilize within an acceptable period time.

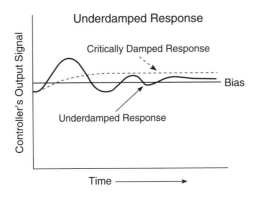

Figure 12.8   Underdamped Response

12.3.5 Unstable with Constant Amplitude Response

The remaining two response categories classify loops that are unstable and there-fore, do not meet any of the three operational objectives of a modulating control loop. Unstable loops have so little damping that they constantly oscillate in reaction to a change in the process load or set point. This response cannot meet the operational objectives of control loop because the transition period is excessively long and the control point never stabilizes.

As the gain of an underdamped loop is increased, the number of oscillations that appear in its transition period also increase. If the gain is increased suffi-ciently, the response reaches a point where it will never achieve a steady state operating mode. This is the point where the response crosses over from under-damped to unstable. Figure 12.9 depicts an unstable response where the oscilla-tions are of the same relative size.

12.3.6 Unstable with Increasing Amplitude

When the gain of an unstable loop is increased even further, the controller's response oscillates more aggressively which each successive oscillation having a larger amplitude. Eventually, the amplitude exceeds the output range of the con-troller and the controller responds like a short cycling two-position loop. Like the previous unstable response, this response cannot meet the operational objectives of a control loop and will cause excessive wear and tear to the control loop com-ponents. Figure 12.10 shows this response.

12.3.7 Responses That Meet the Operational Objectives of a Loop

Only those controller responses that fall within the critically damped and under-damped categories successfully meet the operational objectives of a modulating control loop. Overdamped loops respond too slowly, producing undesirable

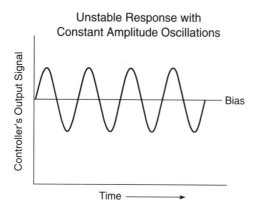

Figure 12.9   Unstable - Constant Amplitude

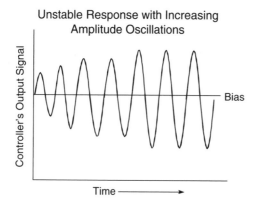

Figure 12.10   Unstable - Increasing Amplitude

steady state errors and long transition periods. Unstable loops cannot meet the requirements of an acceptable response because they react too quickly and cannot maintain the control point at a steady state condition.

In any application, there is only *one* value for the controller's overall gain (P, PI, PID) that produces a critically damped response. If the value of this gain is slightly lower than the critically damped gain, the response is classified as overdamped. If the controller's gain is slightly higher than the critically damped value, the response is classified as underdamped. Underdamped responses can have any value for its gain parameters that keep the controller's response somewhere between critically damped and unstable and still meet the response characteristics of the category. Underdamped responses range from those with slight oscillations (nearly critically damped) to those with many oscillations (approaching unstable). Although a response that is approaching unstable has a long transition period and no longer fulfills the operational objectives of a loop, it still belongs in the under-damped category if it finally reaches a steady state condition.

12.4

CRITERIA FOR EVALUATING CONTROL LOOP RESPONSES

A modulating controller's response must fall into the critically damped or under-damped categories if it is going to meet the operational objectives of a control loop. Although it is easy to determine if the response is critically damped, there are a whole range of gain values that can be used that meet the underdamped response characteristics. To determine if an underdamped response effectively meets the three operational objectives of a modulating control loop, criterion have been developed that assist the technician in determining whether a controller is properly tuned. These criterion are used to evaluate whether a response efficiently balances the opposing requirements of the three objectives. The three criteria used for evaluating an underdamped modulating loop response are:

1. Critically damped
2. Quarter amplitude decay
3. Minimize the integral of the absolute error

When tuning a modulating control loop, the technician adjusts the value of the controller's gain parameters until its response meets any *one* of these three evaluation criteria. When the loop's response matches one of these criteria, the operational objectives of a modulating control loop have been met and the loop will respond in an efficient manner.

In order to calibrate a modulating control loop, the technician must be able to evaluate its dynamic response. One of two conditions must occur to provoke a control loop to respond, either the process must undergo a measurable change in its load or a measurable change must be made in its set point. Since it is difficult to make a finite change in a process load to analyze a controller's response, this option is not used by the calibrating technician. Therefore, to initiate a response of a control loop, the technician makes a *step change* in the set point of the process. A step change is an instantaneous change of 5° or 10° in the set point that causes the controller to respond to what it senses as a finite change occurring in the process load. Since the response must be evaluated several times, the usual procedure is to increase the set point the first time, decrease it the second time and keep alternating the direction of change until the loop is calibrated. The process must be allowed to stabilize before the next change in set point is made.

To evaluate a controller's calibration, the technician inputs a step change in the loop's set point. The output signal of the controller is monitored to determine its current response category. After the loop has stabilized, the proportional gain is adjusted in a direction that moves the response toward the critically damped or slightly underdamped category. Another step change in the set point is initiated and the response evaluated. This change in the set point is made in the direction opposite of the initial change. This procedure is repeated until the response matches one of the three response criteria.

12.4.1 Critically Damped Criterion

As stated previously, a critically damped controller responds to changes in the process as fast as possible *without* generating a single oscillation. To meet this criterion, the technician must be sure the response is not overdamped or underdamped. To determine if the response is critically damped and not overdamped, the proportional gain is increased slightly and a step change is initiated. If the controller responds with a small oscillation in its output signal, the loop was most likely critically damped before the gain was increased. To return the response back to critically damped, the gain is reduced back to its previous value. If the increase in proportional gain did not generate a slight oscillation, the loop is still overdamped and the gain must be increased in steps until the loop becomes slightly underdamped. Once an oscillation occurs, the proportional gain is reduced slightly to achieve a critically damped response. Figure 12.11 depicts a critically damped response.

A critically damped response is slower than an underdamped response and will consume more energy due to its larger steady state error. It is used for applications that cannot tolerate oscillations in the control point during a load change. This criterion is not usually used in HVAC applications because most applications can tolerate a few oscillations without any problems.

12.4.2 Quarter Amplitude Decay Criterion

Quarter amplitude decay is the criterion of choice for HVAC processes. Quarter amplitude decay is used in the evaluation of underdamped responses. This criterion optimizes the number of oscillations allowed in a properly tuned underdamped loop while minimizing the length of the transition period. Quarter amplitude decay is widely used by HVAC calibrating technicians because of its ease of use in the field.

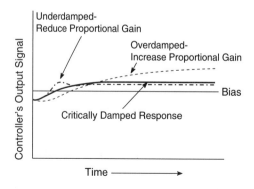

Figure 12.11   Critically Damped Criterion Response

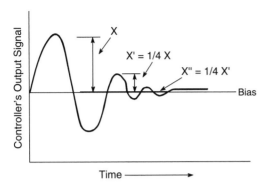

Figure 12.12 Quarter Amplitude Decay Criterion

Quarter amplitude decay limits the amplitude of each successive oscillation to a magnitude that is equal to approximately *one fourth* of the size of the previous oscillation. This reduction in the size of approximately 75% of the previous magnitude causes the controller's output signal to stabilize within three or four oscillations thereby optimizing the length of the transition period. A quarter amplitude response is shown in Figure 12.12.

This evaluation is most often used because it is easy to observe the decrease in the size of the successive oscillations in the controller's output using a pressure gauge or electric meter. Generally, if the signal stabilizes within three or four oscillations the response is considered to meet the quarter amplitude decay criterion. If the loop oscillates more than four times, a slight reduction in the proportional gain will increase the damping and slow the response, reducing the number of oscillations. If fewer than three oscillations occur, the gain can still be increased slightly to improve the efficiency of the process without producing an unstable response.

12.4.3 Minimize the Integral of the Absolute Error Criterion

Minimize the integral of the absolute error criterion is also used for evaluating underdamped controller responses. It is a mathematical based criterion that solves for the algebraic sum of the area under the response curve using integration. Each separate area within the limits defined by the response curve and the controller's bias line is calculated and added to the other areas. By reducing the total area of the response, the loop is tuned to respond as quickly as possible, minimizing the transition period and maximizing the efficiency of the process.

This criterion is mathematically more accurate than quarter amplitude decay in the evaluation of underdamped responses. It is also unrealistic for the calibrating technician to perform this procedure in the field. This criterion lends itself nicely to control loops that use microprocessors for controllers because computers are

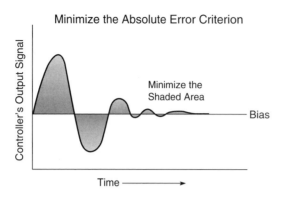

Figure 12.13   Minimize Integrated Error Criterion

very adept at performing mathematical calculations. Many digital control systems employ self tuning algorithms that employ this criterion to tune the loop automatically. In self tuning applications, the controller will initiate a step change in the set point and calculate the area under the response curve, then it will alter the proportional gain and repeat the procedure until the area under the response curve is minimized. Figure 12.13 depicts a minimized error response.

12.5

SUMMARY

Closed loop control systems are designed to economically maintain a process control point within its throttling range. The presence of the feedback path permits the controller to become self-correcting, altering its output signal until the energy being transferred into the process balances with the load. This characteristic of closed loops generally produces oscillations in the controller's output signal whenever a load change occurs. To regulate the number and size of these oscillations, the controller's damping factor is varied by adjusting its gain parameters. As the gain is increased, the damping applied to the response decreases, increasing the speed of the change in the output signal and the number of oscillations that occur. Conversely, as the gain is decreased, signal damping is increased, reducing the speed and number of oscillations of the controller's output signal.

Modulating loops are designed and calibrated to fulfill the following operational objectives:

1. Minimize the size of the initial change in the control point.
2. Minimize the length of time of the transition period.
3. Minimize the size of the remaining steady state error.

Meeting these objectives insures the loop is tuned efficiently and will not become unstable in response to normal changes in the load.

The damping factor determines which of the five categories a modulating response belongs to. Each category is based upon the size and frequency of the oscillations that occur during the transition period of the response. The five response categories are:

1. Overdamped
2. Critically Damped
3. Underdamped
4. Unstable with Constant Amplitude
5. Unstable with Increasing Amplitude

An overdamped loop has the lowest controller gain, slowest response and highest damping of the five responses. Increasing the controller's gain reduces its damping, shifting its response toward critically damped. Further increases in the controller's gain moves the response through the underdamped, unstable with constant amplitude and unstable with increasing amplitude responses.

When the controller's gain is too small with respect to the process characteristics, its response will be overdamped. This response is too slow to properly meet the first and third operational objectives of modulating control loops. It also wastes energy and fails to adequately maintain the control point of the process within a reasonable throttling range. As the gain is increased to speed up the loop's response without producing any oscillations, the loop responds in a critically damped manner. Its controller's output signal changes as fast as possible without creating any oscillations. This response is most efficient at meeting the second operational objective at the expense of producing less than optimal response to satisfy the first and third objectives. This response is used for processes that cannot endure oscillations during periods of load changes.

Increasing the controller's gain of a critically damped loop shifts the response to underdamped, which is characterized by oscillations in its output signal. The number of oscillations in the transition period must be optimized to produce an efficient loop response that meets all three operational objectives.

The three criteria used to evaluate a response to be sure its operating efficiently are:

1. Critically damped
2. Quarter amplitude decay
3. Minimize the integral of the absolute error

These criteria are only used to evaluate underdamped and critically damped responses. The other three categories, overdamped, unstable with constant amplitude and unstable with increasing amplitude fail to meet the operational objectives of a modulating control loop. A properly tuned modulating control loop will match one of these criteria. Quarter amplitude decay is most often used because it is easily performed in the field. Minimizing the absolute error is often used by digital control systems self-tuning algorithms.

12.6

EXERCISES

Determine if the following statements are true or false. If any portion of the statement is false, the entire statement is false. Explain your answers.

1. Minimizing the integral of the absolute error is a criterion for good control.
2. The first objective of a control loop is to minimize the amount of change in the control point that occurs in response to a load change in the process.
3. Critically damped control responses meet the first and second objectives of a properly tuned control loop.
4. The second objective of a control loop is met when the loop is tuned with a quarter amplitude decay response.
5. Critically damped loops using a proportional only controller will have a larger steady state error than a proportional controller tuned for quarter amplitude decay.
6. Increasing the gain decreases the transition period of an overdamped response.
7. An unstable response minimizes the initial change in the control point.
8. Decreasing the gain of a critically damped loop changes its response to underdamped.
9. As the proportional gain is decreased, the damping factor of the controller decreases.
10. Criteria for proper tuning only applies to overdamped and critically damped responses.

Respond to the following statements and questions completely and accurately, using the material found in this chapter.

1. Why is an open loop controller easier to tune than a closed loop controller?
2. List the steps required to calibrate an open loop controller.
3. Describe the characteristics of a closed loop configuration.
4. What is dynamic analysis and why is it used?
5. Define the transition period of a response.
6. What effects do process lags have on the length of the transition period?
7. What effects do process lags have on the phase shift between the process conditions and its controller's response?
8. How do changes in the proportional gain of a controller affect the phase shift of its process?
9. Describe in detail, the characteristics of each of the operational objectives of a modulating control loop.
10. How does the magnitude of the proportional gain affect each of the operational objectives of a control loop?

11. Describe the damping and gain characteristics of each of the five response categories.
12. How do changes in the controller's proportional gain determine which category its response belongs to?
13. Define controller damping.
14. What effect does changes in the controller's damping factor have on the category the response belongs to?
15. Which of the five possible responses meet the three operational objectives of a control loop?
16. Describe the characteristics of each of the criterion used to evaluate a modulating loop.
17. Which evaluation criterion is most often used by calibrating technicians? Explain your answer.
18. List the steps used to calibrate a modulating control loop.
19. What causes too many oscillations to be produced by an underdamped controller? Describe a solution that will correct this problem.
20. How can a calibrating technician determine if a loop is overdamped?

CHAPTER 13

Microprocessor-Based Control Systems

The information presented in previous chapters is equally applicable to pneumatic, electric, analog electronic or digital electronic control systems. They all respond the same regardless of the signal types used by each system. In recent years, digital electronic systems have grown to be the most common platform specified in new and retrofit HVAC applications. Chapter 13 reviews the design and operational characteristics of these control systems.

OBJECTIVES

Upon completion of this chapter, the reader will be able to:

1. Describe the purpose of the major electronic components found in microprocessor based control systems.
2. List the operational characteristics of the common point types used in HVAC applications.
3. List the installation guidelines for DDC equipment and signal lines.
4. Describe function of a point database and the application program.
5. Describe the characteristics of the Energy Management Strategies.

238

13.1

WHAT ARE MICROPROCESSOR-BASED CONTROL SYSTEMS?

Microprocessor-based control systems combine computer hardware technology with user generated software programs to perform binary, modulating, energy management and logic based control strategies for the operation of building systems.

Digital electronic control systems are designed around the capabilities of an electronic integrated circuit chip called a *microprocessor*. This chip performs mathematical, logic and control functions using programs stored in the system's memory. It can create responses that emulate loop controllers, relays, signal converters, time clocks, signal selectors, signal limiters and any other component used in pneumatic, electromechanical and analog electronic control systems. Microprocessor-based control systems are known by many acronyms, the most popular being EMS, DDC, EMCS and BMS or BAS. The following sections outline the significant differences between the operational characteristics of these systems.

EMS—Energy Management System

An energy management system can only perform binary control strategies. A microprocessor chip located in the panel controls the operation of small, onboard control relays. Sophisticated strategies that were previously impossible to perform with a simple time clock are used to reduce electrical consumption and demand. The control relays cycle electrical machinery on and off based upon the time of day, equipment operating schedules or the current electrical demand. These systems could not perform closed loop modulating control. They were popular in the early 1980s and are no longer being installed because current technology allows systems to perform all the functions of EMS systems, along with modulating control and more advanced energy efficient operating and maintenance strategies.

DDC—Direct Digital Control or Distributed Digital Control

DDC is an acronym for the phrase *distributed digital control* or *direct digital control*. The terms distributed and digital emphasize the differences between these control systems and analog electronic or pneumatic control systems. *Digital* characterizes control systems that use *binary* signals to represent all the information (temperatures, humidities, errors, output signals, etc.) that is stored or manipulated by the microprocessor in its panel. This characteristic differs from analog electronic and pneumatic systems that use continuous or modulating signals to represent data.

The term *distributed* underscores the ability of these systems to increase system reliability, speed, computing power and responsibility by dividing the control functions of a facility among a number of smaller microprocessor based panels. Each of these panels can operate alone or as part of a larger integrated facility management system. This differs from earlier digital control system architecture that used one central computer to perform all system functions.

DDC systems perform all the functions of an EMS system along with the ability to perform modulating loop control, advanced energy management strategies, self-tuning, custom programming, advanced communications and other functions.

EMCS—Energy Management and Control System

Energy Management and Control Systems have the same operational features as DDC systems. The name indicates that they can perform the energy management function of early microprocessor based systems along with the ability to perform modulating control.

BMS and BAS—Building Management System or Building Automation System

BMS and BAS systems combine the functions of DDC systems with fire, security, lighting and other types of facility management systems into one integrated control system.

13.1.1 Operational Characteristics of the Components Found in a DDC Panel

DDC systems can have from one to hundreds of control panels located throughout the buildings of a facility. The sensors and actuators of each control loop are connected by wires or tubes to termination strips found in these panels. The microprocessor on the panel's circuit board performs all of the control strategies required to operate the connected equipment. Most installations also have a personal computer located in a maintenance office that provides a means of accessing the entire system from a single location. This work station computer stores the programs that are operating in each panel, characteristics of all devices connected to the field control panels, system graphics, programming tools and other productivity enhancing software. It also provides a modem for external communications along with a mouse and a keyboard to access system information. Figure 13.1 shows a photograph of various DDC panels.

DDC hardware and other computer based equipment are known as *digital* systems. Digital electronics is a field of technology that applies low voltage DC circuits to generate and respond to binary (two-state) signals. Digital systems can only recognize two signals that must be used to represent all data. These two signals are called **logic 1** and **logic 0.** Groups of one's and zero's represent numbers, words, signals and all other data in a digital system. Each 1 or 0 is called a *bit*, which is an acronym for *BI*-nary digi-*T*. Bits are electrically represented with low voltage DC signals. A logic 1 is typically represented by a 5 volt signal and a logic 0 is characterized by a 1.5 volt signal.

A logic 0 bit is not represented by a zero volt signal because that signal can be produced by a system fault such as an open or short circuit. Consequently, if a digital electronic device received a 0 volt signal it would not be able to differentiate

Figure 13.1 DDC Control Panels

between real data and a fault in the input circuit. Raising the 0 bit voltage level to 1.5 volts permits the circuits to differentiate a true logic 0 bit from corrupted data. Figure 13.2 shows the range of voltages that are representative of logic 1's and 0's. Note, any signals that fall within the 2 to 3 volt range are discarded because they could have been produced by a slight voltage variation in the logic 0 or logic 1 circuit. The following sections briefly outline the major components in a digital microprocessor-based DDC system.

13.1.2 The Microprocessor Chip

The microprocessor is a powerful integrated circuit chip that performs the data manipulation, system diagnostics, math, logic and other control functions within digital computer systems. Each microprocessor chip is constructed of hundreds of thousands of components *etched* onto a piece of silicon substrate that is about 1 square inch in size. Transistors, diodes, resistors, and capacitors are configured into various circuits on the chip. Each circuit responds to the binary instructions and data being transferred to and from the chip's terminals. Current DDC technology incorporate microprocessor chips comparable to the Intel® 386 and 486 chips found in personal computers. There are also a number of smaller and slower microprocessors used in single function DDC panels to handle specific data manipulation procedures. This frees the main processor from performing simple routine chores.

Data is transferred within a microprocessor in strings of 1's and 0's called *words*. Word size is a function of the microprocessor. They range from 4 bit words used in older systems to 64 bit words used in some current system components. Each word contains the data requested by the microprocessor to complete a required instruction.

Whenever data is called from a memory storage location, it is stored in registers. A register is a block of memory located on the microprocessor that temporarily holds one or more binary words. A 16 bit register can hold one 16 bit word or two 8 bit words while a 32 bit register holds one 32 bit, two 16 bit or four 8 bit words.

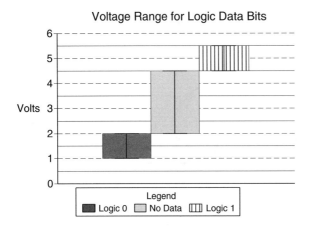

Figure 13.2 Voltage Levels for Logic Data

The length of the word has a direct bearing on the operating speed of the system. The main microprocessor chips used in DDC panels move data in 16 bit words. A microprocessor that moves information in 16 bit words is twice as fast as one that manipulates data in 8 bit words. The personal computers used for the DDC system's work station move data in 32 and 64 bit words.

The circuits on a microprocessor chip are configured into various sections and performs a related set of tasks. One of these sections is called the *arithmetic logic unit* (ALU). The ALU executes mathematical calculations and also performs *relational* and *logical* comparisons between data stored in its registers. There are two registers used by the ALU to hold data. One is called the *accumulator* and the other is called the *data* register. The accumulator holds the mathematical *operand* before the ALU performs an operation with it using the word stored in the data register. At the completion of a math or logic operation, the *result* replaces the word in the accumulator. The accumulator is now ready for the next calculation. The data register also holds instruction codes needed by the microprocessor, data that is called from a memory location and data that is to be transferred back to a memory location.

Other registers are located in the microprocessor to facilitate the movement of data. An address register holds the memory location of the word being processed by the ALU. Another register is a program counter that controls the sequence of the execution of the program instructions. Additional information on the operation and construction of microprocessors and integrated circuits can be found in electronic textbooks.

13.1.3 The Data Bus

A *bus* is a circuit term that describes a group of parallel conductors used to transfer data between memory, registers and other integrated chips on a circuit board. A bus is designed to transfer either 4, 8, 16, 32 or 64 bit words between locations within the computer. The larger the bus, the faster the data movement and the computer's performance. The two primary buses in the computer are the *data* bus and the *address* bus. The data bus is used to transfer data requested by a chip from its memory location and to return data back to memory. The address bus informs the various chips on a circuit board when it is their turn to transmit or receive information off the data bus. Access to these buses is coordinated by the microprocessor and the computer's system clock to prevent more than one chip from accessing the data on the bus at the wrong time, corrupting the data on the bus.

13.1.4 The System Clock

The system clock is an integrated circuit chip that generates a square wave pulse train that is used to synchronize all the other chips in a DDC panel or computer. The clock and its associated control lines sequence the individual systems so only one chip has access to the data bus at one time. If the clock were not present, the individual systems would operate independently. It would be impossible to perform any functions with a computer that lacked a clock. The clock's speed sets the speed of the microprocessor. Faster speeds allow the processor to perform more functions within a given period of time. High speed processors currently available for personal computers operate with clock speeds over 150 megahertz or 150 million cycles per second. Current DDC technology incorporate main microprocessor chips operating at 8 to16 megahertz. Figure 13.3 shows the shape of the 10 megahertz square wave.

13.1.5 System Memory

DDC systems store operating instructions and data in memory locations etched on integrated circuit chips. Data is stored as the presence or absence of a voltage at the appropriate bit address. A 5 volt signal stored at a bit location on the chip represents a logic 1 stored at that location. A 1.5 volt signal would represent a logic 0 stored at that memory location. There are two classes of memory used in computer systems, *Random Access Memory* (RAM) and *Read Only Memory* (ROM).

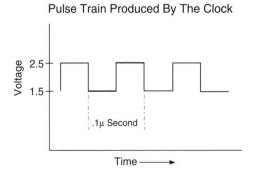

Figure 13.3 Square Wave Pulse Signal Created by Computer Clock

ROM is *permanent memory*. Information can only be read from the ROM chip memory, that is why it is called *read only* memory. Once the instructions and data are loaded or "burned into" the chip, they cannot be altered by the loss of power. Any changes that need to be made in the stored data require the removal and replacement of the chip. Since ROM is permanent memory, the instructions stored in ROM are called *firmware*. Information stored in ROM is retained in the chip without the need of an external power source. If data must be stored or altered in memory a different type of memory must be used.

ROM contains the basic instructions required by the microprocessor to start up the system, perform self diagnostics on its internal systems and to interpret operational instructions. The programming code needed to develop an application's operating program and its point database is also stored on a ROM chip so it is available when the panel is installed. In earlier digital control systems, the circuit boards that contained the ROM chips had to be replaced if the instruction set for the DDC panels was to be upgraded to a newer revision. To reduce the expense associated with these periodic upgrades, boards are now designed with the ROM chips installed in chip sockets instead of being soldered onto the boards. When an upgrade is performed, the old chips are pulled from their sockets, and new chips inserted in their place. A typical DDC panel has about 256 to 512 Kb (kilobytes) of ROM on its main controller board.

Different types of ROM have been developed that function as a semi-permanent storage media that retains data without the need of applied power but can also be erased and rewritten using the proper equipment or circuitry. To retain the advantages of non-volatile memory while increasing the flexibility of the chip to accept changes in its firmware, *EPROM* chips were developed. EPROM is short for erasable, programmable ROM. EPROM memory chips are erased by an application of ultraviolet light to a window located on the top surface of the chip. When the UV light has been applied for a 15 minute period, the energy in the light resets all the memory bits to logic 0s. A special machine is used to reprogram the chip with its revised instruction set. EPROM chips can be reprogrammed about 40 times before they must be replaced.

Another non-volatile ROM chip that can be reprogrammed is called *EEPROM* or *double* E-PROM. EEPROM is the acronym for *electrically* erasable, programmable ROM. EEPROM is erased by the application of an electrical signal to a particular circuit on the chip instead of an ultraviolet light. EEPROM chips are reprogrammed right on their board without the need of special equipment. These chips can be reprogrammed up to 10,000 times, making them the more versatile ROM chip used in DDC applications.

RAM is an acronym for *random access memory*. Although this phrase is somewhat dated, it's still used to describe the volatile memory used in computers. When power is removed from a RAM chip, all the data stored within the chip is erased. A more appropriate description of RAM memory is *read/write* memory. This phrase accurately depicts the difference between ROM and RAM. Data can be read from or written to RAM but only read from ROM. Memory locations in RAM are constantly being rewritten in response to directions from the microprocessor. This is the reason that RAM memory is designed to be volatile. Data and instructions stored in RAM are called *software* because they can be easily erased or changed.

DDC application programs and databases are stored in RAM. Batteries or capacitors are wired to the memory chips to prevent the loss of the data during short duration power interruptions. When power to the panel is interrupted, the chip draws power from the capacitor or battery to maintain the information stored in RAM. If the capacitor or battery loses its charge before regular power is restored to the panel, the data in RAM is erased and will have to be reloaded once power is restored. A basic DDC panel is equipped with 512 kilobytes of RAM. The newer DDC panels have slots available to increase the onboard RAM to 8 Mb (megabytes) to handle larger programs and more point information.

13.1.6 Input and Output Interfaces

The input and output interfaces allow the field devices of the control system to be connected to the microprocessor and other DDC panel circuitry. The interface consists of terminal strips that provide a means of connecting sensor and actuator wires to the circuit board. In pneumatic signal applications, termination ports are available for supply air and output signals to the actuators. Each location on the termination interface has its own unique address to allow the microprocessor to differentiate it from the other points connected to the system. The addresses are stamped on the circuit board or termination strip so the technician knows where to properly connect the signal lines of a device. The locations are entered into a program called the database which acts as an address book for the microprocessor.

The mechanical and electrical systems controlled by DDC equipment operate in an analog environment. The binary DDC computer must correctly interface with the analog world in order to perform the desired control of the systems. To accomplish this interface, a DDC panel contains circuits that convert *analog input* signals into binary words so they can be processed by the microprocessor and stored into memory. An integrated circuit chip has been designed whose only function is to perform the conversion between analog signals and binary words. This chip is called an *analog to digital converter* (ADC). The signal arriving at the input terminals of the DDC panel is converted into an 8 or 12 bit word by the ADC.

The number of bits used in the converter's design determines the resolution of the input signal. An 8 bit converter divides the sensor's range into 256 parts (2^8). Therefore, an 8 bit converter can measure changes of 0.4° (100/256) for a sensor with a 100° span or 0.8° with a sensor having a 200° span. A twelve bit converter divides the sensor's range into 4096 (2^{12}) parts. If the same sensors listed above were connected to a 12 bit converter, the panel is able to record changes of 0.02° and 0.05° respectively. Therefore, the larger the word size of the converter, the better the resolution of the connected analog device. DDC panels are currently available with 8 or 12 bit ADC converters.

DDC panels also contain devices that convert the computer's binary words into analog output signals that operate actuators. A *digital to analog converter* (DAC) is used on the *analog* output terminations of a DDC panel. These chips convert a binary word into an analog voltage or current signal that represents the output signal of the digital controller. Typical output ranges of electric signals are 0 to 5 V dc, 0 to 10 V dc, 0 to 20 mA and 4 to 20 mA. Most DDC panels have an 8 bit DAC that generates changes in the output signal of 0.4% of the output range. In applications incorporating pneumatic actuators, a device called an *electric to pneumatic converter* is used to convert a binary word into a 3 to 15 psi signal.

Most DDC field panels do not have an individual ADC or DAC chip for each termination address located on their I/O interface terminal strips. In most designs there is one ADC and one DAC that are shared between all of the analog devices. Sharing the converter chip is accomplished through an electrical wiring/switching scheme called *multiplexing*. Multiplexing is a procedure where each terminal address is momentarily connected to the converter chip's input terminals. Once connected, the chip performs the required conversion between the analog signal and the binary word. In an ADC application, the resulting binary word is sent to its assigned address in RAM memory. After the converter chip updates the value of the input or output point, the next termination address is connected to the converter and the process repeats itself. The multiplexed input and output addresses are constantly scrolled, one after another in a continuous loop. Multiplexer circuits are constantly polling the points and updating the information in the computer's memory. A typical DDC system can update the values of all its input and output points within a few seconds.

When the application program requires information on the value of a measured variable, it looks up its latest value stored in its memory location. It does not read the signal on its termination address. The output signal multiplexer updates in a slightly different manner than the input procedure. Once the microprocessor and application program command an output point to a new position, the new binary word is placed in the memory location assigned to that point's address. It is not sent directly to the output device. The DAC multiplexer circuit scrolls through the memory addresses of the output points and updates their output signals one at a time. The output multiplexer connects the DAC converter chip output port to an analog point termination address, draws the associated 8 or 12 bit word from the proper memory location and converts the word into the appropriate analog signal for that address. The output circuit of each termination address samples the data from the converter and holds its output signal at that value until it is multiplexed again. Multiplexers make it possible for many points to share the same converter, reducing the cost of the DDC system while still maintaining control, comfort, and safety.

13.1.7 Communications Techniques

Computers communicate with other peripheral devices using strings of 1's and 0's sent through a pair of wires. A peripheral device is any piece of equipment that sends or receives data through connecting wires. Keyboards, display monitors, printers, DDC panels, and exterior modems are some of the peripheral devices used in building automation systems.

Data can be transmitted using two possible modes of communication, parallel or serial transmission. The *parallel* mode transmits binary words over an exterior bus that connects the two devices. The number of conductors or data lines in the bus corresponds to the length of the binary word being transmitted. Parallel transmission cables that connect computer peripherals typically use 8 data conductors along with a couple of control wires. Additional conductors are present in the cable to control the transmission of data between the components. Parallel transmission mode can only be used over short distances (usually less than 10 feet) unless a signal amplifier is installed to boost the strength of the signals. Work station printers typically incorporate parallel mode transmission methods to speed up the printing process.

Serial transmission transfers a string binary information through a single pair of wires. The binary string typically consists of an eight bit word with a stop bit that separates each word and in some protocols, another bit that is used to determine if the word was corrupted during its transmission. Each string of data is deciphered into its 8 bit word and the control bits before the data is stored in RAM. DDC systems use the serial transmission methods to exchange information between other DDC panels, expansion boards, single unit panels, and the work station computer.

The rate of transmission between components is measured in *bits per second* (BPS) and *baud* rate. BPS is a measure of the total number of bits (data plus control) transferred through the wires within a one second period. Baud rate is the rate of transmission measured by the number of data words or unique codes being transmitted each second. Baud rate defines how many packets of information are being sent whereas BPS indicates the number bits (packets x bits per packet) being transferred between devices. Although both methods of describing serial data transmission rates are used, DDC communications are typically defined in terms of baud rate. The common baud rates used by DDC systems are 4800, 9600, and 19,200 baud between peripherals over a twisted pair of communication wires.

The electrical communication standard used by the computer industry that defines the connections and signal levels for serial communication is the *RS-232* standard, written by the Electronics Industries Association. This standard specifies the necessary electrical, mechanical and functional characteristics required for serial data exchange between two pieces of equipment. RS-232 converts the low voltage binary string generated by the computer into higher level signals in preparation for transmission. Increasing the signal level reduces the effects of other electrical noise and interference that may be injected into the information while it's being transmitted. RS-232 also defines the pin configuration of the DB-9 and DB-25 pin connectors found on computer interface boards.

A *modem* is an electronic device used to transfer data over telephone lines. The modem provides an interface between the analog signal used on telephone lines and the binary signals used by computers. Modem is an acronym for *MOdulator/DEModulator*, which accurately describes the device's function. Two modems are required to transfer data across phone lines. One modem is installed in the sending computer and the other is installed in the receiving computer. The transmitting modem converts a binary string of data from the computer into a *modulating* analog signal, similar to an FM radio signal. A typical modem converts a logic 0 bit into a 1070 Hz signal and a logic 1 into a 1270 Hz signal for transmission. Others use 2025 (logic 0) and 2225 Hz (logic 1) frequency modulated signals. These modulated signals are nearly unaffected by the normal electrical noise found in telephone switch systems. The modulated signal is transmitted over telephone lines to the receiving modem. The receiving modem *demodulates* this analog signal into binary strings using the RS-232 protocol. The demodulated bit string is in a format understood by the receiving device so it can be displayed on its monitor or stored in its memory.

Modems can transmit using half or full duplex modes. Half duplex modems can transmit and receive data, but not at the same time. Full duplex modems transmit and receive at the same time by using different frequencies for the transmitted and receiving data strings. These two frequencies do not interfere with each other on the telephone line so they can travel in different directions at the same time, over the same two lines. Most modems used in DDC applications are full duplex devices.

New generations of DDC panels are equipped with a modem communication system that can automatically answer a phone call from another computer or dial a remote computer in response to a system event. The automatic dial function is used to alert the proper people or equipment whenever a predefined condition occurs within the system. In larger DDC applications, the work station computer contains a modem and the software to operate it. When an alarm condition is detected within any of the processes throughout the entire control system, the work station computer automatically dials the phone number(s) programmed into the list of connected computers or personal pagers. The program can initiate the call to another computer directly or send an alpha-numeric message to a pager. Typically, the building operator or service company has a computer operating or a pager to receive these alarm messages. The receiver of the message can call up the system with another computer and read the information associated with the alarm. In many cases, the problem can be temporarily corrected over the phone. This technology facilitates system maintenance through remote access of systems from home or the office.

13.2

DDC HARDWARE POINTS

The previous section presented information on the important components found inside a DDC panel or work station computer. DDC *hardware* points are those devices located external to the panel which couple the HVAC process to the DDC computer. All of the *physical* devices, sensors, actuators, switches, etc. connected to a DDC panel's input and output termination strips are called *hardware points*. The term *physical* denotes a control device that has been installed in the field and connected to a DDC panel's input or output termination address. There are other points in a DDC system that are created by the DDC system's firmware. These points are called *software* points. They only exist in the memory circuits of the DDC panel and are entered into the system by the application engineer, control technician, or building operator. Software points are typically used to represent set points, gain parameters, results from calculations, the current state of a process, or any other information that is useful to the operator. The following sections highlight the characteristics of physical points commonly used in DDC systems.

13.2.1 Binary Input Points

Binary input points are connected to the input termination strips labeled as *binary* or *digital input* (BI, DI). Devices terminated on BI or DI addresses generate a two-state input signal that indicates a change has occurred in the state of a process variable. Binary input points are commonly used to show the operating status of pumps, fans, drives, occupancy, damper position, safety conditions, and other two-state information. Binary addresses can also be configured to accept signals that indicate the current status of fire, lighting, access, or other systems integrated into a building automation system.

A change in the state of a binary input is generated by the opening or closing of a set of *dry* contacts located in the field. A change in state of a binary input is read by the microprocessor by the presence or absence of a low voltage DC signal across the contacts terminated on the binary input address terminals. This small voltage is generated by circuitry on the termination board. When the field contacts close, the potential difference across the terminal screws shorts to zero volts. When the contacts open, the voltage across the point's terminals changes state to the magnitude being generated by the termination board.

Figure 13.4 shows a section of the binary input termination strip. Each termination address has three screws, two are used for the contacts and the remaining one is a grounded screw used to ground the shield of the wire that connects the field point to the panel. Dry contacts have *no* external voltage applied across their terminals. The only voltage across the contacts is generated by the termination

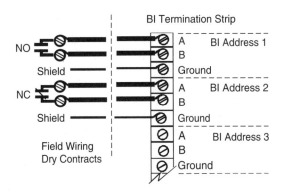

Figure 13.4   Typical Termination Detail for Binary Input Points

board. If another voltage source were applied to the contacts it could damage the electronics on the termination board.

13.2.2 Pulse Accumulator Points

Gas, water, condensate, electric, and other meters can be connected to a DDC system using binary input addresses. These meters can be configured to indicate flow by pulsing a set of dry contacts located in the meter head. The pulse rate is proportional to the volume flowing through the meter. Each meter has a multiplier called a *meter constant* which indicates the amount of flow that is required to generate a pulse of the meter's dry contacts. When the number of pulses is multiplied by the meter constant, the consumption of the utility can be recorded by the DDC system.

When a meter with pulse accumulator points is connected to the binary input address, the technician programs the DDC panel so it knows the device terminated at that address is a pulse accumulator point. The technician also enters the meter constant for that input point. The DDC panel's microprocessor maintains a count of the number of contact closures and multiplies them by the meter constant to calculate the totalized flow. Most DDC panels have a maximum limit on the number of binary input addresses that can be configured as pulse accumulator points because these points require more of the system's resources to function.

13.2.3 Binary Output Points

Binary or *digital output* (BO, DO) points are used to cycle electric loads on and off in response to the input conditions and the application program. Each binary output address is connected to an on-board relay having normally open, normally closed and common contact termination screws. These relays have a low voltage DC coil that switches the position of their contacts in response to a signal from the microprocessor. These contacts are limited in the amount of voltage and current they can switch. Threfore, the panel's control relay is typically used to switch an auxiliary 24 volt AC relay that is located in the field. The auxiliary control relay isolates the sensitive electronic circuits of the DDC panel from the higher voltages used by the control circuits of the field contactors and motor starters. These higher voltages and currents are switched outside of the DDC panel to limit the possibility of transient electrical spikes from harming the panel's electronic components. Figure 13.5 shows a termination layout for binary output points.

Binary output relays can be configured to operate as momentary (pulse) outputs or as latching outputs. A *momentary* output pulses its relay coil for one second whenever the point is commanded on. The control circuit of the load is responsible for sealing the control circuit so the load remains in operation after the pulsed contacts open. Momentary configured outputs are used to emulate a momentary push-button on/off switch. An additional binary address is required to pulse the equipment off. A *latching* output relay configuration energizes the DDC relay's coil and maintains the position of the contacts until the microprocessor issues a command to change their state. At that point, the relay coil is de-energized and the contacts return to their normal position. Most HVAC applications use the latching binary output point configuration.

Several output relays can be configured with software to operate together to provide slow-fast-stop and on-off-auto switching strategies. They can also be coupled with a binary input point to provide a feedback signal that confirms that a command issued to a binary output point has performed its intended function. In these configurations, the binary output point is commanded on and a set of status contacts located in the field change their state when the load is operating correctly. If the status contacts do not change their state, the microprocessor will issue an alarm message stating a fault condition exists. Figure 13.6 shows a differential pressure switch commonly used to prove air flow in a duct, verifying the fan's operation. The contacts for this switch are terminated on the binary input termination strip. Termination diagrams for these and other configurations are shown in the installation and engineering manuals for the DDC system.

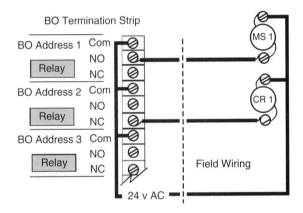

Figure 13.5   Termination Diagram for Binary Output Points

Status Contacts for Binary Input

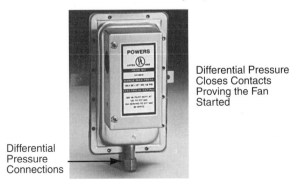

Figure 13.6   Binary Input Status Hardware Point

13.2.4 Analog Input Points

Analog input (AI) addresses are used to terminate analog sensors to the DDC panel. Analog sensors generate continuous signals that must be converted into their equivalent binary word by the DDC panel's analog to digital converter. DDC panels can accept 4 to 20 milliamp current signals, 0 to 5 and 0 to 10 volt signals, variable resistance signals and 3 to 15 psi pneumatic signals (through a signal converter).

Thermistors are solid state devices that generate a change in electrical resistance in response to changes in the measured temperature. These are *passive* devices, which means they do not need a voltage supply to operate. They are connected by two wires to the signal and common terminals of an analog input termination address. Metal RTDs and other *active* sensors that generate a current or voltage signal use a signal converter that is located near the sensing element. The signal converter must be supplied with power in order to generate an output signal. The necessary power is supplied by terminals on the AI termination board. One connecting wire of an *active* sensor is terminated to the AI terminal labeled *(+24 v)* or *voltage*. The remaining wire is connected to the *signal* terminal. Address 2 in Figure 13.7 depicts the termination detail for a sensor having a signal converter. The technician programs the DDC database to recognize the characteristics of the terminated sensor so the ADC can properly convert the input signals into useful data. Figure 13.8 shows an RTD sensor that would be connected to the AI terminals of a DDC panel.

13.2.5 Analog Output Points

Analog output termination addresses are used to connect analog actuators to the DDC panel. The analog output signals generated by the *sample and hold* circuit of the output address are sent to the actuator. DDC panels can generate 4 to 20 mA, 0 to 5 or 0 to 10 volt signals as shown in Figure 13.9. They can also produce pneumatic signals using an electric to pneumatic signal converter. As in all the other applications, the controls technician programs the DDC database to produce the signal required by the connected actuator. Figure 13.10 shows a device that converts the signal from the digital computer into an analog pneumatic signal.

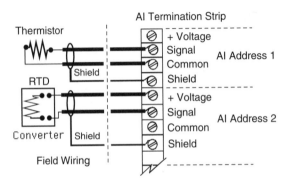

Figure 13.7   Typical Termination Detail for Analog Input Points

RDT Duct Sensor and Signal Converter

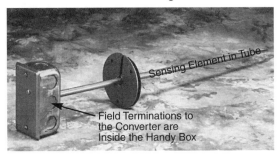

Figure 13.8 RTD Sensor and Signal Converter

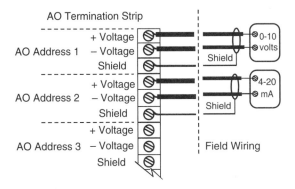

Figure 13.9 Typical Termination Detail for Analog Output Points

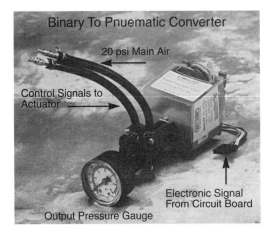

Figure 13.10 Electronic to Pneumatic Converter

13.3

ELECTRICAL MAGNETIC INTERFERENCE

All field points that send or receive electric signals from the DDC panel are connected to the panel by lengths of 18, 20 or 22 gauge insulated wire. These conductors are run through mechanical rooms, up and down pipe and duct chases, within ceiling spaces and wall cavities. Signals that travel on wires located within areas that contain high voltage conductors, high power motors or transformers are prone to corruption by electrical noise generated by these sources. Electrical and magnetic noise can be injected into the signal wires, altering the information they carry. If this occurs, the DDC panel will receive incorrect information on the current condition of the process. Consequently, the microprocessor will also generate incorrect responses. *Electrical magnetic interference* (EMI) is the phrase that describes the corruption of control signals that results from capacitive or inductive coupling.

Capacitors are constructed by separating two foil plates by a dielectric or insulating material. *Capacitive coupling* occurs whenever two conductors are separated by an insulator. The insulating material prevents conduction between the plates but allows the electrostatic charge of the electrons to induce current flow in the other plate. Similarly, whenever two wires of different voltages are positioned near each other, and separated by a dielectric (air), the higher voltage conductor induces a signal into the lower voltage conductor. The amount signal (noise) induced by the higher voltage wire is proportional to the difference between the voltages of the two wires and inversely proportional to the square of the distance between the conductors. If signal lines are installed close to or within conduits that contain higher voltage (over 100 volts) wires, the capacitive noise injected into the signal wire corrupts the information contained in the signal.

Electrical noise is also injected on signal wires by *magnetic induction*. Whenever a wire is cut by a moving magnetic field, a current is induced in that wire. The stronger the magnetic field, the greater the induced current. Consequently, when large alternating currents are present within the area of the signal wires, they produce strong 60 hertz oscillating magnetic fields. These fields inject noise into signal wires that are located near the electromagnetic sources. The amount of current induced in the signal wire is proportional to the strength of the magnetic field and its frequency.

Several installation guidelines have been developed that reduce the effects of signal corruption by capacitive coupling and magnetic induction. The most important guideline followed during the installation of control signal wires is: *Never run signal wires within the same conduit or trough as power wires.* Signal wires are always run in separate conduits or toughs that are separated by at least four and preferably, six inches. This minimum distance decreases the effects of EMI. Another installation guideline calls for signal wires and their conduits to be installed as far away from high EMI sources as realistically possible. Never install DDC panels or run their signal wires near high voltage feeds, large electric motors or transformers. To further minimize EMI affects on signals, special control wire has been developed to connect physical points to the DDC panel termination strips.

One of the more popular cables is called *twisted shielded pair* (TSP) wire. It is constructed with a *pair* of 18, 20 or 22 gauge insulated wires that are *twisted* around each other in the axial direction, throughout the entire length of the wire. The two conductors are also wrapped with a conductive foil *shield* that extends the length of the wire. In some TSP configurations, the shield is held in place around the twisted wires with an additional non-insulated wire that is spiraled around the twisted pair. This extra conductor provides an easy method of connecting the shield to the termination strip. The entire TSP wire may be surrounded by a plastic insulating cover.

The shield around the twisted pair of conductors is used to reduce the effects of capacitive coupling. The shield acts as one of the plates of a capacitor while any nearby high voltage wire acts as the other plate. Consequently, the high voltage plate couples with the shield, injecting any noise into the foil conductor which in turn passes the stray signal directly to an electrical ground. Coupling does not occur between the signal wires and the shield because of the small difference in voltage (10 volts) that exists between the control signal and zero volt (ground) potential of the shield.

Proper termination of the shield is imperative to its success in minimizing capacitive coupling. To be effective, the shield is only terminated to earth ground *at the DDC panel.* The field end of the shield is stripped from around the wire and insulated with electrical tape to prevent it from making contact with earth ground at the sensor's location. Grounding a shield at both ends may result in the generation of circulating currents within in the shield. These currents raise the voltages on the DDC board and may damage the control system.

Twisting the signal wires reduces the injection of noise produced by magnetic fields that cut through the wire. The twists in each conductor cause the magnetic field to cut the wire in opposite directions. This produces opposing currents of *equal magnitude* on the same wire. The two opposing currents cancel each other, eliminating the effects of magnetic fields on signal carried by the wires.

Additional electrical noise can be produced by normal control relay operation. Whenever the flow of current through a coil of wire is interrupted, the magnetic field it produced collapses rapidly, generating a momentary high voltage spike across the coil's terminals. This spike can destroy static sensitive integrated circuit chips found in DDC panels. To minimize the possibility of injecting a voltage spike into the circuit board, all relay coils have a device called a *varistor* installed across their coil leads and contacts. A varistor is a solid state component that has a high resistance to normal voltages but shorts its leads together whenever the applied voltage exceeds its design threshold (400 volts).

When the coil of a binary output address is commanded open, the high voltage spike created by its collapsing magnetic field is shunted to 0 volts before it can harm the components on the circuit boards. Varistors are also added to the binary input terminations to prevent field generated spikes from being transferred into the input termination board. EMI is also the driving force behind the control specification that prohibits the installation of other relays, transformers, E/P's or other like devices inside or near the DDC enclosure. These devices would be too close to the chips to shield them from the negative effects of their magnetic induction. Control system engineering and installation manuals outline the precautions that should be followed to minimize the effects of EMI in DDC installations.

The integrated circuit chips on the circuit boards are also sensitive to stray magnetic fields. A discharge of static electricity in the vicinity of a chip can easily destroy it. To minimize the dangers of static discharge when servicing a DDC panel, the technician wears a grounding strap that bleeds any excess electrons in the body to ground. Since the DDC panel is already grounded, no static discharge can take place. If a strap is not available, always make a habit of touching the grounded metal cabinet to drain any excess charge before touching a board. Never disconnect a wire or pull a board from its connectors when the panel is powered up. Removing a powered board will usually render it useless.

13.4

DDC PROGRAMS

DDC systems require software programs to customize the panel for a particular installation. These programs contain the instructions needed by the microprocessor to interpret the input data from the sensors and generate the correct output responses to changes in the processes. The program is developed using the firmware stored in ROM software and the programming instruction manuals available from the system's manufacturer. Figure 13.11 shows the software disks and manuals used to develop the DDC program.

Different methods of writing the program code for DDC panels have been developed. Each succeeding generation of programming software becomes more

Figure 13.11   Software and Manuals Used to Develop DDC Programs

user friendly. Earlier generations of code were more symbolic and cryptic to write, making troubleshooting a cumbersome procedure. Current generation firmware produces menu driven templates and standardized code that prompts the programming technician to fill in the blanks with required data. The program checks the data entered into each field to be sure its appropriate for that parameter. If the data was incorrect, the program flags the programer, requiring changes to be made before the data is stored. This minimizes the number of start-up problems associated with the software. To increase the versatilely of the system, custom routines or strategies can be written to enhance its operation.

13.4.1 System Database

Two related software programs must be input into the DDC panel to create a functional control system. One program is called the *database,* which stores point and system configuration information. The second program is called the *application program,* which instructs the DDC system on the operational requirements of the processes. The database is an electronic description of the DDC system's hardware configuration. It includes a listing of all the DDC panels and peripheral devices attached to the system's communication busses. Each DDC panel and the points within each panel are identified by their own unique address. These addresses are stored in the database which resides in the panel's RAM memory and at the central computer. When an application program needs information from a point located somewhere else in the system, it reads the point's address from the database and requests the information from the panel that the point is terminated on.

Every point, whether software or hardware, is identified in the database. It holds a record of all the operational characteristics pertaining to every point in the system. This information is used by the microprocessor to identify the point and update its value or status. The operator can call up any point that is attached to any panel on the communications bus from the work station computer or with a hand-held programming instrument. The point's current condition can be read, adjusted or overridden to analyze the response of the process. Table 13.1 lists some of the point characteristics found in a typical DDC database. A brief description of the characteristics follows the table.

13.4.2 Application Program

The application program is a list of instructions or code that is used by the microprocessor to control the connected processes. The DDC application program is generated by the programming engineer or technician after the database has been entered into RAM memory. The program instructs the system on how to operate equipment during occupied and unoccupied periods along with the system's response during emergency conditions. It also describes start-up and shut-down sequences, loop modulation, equipment starting and stopping, temperature reset schedules, energy management routines and any other code required to meet the operational specification.

Table 13.1   POINT CHARACTERISTICS IN TYPICAL DDC DATABASE

Point Characteristic	Analog Input	Analog Output	Binary Input	Binary Output
Point Name	x	x	x	x
Point Type	x	x	x	x
Description	x	x	x	x
Gain	x	x		
Bias	x	x		
Normal position			x	
Alarm Limits	x	x		
Engineering Units	x	x		
Initial Value		x		

Point name is an alpha-numeric code (usually 8 characters long) that identifies the point to the operator and the system. The name is used by the operator to access point information and to write the application code.

Point Type describes the point's general characteristics, analog input, analog output, binary input, binary output, etc. The point type is used by the firmware to request the correct information from the programmer during the development of the database.

Description augments the point name. It is used to identify the point or its location in easily recognized words (president's office, Room 142, Chiller no. 3, etc.)

Gain & Bias are used by the microprocessor to develop the transfer function of analog points so their magnitudes can be calculated and stored in memory.

Normal Position is used to indicate the normal position of binary input contacts, NO or NC.

Alarm Limits are used to indicate the permissible high and low limits of an analog point. When the magnitude of a point exceeds its limit, the database generates an alarm message.

Engineering Units are used to display the units of a value of an analog point when it is displayed or printed.

Initial Value is the value of an analog output point that is used when the system is initially started. It insures a fail-safe condition exists at start-up.

The program is developed using either standardized programmed routines or from scratch. Standardized routines are available from the control company's library of proven strategies. They are used to simplify programming of new systems. A strategy is called up and the appropriate point names are written into the code to customize the program for the new application. Sensor names, actuator names, loop gains and operating schedules are added to the generic code and loaded into the panel's memory. After the system is started, the gains are adjusted to tune the loop using the procedures described in Chapter 12. Using the pre-programmed code establishes a standard throughout a facility where all similar processes operate with the same sequence of operation. This simplifies analysis and maintenance of the systems because they all operate with the same strategy.

In those applications where a standardized strategy is not available to meet the requirements of the process, customized software code can be written by the programmer. The format of the program differs between manufacturers. The programming manuals list all the available commands, their proper syntax and examples of their application. Custom programs can be very powerful, but increase the chances that the technician will have difficulty in following the logic if they are not properly documented.

The sequence of execution of the application program's instructions is based upon the existence of certain conditions within a process. For example, if the outside air temperature is below 60 °F, the economizer cycle can be used or the heating system can be enabled, etc. To determine whether the proper process conditions exist, *relational* and *logic* operators are available in the programming code. Relational operators determine if one value is greater than, less than or equal to another condition. Logic operators compare two conditions and generate a binary, True/False response. The logic operators are AND, OR, NOT and Exclusive OR. A summary of logic operators appears in Table 13.2. Other program statements are available to change the characteristics of points, the programs sequence, generate emergency responses, perform energy management strategies and numerous other functions. Programming documentation is available from the manufacturer to assist the technician in reading and writing DDC programs.

Table 13.2   SUMMARY OF LOGIC OPERATORS AND THEIR FUNCTION

Operator	Description	Example
AND	If both conditions are true, the result is true	If time > 8:00 AND day = Monday then fan = on
OR	if either condition is true, the result is true	If outside air temp > 65 or heat = on then dampers = minimum position
NOT	Inverts the state of a binary command or operator	NOT on = off; NOT off = on NOT 1 = 0; NOT 0 = 1
NOT AND = NAND	If both conditions are true, the result is false	
NOT OR = NOR	If either condition is true, the result is false	
Exclusive OR	If only one condition is true, the result is true	

13.5

ENERGY MANAGEMENT STRATEGIES

Energy management strategies are special HVAC routines that are used to improve the operating efficiency of equipment and reduce the cost of their operation through flexible scheduling, limiting operation and altering set points. These strategies were designed to reduce electrical consumption and demand. *Consumption* is a measure of electrical energy used by the equipment in a facility. Consumption is measured in units of kilowatt-hours. Utilities charge between $0.02 and $0.15 per kilowatt-hour. To reduce consumption charges, equipment can be scheduled to operate only when needed, using a DDC system.

Demand is a measure of the rate of electrical usage by a facility. It is measured in units of kilowatts and appears as a separate charge on commercial utility bills. Demand charges fall within a range of $5.00 to $10.00 per kW. These charges offset the costs incurred by utility companies to maintain a capability of supplying the momentary requests for power. Commercial customers are charged a capacity fee that is based upon the *highest* demand (kW) that they incurred within a 15 minute window during the last billing period.

Customers may also be charged an additional demand related fee based upon the highest demand that has occurred over the last 12 billings. This fee can cost about $2.00 per kW and must be paid for the next 11 months or until a larger demand has been set by the facility. These two demand related charges can account for a large percentage of a customer's monthly utility bill. The following energy management strategies can reduce electrical consumption and demand charges and are available in the firmware of DDC systems.

13.5.1 Time of Day Scheduling

Time of Day scheduling reduces electrical *consumption* by commanding equipment off when it is not needed. Scheduling reduces the cost of electrical consumption by reducing the operating hours of electrical equipment. When the equipment is off, it does not draw power from the utility, reducing the total kilowatt hours consumed during the billing period.

Time of day operation was the first energy management routine used to reduce the cost of electrical consumption. Time clocks were installed in systems to command equipment off during unoccupied periods. Time of day programming has advanced from the time clock era by incorporating an electronic calendar and clock into DDC panels to schedule the operation of equipment controlled by the system. The time of day function allows the user to program different operating schedules for each day, week, month or season. Equipment can also be programmed to operate with different schedules during the same day.

Each schedule should include the extra time needed to pre-condition the zone before occupants arrive. For example, if the set point of a zone is reset to 60° during unoccupied intervals in the winter, the zone may require an additional hour to bring its temperature back to 72° before the occupied period begins. The programmed time of day schedule must account for the extra time necessary to precondition the zone.

Temporary schedules can be added to the system to accommodate transient or short term utilization of equipment. Holiday dates can also be programmed along with a holiday schedule to optimize time of day related savings. Schedules are entered and modified by the building operator. Any scheduled point or system can be manually overridden by the building operator to accommodate temporary schedule requests from building occupants.

13.5.2 Optimum Start/Optimum Stop

Optimum Start and *Optimum Stop* are strategies that further reduce electrical *consumption* in comparison to time of day scheduling. They *vary* the scheduled start and stop times of the equipment based upon current environmental conditions. The time of day program confines equipment to a fixed operating schedule. On extreme weather days, the zone may take 60 minutes to reach comfort levels while during milder weather it may only take 20 minutes. The time of day program schedule must reflect the earliest start time to be sure the zone is at comfort set point before the occupants arrive. Any additional time the equipment operates at the occupied period set point, before the occupied period begins, wastes energy.

Optimum *start* is a sophisticated energy management algorithm that varies the start time of the equipment based upon a number of related process variables. The optimum start algorithm looks at the zone's present conditions, the outside air temperature, the room's thermal resistance and thermal capacitance, the earliest and latest allowable start times and other related variables. The program keeps a log of the amount of time it took for the zone to reach comfort set point over the past starts and uses all this information to determining when to start the equipment for the next occupied period. The optimum start strategy starts the equipment as *late* as possible while still insuring the zones will be at comfort levels by the start of the occupied period.

Optimum *stop* uses the *time constant* of a zone's thermal characteristics to stop the unit before the end of the occupied period. The strategy makes sure the zone temperature stays within its throttling range after its HVAC equipment has been commanded off. Zones with large thermal capacitances can be shut off earlier and allowed to drift knowing they will not deviate out of their comfort range before the occupants begin to leave. This strategy can shave additional kilowatts off the utility bill when compared to a fixed time of day schedule.

Optimum stop is not as widely implemented as the optimum start strategy. When the fans are commanded off before the unoccupied period begins, ventilation air flow into the zone stops, degrading the indoor air quality of the zone. The loss of air movement and background noise has been found to be more disruptive to the occupants daily routines, reducing their end of the day productivity.

13.5.3 Duty Cycling

Duty cycling aggressively cycles equipment on and off based upon its elapsed operating time or zone temperature to reduce electrical *consumption*. This cycling occurs during the occupied period when the equipment would have previously been operated continuously. When properly applied, duty cycling can be used to improve the overall operating efficiency of large systems that have been oversized for the current operating conditions. There are two methods used to implement a duty cycle strategy. One is based on time and the other is a function of the zone's temperature.

Time based duty cycling allows a load to be configured to operate on a fixed frequency time schedule. For example, a load can be programmed to run for ten minutes and stay off for the next ten minutes, repeating this pattern throughout the occupied period. The amount of time the unit is scheduled to operate is variable. It is based upon the maximum cycling frequency determined by the equipment manufacturer. A schedule of equal on and off intervals would avoid using one half of the electricity that would have been used if the equipment operated the entire occupied schedule.

There are drawbacks to using the time based duty cycle strategy that may outweigh the savings in electrical consumption. Larger motors experience an increase in the amount of belt and bearing wear when aggressively scheduled. Large motors also generate excessive heat in their windings when they aggressively cycled. Their high starting torques and multiple starts also induce excessive torque that can lead to premature shaft failures. The rapid pressurization and depressurization of the connecting ductwork also generates noises that are distracting to building occupants. All of these reactions to duty cycling must be evaluated against the projected savings before implementation.

Duty cycling based on temperature can be safely applied in large zones that are served by multiple rooftop units such as those found in department stores and supermarkets. All of the units serving the zone can be duty cycled based upon the average zone temperature. This strategy allows units to come on line only after those already operating are fully loaded. As the load decreases, some of the units are cycled off, allowing those remaining on line to operate at full load, where they are more efficient. Of all the energy management strategies, duty cycling is the least implemented because of its limited applications and associated disadvantages.

13.5.4 Demand Limiting

Demand limiting cycles off electrical equipment to prevent a facility's electrical demand from exceeding a predetermined set point. Demand limiting is the only energy management strategy that is designed to directly limit a facility's electrical *demand*. It prevents a facility from drawing too much energy by shedding (turning off) equipment when the current demand approaches the demand limiting set point. After the demand has decreased sufficiently, loads are restored (turned on) based upon the amount of time they have been off and their priority in the system.

Loads programmed with the demand limiting feature need certain parameters entered into the database that define how that piece of equipment must be shed and restored. The load's priority, minimum operating time, minimum off time and maximum off time are entered into the database to prevent the load from overheating or the process from reaching unacceptable conditions whenever demand limiting is in use. The priority defines each load involved in the strategy as either non-critical, critical or very critical. Minimum operating time allows the load to operate a fixed quantity of time before it is available for shedding. This allows sufficient time for large motors to dissipate the heat associated with starting. The minimum off time prevents equipment from short cycling.

When a building's electrical demand approaches the strategy's set point, the loads with the lowest priority are shed first in an attempt to reduce the electrical demand. If the demand continues to rise, higher priority loads are shed, one at a time, to keep the demand below the set point. Loads continue to be shed until the demand drops below a predefined deadband. If any load is shed longer than its maximum off time, it is restored into operation and another load is shed. All equipment must operate longer than its minimum on time before it becomes eligible for shedding. When the demand falls below its deadband, higher priority loads are restored first, in a manner that will not allow the demand to exceed its set point as the load is restored.

Demand limiting reduces kilowatt associated demand charges at a cost of periodic reductions in production or comfort. Every piece of equipment that is shed reduces the number of kilowatts being drawn from the utility. If, after all possible loads are shed, the demand continues to rise, a new maximum demand will be set. When this occurs, the demand set point should be evaluated to determine if it is realistically set or whether it must be raised. This strategy can have a dramatic effect on the utility bill because the costs associated with the demand portion of the utility bill can be one hundred times greater than those associated with consumption costs during peak operating periods.

13.5.5 Temperature Reset

Temperature set points can be reset to reduce the load on HVAC equipment, thereby reducing their electrical consumption and the facility's electrical demand while continuing to maintain a level of comfort within the zone. During summer months, occupied temperature set points can be raised during periods of high electrical demand to reduce equipment load. If a temperature difference of ten to 15 degrees is maintained between the zone and the outside temperature, occupants still remain relatively comfortable as long as the relative humidity is kept below 60% and air movement in the zone is maintained. When the electrical demand increases toward its set point, the DDC system can be programmed to raise the discharge air temperature and chilled water set points to reduce the equipments load.

During unoccupied hours, set points can be automatically reset to limit equipment operation, thereby reducing electrical consumption. During summer months, unoccupied zones are allowed to drift up to 85 °F before equipment is allowed to run. During winter months, set points are lowered to 60 °F during unoccupied periods to reduce the amount of heat transferred into the zone. If the temperature exceeds an unoccupied (reset) set point, equipment is commanded on and remains operational until the zone temperature changes 5°. After the temperature comes back into acceptable range for an unoccupied zone, the equipment is commanded off and the zone temperature allowed to drift.

13.6

SOFTWARE ANALYSIS FUNCTIONS

DDC software packages provide methods for monitoring the system's operation. A procedure called *point trending* is available that instructs the work station computer to read and store the value of a point at a desired time interval so it may be analyzed in the future. Analog point trending is used to monitor temperatures within a zone or process to help determine the cause of a problem. Analog output points can also be trended to analyze their dynamic response curves to determine if calibration is required. Trends can be established for binary points to indicate the start and stop times of equipment along with its cycling interval. The information contained in a trend assists the technician in analyzing processes that are experiencing intermittent problems that are difficult to observe in the field.

DDC systems also have software that allows the operator to override the status and values of the points in the system. Point commanding routines provide a means of changing set points, control points, cycling equipment and evaluating control strategies from the work station location. The work station can be configured so the values of any points being manually altered are automatically printed out on the report printer to create a hard copy of all the system changes. This is another tool available from analyzing systems that is not available in pneumatic and analog control systems.

13.7

SUMMARY

DDC systems are the most common control systems currently being installed in the HVAC industry. They use computer based hardware and software technologies to maximize the operating efficiency of the connected equipment. DDC system architecture *distributes* the responsibility for control by installing microcomputers in panels located throughout the facility. A twisted pair of communication conductors connects all the DDC panels in a system to a data acquisition and programming computer called a work station or head end computer. The communications bus makes it possible for the panels to distribute the work among many computers, share data and share system resources. The personal computer at the work station is used to collect and store operating data, archive programs, analyze system operation, develop programs, override local panel control, dial out during emergencies, and generate reports. The work station also allows complete access to the connected systems from one convenient location using a text or graphic interface.

Energy management functions are used to reduce the electrical consumption and demand charges within a facility. Time of day scheduling incorporates regular, temporary, and holiday schedules to control the operation of equipment so it only operates when needed. Optimum start and optimum stop strategies generate greater consumption savings than those produced by time of day scheduling by using dynamic schedules that alter the start and stop times based on the thermal characteristics of the zones, the outside air temperature, and a history of previous cycles. Duty cycling aggressively cycles oversized or duplicate equipment to force the operating equipment to operate closer to their full load capacity. Demand limiting strategies shed loads as the facility's demand approaches its set point to limit demand associated utility costs and the possibility of establishing new demand peaks. Temperature set point reset strategies are also used reduce the load experienced by operating equipment to reduce energy consumption.

13.8

EXERCISES

Determine if the following statements are true or false. If any portion of the statement is false, the entire statement is false. Explain your answers.

1. DDC systems control analog systems with binary computers.
2. The microprocessor in a DDC panel performs math and logic functions.
3. RAM is non-volatile memory used to store the user program and database.
4. Firmware is stored in ROM.
5. Sensors are terminated on the analog input termination strip.
6. Pulse accumulator points are a form of binary output point that are used to connect various flow meters to the DDC system.
7. Demand limiting reduces the monthly charges associated with the rate of energy utilization.
8. Varistors protect DDC circuit boards from stray voltage spikes.
9. Time of day scheduling reduces consumption more than optimum start/stop strategies.
10. Analog sensors that generate voltage or current signals have one wire connected to the voltage terminal screw of an analog input address.

Respond to the following statements and questions completely and accurately, using the material found in this chapter.

1. Describe the characteristics of a DDC system.
2. Describe three methods of minimizing the effects of electromagnetic interference on signal wires.
3. What is a microprocessor and what is its function in DDC systems?
4. What are the major differences between RAM and ROM?
5. What is firmware and where is it stored?
6. Describe the function of a digital to analog converter and an analog to digital converter.
7. What is multiplexing and why is it used in DDC panels?
8. What are dry contacts and why are they specified in binary input applications?
9. Why is the shield of a signal wire only terminated at the DDC panel?
10. What is a point database?
11. What is the application program in a DDC panel?
12. Describe the OR logic function and give and example of its use in an HVAC application.
13. Describe the AND logic function and give and example of its use in an HVAC application.
14. Describe the Exclusive OR logic function.
15. What is the difference between electrical consumption and demand?
16. Describe the advantages and disadvantages of the time of day energy management strategy.
17. Describe the advantages and disadvantages of the optimum start energy management strategy.
18. Describe the advantages and disadvantages of the optimum stop energy management strategy.
19. Describe the advantages and disadvantages of the demand limiting energy management strategy.
20. Describe the advantages and disadvantages of the duty cycling.

PART 3

HVAC Applications

CHAPTER 14

Documenting Control Processes

Chapter 14 introduces the reader to the characteristics and features of some of the methods used to present detailed information about the design and operation of control systems. These techniques will be incorporated throughout the remaining chapters in the book to present the characteristics of the common control strategies used in HVAC systems.

OBJECTIVES

Upon completion of this chapter, the reader will be able to:

1. Describe the characteristics and uses of a system drawing.
2. Describe the characteristics and uses of a sequence of operation.
3. Describe the characteristics and uses of a flow chart and its common symbols.

14.1

SYSTEM DRAWINGS

A system drawing is a collection of lines and symbols that illustrate the mechanical, electrical and control components that describe the installation of an HVAC system.

These drawings are used to guide the technicians locating and connecting the system's components. After the project is complete, a revised set of drawings called *as built's* or *record* drawings are developed that reflect any changes that were made during the installation. *As built* drawings are used as a diagnostic resource for service technicians and building operators to analyze control system problems.

System drawings show each control device and the associated mechanical equipment using symbols that are similar in appearance to the device's two-dimensional outline. Figure 14.1 shows a symbolic representation of a voltage to pneumatic signal converter. Note the similarity in their shape and labeling. Each symbol is placed on the drawing in the location and position that represents its actual location in the field.

Currently, there are no standard symbols adopted by the HVAC industry to depict mechanical and control devices. Each manufacturer has their own set of symbols used to develop their system drawings. Although they appear similar, they are not derived from a standard, but from the general shape of the component.

Straight lines are used on system drawings to represent the signal paths that connect the control devices together. Control lines can be drawn using different line types (dashed, solid, dots) to differentiate between different signal types. Geometric symbols (square, circle, pentagon, hexagon) are often used to identify different types of terminations, simplifying the control system drawing. A single line is often used to represent more than one tube or wire when there is little chance of

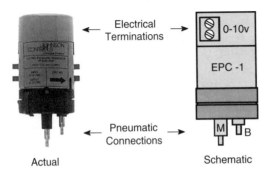

Electric to Pneumatic Converter
Converts 0 - 10 V to 0 - 20 psi

Figure 14.1   Schematic Symbol of an Actual Device

confusion over the intent of the line. All the control symbols and line types that are shown on the drawing are summarized in a legend and identifiable by their tag (EPT-1) which will be listed on the bill of materials.

Figure 14.2 depicts the system drawing of an air handling unit and its associated temperature control system. The ductwork, dampers, filters and fans are also schematically shown using a realistic representation of their actual shape. The two-dimensional representation depicts the configuration of the ducts and the relative location of the dampers, fans and sensors.

Proper labeling of the system drawing is essential if it is to be used as a reference source. Each control device is labeled with an acronym that uniquely identifies the device in the system. The acronym typically uses the first letters of the component's description along with a sequential number. In Figure 14.2, the damper motors are labeled DM-1, DM-2 and DM-3, the temperature sensors are designated TS-1 through TS-4 and the valve is tagged V-1. A legend is required on each drawing to indicate the meaning of the acronyms and symbols used on the drawing.

14.1.1 Single Line Drawing

A single line drawing uses one line to represent the ductwork, with the symbols for the control devices and mechanical equipment superimposed upon the duct lines.

A single line drawing is becoming the preferred method of representing HVAC systems and their controls. This type of illustration fulfills the purpose of the system drawing using a simplified format. Figure 14.3 depicts the same mechanical system that was shown in Figure 14.2. It differs in that it shows the ducts as a single line and the field control devices to the DDC panel. The type of illustration method is based solely on the preference of the customer and the control company.

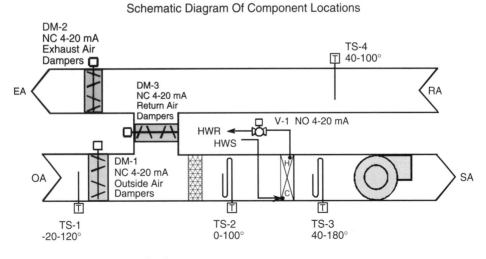

Schematic Diagram Of Component Locations

Figure 14.2   AHU Control Schematic

Line Diagram of System Installation

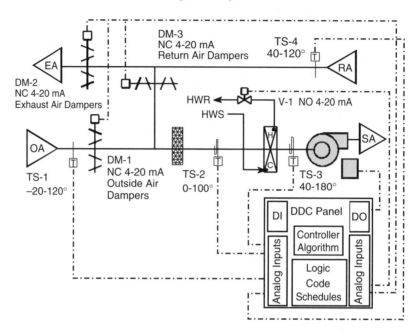

Figure 14.3 Single Line Drawing of an AHU and its Control System

14.2

SEQUENCE OF OPERATION

A sequence of operation is a concise summary of the operational attributes of the mechanical equipment in a system and their control loops.

The sequence of operation is developed by the mechanical engineer in conjunction with the control system's application engineer. It is another reference document used by installation, commissioning, operations and service personnel that thoroughly describes the intended operation of the control system. The sequence of operation is placed on the system drawings, beside the diagram of the system it describes. Placing the sequence on the drawings makes it easier for the reader to reference the components and interactions that are being described in the sequence of operation. The sequence of operation is also called the *description of operation*.

Writing a complete sequence of operation requires practice. Once complete, it must summarize all the necessary factual information in clear, grammatically correct statements. After reading the sequence, a mechanically inclined person that was not familiar with the application described in the sequence should be able to determine whether the system is operating correctly. Many control companies have standard sequences written for the common HVAC processes. These sequences are copied onto the system drawings and modified as needed. The following sections describe some of the characteristics of a well written sequence of operation.

14.2.1 The Organization of a Sequence of Operation

A sequence of operation is easier to read if it follows a logical format. Well written sequences are organized in logical sections that describe how the processes respond during the pre-occupied, occupied, unoccupied and emergency periods of operation. Each section is further separated into subsections that describe the operation of each process control loop found on the unit, (mixed air temperature, discharge air temperature, humidity control, static pressure, etc.). Additional sections can also be added that describe the operation of zone terminal devices, primary heating and cooling equipment, etc. Each section of the sequence of operation is written as a group of related paragraphs that can stand alone in their description. This method of presenting the operating details of a system allows a technician to zero in on the operational characteristics of the portion of the system that is presently being analyzed instead of having to read the entire sequence to try to find the necessary information.

The number and type of sections used in a sequence are based on the equipment being controlled and the characteristics of the application. A model sequence of operation consists of some or all of the following sections.

System Overview

The system overview briefly describes the mechanical equipment and processes being maintained by the control system.

Energy Management Overview

The energy management overview briefly describes the energy management strategies that are incorporated into the system's operation. Start/stop optimization, demand limiting, time of day scheduling, set point reset, pre-occupied warm-up, pre-occupied purge, and duty cycling are energy management strategies mentioned in this section if they are incorporated into the sequence of operation.

Pre-conditioning Operation

Pre-conditioning mode describes how the equipment operates during the period preceding the building's occupied period. The section describes morning warm-up and purge strategies that are incorporated into the sequence of operation.

Occupancy Operation

The occupied section of the sequence describes how the processes and their control loops operate during the period when the building is occupied. This section is further broken down into subsections for each process.

1. Fan operation
2. Static pressure control

3. Temperature control
   Mixed air
   Supply air
   Zone
4. Humidity control
5. Zone pressurization
6. Lighting control
7. Primary Equipment Control
   Boilers
   Chillers
   Cooling towers

Unoccupied Operation

The unoccupied section of the sequence describes how the control loops respond during the period following occupied time and ending when the pre-condition mode begins. Each subsections describe how a process and its control loops respond when the building enters its unoccupied period of operation. Each of the subsections (1 through 7) listed under the *Occupied Operation* mode are also described in this section, indicating how the loops respond when equipment is commanded off.

Emergency Strategies

This section thoroughly describes the response of the affected control loops in the event an emergency condition such as smoke, fire, freeze or fault (failure to prove) occurs. Depending upon the particular condition, all or some of the following sub-sections (1 through 7) listed under the Occupied Operation mode are included in the section.

Temperature Set Point and Occupied Schedules

The sequence also contains a section that details any tables or schedules referenced in the other sections. Some possible schedules are:

1. Mixed Air Reset Schedule—Schedule " A "
2. Supply Air Reset Schedule—Schedule " B "
3. Hot water Reset Schedule—Schedule " C "
4. Occupancy Schedule " D "
5. Unoccupied Set point Tables
6. Valve Sequencing Schedules

Other Sections

If a specification calls out for information that may be valuable to the operators of the system but do not fall under any of the above sections, they may be added to the end of the sequence of operation.

Keep in mind, the sequence of operation found on a drawing or print may not include all of the changes that have been made in the field after the system was installed and commissioned. In these cases, the original sequence of operation becomes somewhat obsolete and the persons relying on the information must use their analytical abilities to determine what the actual sequence of operation may be. It is the responsibility of the service technician to note any changes that were made to the operating characteristics of a system on the drawings along with the date of the changes and the initials of the person who made the change. This policy will ensure that a current sequence of operation is always in the file.

14.2.2 Information Included in the Sections of a Sequence of Operation

There are certain formatting procedures that can be used to simplify the writing and reading of each section in a sequence of operation. Observing these procedures insures all the information required by a technician to commission or troubleshoot a control system is on the drawings. Include the following information in a paragraph format when describing the operation of each control loop:

1. The loop sensor's input range.
2. The loop's set point.
3. The controller's action (direct or reverse), mode (P, PI, PID, PD) and output signal range.
4. Final controlled device's operating range and normal position.
5. Include an example of the loop's response to an increase and/or decrease in the control point.
6. The point names that were used on the drawing (TT-1, TC-3, etc.) are indicated in parentheses following a reference of the device within the sequence. For example:

   "... the mixed air temperature sensor (TS-3) sends a proportional signal to ...."

   ".... when outside air temperature is below 55 F, as sensed by the O.A. sensor (TT-1) ....."

   "... the heating valve (V-4) will be modulated ....."

7. Temperature reset schedules and occupied/unoccupied time schedules should be included at the end of the sequence of operation. When describing the operation of a loop that uses a time schedule, make reference to Schedule A or Schedule B within the section. Tables 14.1 and 14.2 illustrate the format used to list reset and schedule data.

Table 14.1   SCHEDULE A, SUPPLY AIR TEMPERATURE

OA Temperature	Supply Air Set Point
10	70
70	55

Table 14.2   SCHEDULE B, OCCUPIED TIME

Sun	Mon	Tue	Wed	Thur	Fri	Sat	Holiday 1	Holiday 2
–	7 A.M. to 6:30 P.M.	7 A.M. to 8:30 P.M.	7 A.M. to 6:30 P.M.	7 A.M. to 6:30 P.M.	7 A.M. to 5:30 P.M.	9 A.M. to 3 P.M.	unoccupied	9 A.M. to 3:00 P.M.

Application 14.1 is an example of a sequence of operation for a small laboratory air handling unit developed using the format described above.

APPLICATION 14.1 SEQUENCE OF OPERATION
P.T. LEWIS COLLEGE OF TECHNOLOGY—HVAC LABORATORY
SEQUENCE OF OPERATION

I. System Overview The HVAC & Controls Laboratory at P.T. Lewis College of Technology incorporates a 20 ton, 5000 cfm VAV air handling unit that serves 4 zones, the computer lab, the equipment lab and two adjoining classrooms. Each zone maintains its temperature using a VAV box with terminal reheat. Mechanical cooling is supplied by a 20 ton reciprocating chiller package. Hot water is supplied by a 50 hp water tube boiler. A DDC control system is used to operate the AHU, enable and disable the primary equipment , enable and disable the lighting system based upon a user adjustable occupancy schedule.

II. Energy Management Functions The DDC control system incorporates the following energy management strategies:

A. Time of day occupancy scheduling is used to cycle equipment on and off to reduce the energy consumption of the facility. Holiday scheduling is also incorporated to command the system into unoccupied operating mode on specified holidays.
B. Temperature reset is used to vary the supply and mixed air temperature set points based upon outside air temperature to improve system energy efficiency while satisfying the process load.
C. Economizer control of the mixed air dampers is used to minimize the amount of mechanical cooling required whenever the outside air temperature is below 65°.

III. Primary Equipment Control

A. **Boiler Control**—The boiler (BOILER-1) is enabled whenever the outside air temperature is below 65°. The primary hot water circulating pump (HWP-1) along with the hot water coil circulating pump (HWP-2) are commanded on to provide zone reheat. The boiler and its associated pumps are commanded off whenever the outside air temperature exceeds 65°. *Note: DDC point names are included in parentheses to assist in the location of the point in the database.*

B. **Chiller Control**—The chiller (CHILLER-1) is enabled whenever the outside air temperature (OATEMP) is greater than 55° and the system is in occupied operation. The chilled water circulating pump (CHWP-1) is commanded on to enable the chiller and disabled during unoccupied periods and whenever the outside air temperature (OATEMP) falls below 55°.

IV. Occupied Operation

A. **Fan Control**—The air handling unit's supply fan (SFAN-1) is commanded on based upon the time of day schedule found in Section VII.

B. **Static Pressure Control**—Whenever the supply fan (SFAN-1) is operating, the static pressure control loop maintains the supply duct static pressure at a set point of 1.2 inches of water at the sensor's location.

The supply air pressure is measured by the static pressure sensor (PT-1) located 2/3 of the distance down the supply duct near the entrance of Room 101. The static pressure controller algorithm generates a reverse acting, 3 to 13 psi, PI output signal that positions the supply fan's normally closed (NC) inlet vane damper (DM-4), to maintain the loop's set point.

As the static pressure in the supply duct increases, the signal to the inlet vane damper decreases, modulating them toward their closed position, reducing the static pressure in the supply duct.

C. **Mixed Air Temperature Control**—Whenever the supply fan is operating, the mixed air dampers are modulated to maintain the mixed air temperature (MASET) set point. The mixed air set point is reverse reset based upon the outside air temperature. As the outdoor temperature increases, increasing the cooling load within the zones, the mixed air set point is decreased in accordance to the reset schedule found in Section VII.

The mixed air temperature is measured by the averaging sensor (TS-2) located in the mixing plenum of the air handling unit. The output signal generated by the direct acting, 3 to 13 psi, proportional only controller algorithm modulates the 3 to 13 psi, normally closed outside air damper (DM-1) and the normally open, 3 to 13 psi return air (DM-2) damper actuators to maintain the mixed air temperature at its set point.

As the mixed air temperature increases, the output signal to the damper actuators (DM-1, DM-2) increases to modulate the normally closed outside air dampers toward their wide open position and the normally open return air dampers toward their closed position.

D. **Cooling Cycle Economizer Strategy**—The mixed air temperature control loop operates using an economizer strategy. Whenever the outside air temperature is below the economizer high limit set point of 65°, the outside air dampers are allowed to modulate between 20% and 100% open to maintain the mixed air set point (MASET). Whenever the outside air temperature exceeds the economizer high limit temperature (65°), the outside air dampers are commanded to their minimum position to maintain the 20% minimum requirement of ventilation air during occupied periods.

E. **Heating Mode Damper Operation**—Whenever the supply air temperature control loop is calling for heat, the mixed air dampers are commanded to their minimum position (20%) to minimize the heating energy required to condition the outside air stream as it enters the AHU.

F. **Supply Air Temperature Control**—Whenever the supply fan (SFAN-1) is operating, the hot and chilled water valves (V-1, V-2) are modulated to maintain the supply air temperature set point (SASET). The supply air set point is reset based upon outside air temperature using the same reset schedule as the mixed air temperature found in Section VII.

The supply air temperature is measured by the supply air sensor (TS-3) located in the supply air duct. The output signal from the direct acting, 3 to 13 psi, PID controller algorithm modulates the 3 to 8 psi, normally open hot water valve (V-1) and the 8 to 13 psi ,normally closed chilled water valve (V-2) actuators to maintain the supply air temperature (SATEMP) at its set point. As the supply air temperature increases, the output signal to the valve actuators increases, modulating the 3 to 8 psi, normally open hot water valve toward its closed position. If the output signal exceeds 8 psi, the hot water valve is completely closed and the 9 to 13 psi, normally closed chilled water valve begins to open.

G. **Lighting Control**—The laboratory lights are enabled and disabled based upon the operating schedule found in Section VII. Enabling the lights permits the room occupants to turn the lights on and off with the wall switches. Disabling the lights prohibits occupants from turning the lights on during unoccupied periods. Emergency lighting provides a constant source of light to permit safe egress.

V. Unoccupied Operation
Unoccupied mode is based on the occupancy schedule found in Section VII. When the building is unoccupied, the supply fan (SFAN-1) is commanded off.

Analog Outputs—Whenever the fan is commanded off, the mixing dampers (DM-1, DM-2), hot and chilled water valves (V-1, V-2) and the inlet vane damper (DM-3) actuators are commanded to their normal positions.

Chiller—The chiller (CHILLER-1) and the chilled water pump (CHWP-1) are commanded off.

Boiler—If the outside air temperature drops below 65°, the boiler (BOILER-1) is enabled and its associated circulation pumps (HWP-1 and HWP-2) are commanded on.

Low Temperature Limit—If any zone's temperature falls below 55°, the supply fan (SFAN-1) is commanded on and allowed to operate until the zone temperature exceeds 58°. After the zone reaches 58°, the supply fan is commanded off and the zone temperature is allowed to drift.

Lighting—The laboratory lights will be disabled based upon their schedule, which allows for an extended operation for the night cleaning crew.

VI. Emergency Strategies
Whenever the air temperature downstream of the heating coil falls below 36°, the low limit thermostat (LL-1) opens the control circuit to the fan motor starter, disabling the fan. The mixing dampers (DM-1, DM-2), hot and chilled water valves (V-1, V-2) and the inlet vane damper (DM-3) actuators are commanded to their normal positions.

VII. Schedules

A. Supply Air and Mixed Air Temperature Reset Schedule

Outside Air	SA Set point
0°	65°
55°	55°

B. Weekly Occupied Schedule

	Monday	Tuesday	Wednesday	Thursday	Friday	Saturday	Sunday
Start	7:30 A.M.	7:30 A.M.	7:30 A.M.	7:30 A.M.	7:30 A.M.	12:01 P.M.	12:01 P.M.
Stop	4:45 P.M.	4:45 P.M.	8:00 P.M.	8:00 P.M.	4:45 P.M.	8:00 P.M.	8:00 P.M.

C. Holiday Dates

December 20 until January 3, May 31, July 4, September 4, November 22—26

D. Lighting

Monday–Friday	On–7:00 A.M.–11:30 P.M.
Saturday	On–12:01 P.M.–11:00 P.M.
Sunday	Off

14.3

FLOWCHARTS

Flowcharts are symbolic representations of a sequence of operation. They consist of a number of interconnected symbols that indicate the logical flow through a program and simplify the analysis of complex application software. Technicians use flowcharts to troubleshoot code because they make it easier to visualize the logical interconnections that exist between the sections of the sequence of operation and their corresponding sections in the application program. Flowcharts can also be used to improve the efficiency of the DDC program by grouping similar strategies in groups called subroutines. Writing the program using subroutines minimizes the amount of system memory required to store the program, makes the program more efficient by eliminating duplication of code, and shortens the time needed for the microprocessor to execute the program.

14.3.1 Flowchart Symbols

Flowcharts can consist of five basic symbols and connecting lines with arrow heads. Each symbol represents a specific *type* of task that is to be performed. The five symbols and their characteristics are described in the following sections.

Diamond

Diamonds are used to represent a *decision* that must be made before one of two possible strategies is performed. A common DDC decision statement is whether or not a building is occupied. If it is occupied, the occupied sequence of operation is followed. If the building is not occupied, the unoccupied strategy will be performed.

The symbol shown in Figure 14.4 shows that the diamond symbol has two output paths labeled YES and NO. If the building is occupied, the logic connected to the YES arrow is performed. If the statement is false, the NO path is followed. The arrow at the top of the diamond connects the previous symbol with the decision statement.

Parallelogram

Parallelograms are used to represent an *Input* or an *Output* function that is to be performed, such as turning a piece of equipment on or off, modulating a final controlled device or receiving a reset signal. In the Application 14.1 a parallelogram would be used to represent the modulation of the mixing dampers, valves and inlet vane dampers. Figure 14.5 depicts the symbol for the input or output statement in the sequence of operation or application program. This symbol has one input and one output arrow.

Square

A square is used to represent the performance of a procedure. The procedure may be a calculation, assignment of a set point or bias, setting of an operating mode or selecting a minimum or maximum value. In Application 14.1, a square would represent the calculation of the reset temperature set points. Figure 14.6 depicts the symbol for procedure statements in the application program. This symbol also has one input and one output arrow.

Decision Symbol

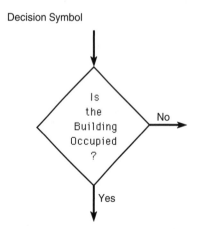

Figure 14.4 Decision Symbol

Input/Output Symbol

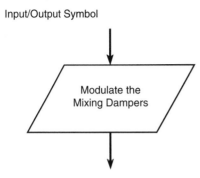

Figure 14.5 Input/Output Symbol

Procedure Symbol

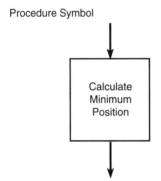

Figure 14.6 Procedure Symbol

Oval

Ovals represent the start and finish statements of a program. Therefore, there are only two ovals found on a typical flowchart, one at the beginning of the logic and the other one at the end. The "end" oval can command the program to stop or to return to "start" and begin over. DDC programs replace the word "end" with "return to start" because they continually loop, performing the same logic throughout their operation. These symbols have only one input or one output arrow.

Circle or Balloon

Circles with letters in their center's are called balloons. They show the continuation of logic that moves to another page or column. There can be many sets of balloons on a single flowchart. Figure 14.7 shows the oval and balloon symbols. These symbols have only one input or one output arrow.

Lines are used to connect the symbols to each other. Each line has an arrowhead that shows the direction of the logic as it moves through the diagram.

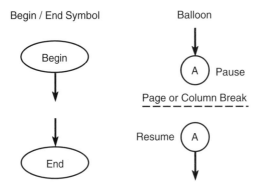

Figure 14.7   Ovals and Balloon Symbol

14.3.2 Flowchart Characteristics

In order for flowcharts to be universally understood, the following rules are used when creating the diagrams.

1. Chart logic always flows from the top to the bottom of the page.
2. Logic (arrows) never flows up from a symbol, it always proceeds down or to the right.
3 Logic symbols flow from left to right, only an arrow can be drawn from right to left to show the direction of the logic returning to a symbol located on the left side of the page.
4. The flowchart MUST follow the logic and sections within the Sequence of Operation.
5. The flowchart shows how alarm conditions cause the logic to bypass the normal operating program by looping around normal operation functions.
6. The last symbol returns the logic back to the start symbol using an arrow or "return to start" oval. This shows that the DDC control system loops continuously.
7. Arrowheads show the direction of logic flow.
8. All logic statements are written inside the flowchart symbol except for Yes and No labels of decision symbol.
9. Flowcharts are to be drawn very neatly.
10. There should be no duplicate logical sequences on the finished flowchart.

14.3.3 Developing the Flowchart

A flowchart is developed after the sequence of operation has been written. The proposed sequence of operation for a project is usually written in the job specifications and found on the system diagram. The control system application engineer verifies and refines this sequence into the actual sequence of operation that will be performed by the control system. The following steps are used for developing a flowchart.

1. Develop a correct Sequence of Operation.
2. Title the flowchart.
3. Draw the "start symbol" (oval) in the upper left corner of the page.
4. Draw vertical line with an arrowhead downward from the start oval to the next symbol.
5. Begin drawing symbols to represent the sections in the Sequence of Operation:
   a. Write the code inside symbol.
   b. Label YES or NO (Y or N) at each decision path.
6. Draw appropriate paths leading from the symbol using arrows.

The first section of logic symbols connected to the START oval are typically the alarm decisions. Placing the alarm decisions at the beginning of the chart indicates the microprocessor evaluates all alarm conditions before the system is allowed to operate in its normal modes. If any system point or process is in alarm, the flowchart shows the program logic looping around the remaining sections of the program and executing the emergency strategy. If the system shows no safety or fault related problems, the flowchart continues into the normal logic routines as stated in the pre-occupied sequence of operation.

After the alarm decision logic symbols have been drawn, the pre-occupancy decision symbols are added. This group of symbols determines if the system is in its occupied or unoccupied operating mode based on the time of day, day of week or holiday. During occupied time, the logic continues straight down. If its unoccupied time, the logic loops around the occupied logic symbols typically leaving toward the right of the diagram. The occupied operation follows the schedule section of the flowchart. The occupied section is the most complex part of the flowchart. It includes all the details of operating of the primary equipment and all of the modulating control loops. The subsections found in the sequence of operation are also visible in this part of the flowchart. The unoccupied operation is usually drawn parallel and to the right of the occupied column of symbols. This is the simplest part of the flowchart where all the necessary equipment is commanded off or to its normal position.

The emergency logic is usually shown to the right of the unoccupied logic column if it differs from the unoccupied sequence of logic. The emergency logic details the response of all the equipment in the system in the event that an emergency or fault has occurred. The flowchart in Figure 14.8 represents the logical flow of the sequence of operation found in Application 14.1. The description inside the symbols only contains key words that are referenced back to the sequence of operation. This allows the text to remain inside the outline of the symbols.

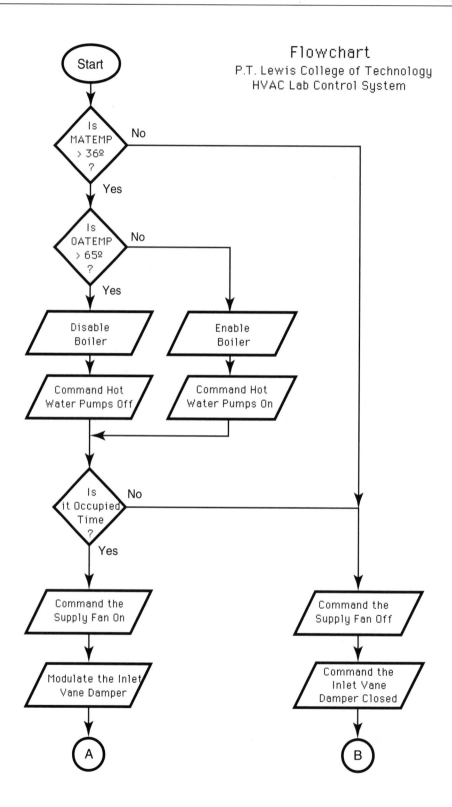

Flowchart
P.T. Lewis College of Technology
HVAC Lab Control System

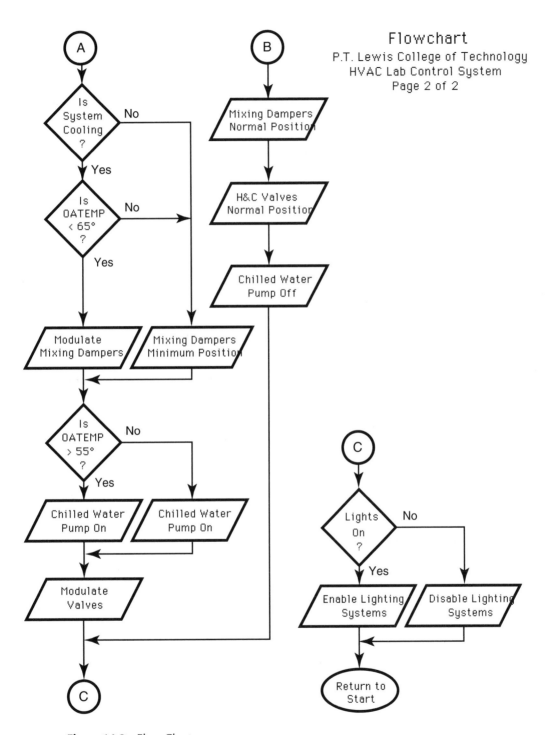

Figure 14.8   Flow Chart

14.4

SUMMARY

The system drawings, sequence of operation and flowcharts provide different tools for describing the layout and operation of a control system. The drawings are used to locate and terminate the components of the control system. A sequence of operation describes how each process is controlled and the interactions that exist between the different processes in the system. Flowcharts are pictorial representations of the sequence of operation that is used to develop a DDC application program and to troubleshoot software and logic problems. When combined, these three resources provide all the information needed to describe and analyze the operation of a control system.

14.5

EXERCISES

Determine if the following statements are true or false. If any portion of the statement is false, the entire statement is false. Explain your answers.

1. System drawings indicate the general location of control devices and the specific locations of their signal paths.
2. Line drawings depict a mechanical system using single lines to represent the ducts.
3. A sequence of operation is a description of the interconnections between control components.
4. The sequence of operation is used in conjunction with the system drawing to analyze a control system's operation.
5. A flowchart pictorially describes the logic used in the DDC database program.
6. A square symbol in a flowchart indicates a decision is being made.
7. A sequence of operation is written in stand alone sections and subsections that describe a particular process or mode of operation.
8. Emergency strategies are drawn at the beginning of the flowchart to indicate that emergency conditions bypass all normal control strategies.
9. An oval is used to indicate the logic of a flowchart continues to another location.
10. A system's drawings, flowcharts and sequence of operation are all reference resources used to install, commission and analyze control system operation.

Respond to the following statements and questions completely and accurately, using the material found in this chapter.

1. An evaporative condenser has a two stage fan that operates based upon the current condensing pressure of the system. The first fan is commanded on when the discharge pressure rises above 170 psi for 5 minutes. The second stage is commanded on after the pressure rises above 185 psi and remains there for 5 minutes. The fans are commanded off in reverse order after the pressure falls 5 psi below their cut in pressure and remains there for 5 minutes. Develop a flowchart for this sequence of operation.

2. A preheat coil operates in sequence with face and bypass dampers. When the outside air temperature is above 40°, the face and bypass dampers remain in their normal position (face open, bypass closed) and the coil's valve is modulated to maintain a leaving air temperature of 55°. When the outside air temperature is below 40°, the valve is commanded wide open and the dampers are modulated to maintain the leaving temperature set point (55°). Develop a flow chart to represent this logic.

3. Develop a flowchart for the following sequence of operation:
   At 7:00 am, the air-conditioning unit will be enabled if the outside air temperature is above 70°. At 6:00 pm the unit will be commanded off. If the room temperature increases above 82° during the unoccupied period, the air-conditioner will be commanded on and operate until the room temperature drops below 78°.

4. Develop a sequence of operation for a residential furnace that operates off a seven day programmable thermostat. Develop your own schedule.

5. Develop a sequence of operation and a flowchart for operating a window air conditioner. Be sure to include the operation of the fan, compressor and overload safety devices.

CHAPTER 15

Mixing Damper Control Strategies

Most commercial HVAC systems use air handling units to condition their zones. These units are built up with different sections that perform a specific process. All the necessary sections are bolted together to create an air handing unit that meets the project specifications by maintaining the correct combination of ventilating, cleaning, heating, cooling, humidifying and pressurizing processes. Each of the remaining chapters in Part 3 describes the purpose and operating characteristics of common air handling unit processes.

15.1

MIXED AIR DAMPER STRATEGIES

Mixed air temperature control strategies modulate outside, return and exhaust air dampers in a manner that fulfills some or all of the following requirements:

1. Maintain the mixed air temperature set point of the air leaving the mixing plenum.
2. Regulate the amount of outside air being brought into the AHU to meet zone ventilation requirements.
3. Pressurize the building by replacing the air that being exhausted from its zones.

The outside and return dampers are collectively known as the *mixing dampers* in all control strategies where both damper actuators are operated in unison using the same controller's output signal. When the mixing dampers are correctly sized and installed, the change in the mixed air temperature will be proportional to the change in the controller's output signal. Figure 15.1 depicts the field components used for a mixed air temperature process. A controller (DDC or single loop) is all that is needed to complete the control loop.

The exhaust air dampers are usually included with the process components of a mixed air temperature control loop. These dampers do not affect the temperature of the air in the mixing plenum but their operation is coupled with the response of the outside air damper. The exhaust air dampers provide a modulating resistance passage between the return air and the exterior of the building. Any return air that is not needed to maintain the mixed air temperature is released outside of the building through the exhaust air dampers and louver. The volume of return air leaving the building is controlled at a value that is slightly less than the amount of outside air being brought into the building. This generates a slight positive pressure in the building to minimize infiltration.

The outside air enters the air handling unit through louvers mounted in the exterior wall. The flow area of the outside air louver is larger than that of the return air duct. This reduces the velocity of the outside air entering the unit to minimize the amount of debris and precipitation that is drawn into the system. If the outside air path, including the louvers, bird screen, duct and damper has a resistance that is much smaller than the pressure drop of the unit's return air path, the nonlinear response of the mixing dampers is increased. Under nonlinear operation, the change in the temperature of the mixed air is not proportional to the change in the

Mixed Air Temperature Process Components

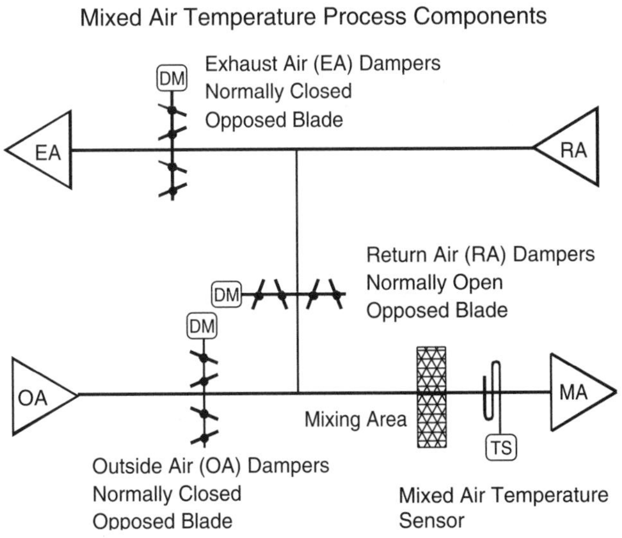

Figure 15.1 Mixed Air Damper Configuration

controller's output signal. Since the dampers and duct size is a function of the equipment design, a control technician must make adjustments in the tuning parameters of the mixed air temperature controller to minimize the nonlinear response.

The following sections explain the purpose and operating characteristics of mixed air processes. Each section contains a system drawing of the process, its sequence of operation and a flowchart of the control logic.

15.1.1 Mixed Air Temperature Control Strategy

A mixed air temperature control strategy modulates the position of the outside and return air dampers to maintain the temperature of the air leaving the air handling unit's mixing plenum at its set point (50 to 65 °F). The temperature is controlled by varying the *mass flowrate* of the outside and return air streams entering the mixing plenum of the air handling unit. During most of the unit's operation, the supply and return air streams are at different temperatures. The temperature of the mixture is proportional to the volumes of outside and return air combined in the mixing plenum.

A response in the mixed air temperature control loop is initiated by a change in the temperature of either of the air streams. As the temperature of one of the air streams decreases, the temperature of the air mixture also decreases. For example, when the outside air temperature decreases, the mixed air temperature control loop modulates the mixing dampers to reduce the amount of cooler outside air in the mixture while proportionately increasing the quantity of warmer return air entering the plenum. A direct acting controller response is used to modulate the normally closed outside air damper's closed as the mixed air temperature decreases. The mixing dampers are allowed to modulate whenever the *outside* air temperature is:

1. *Below* the mixed air temperature set point.
2. *Above* the temperature that would cause the outside air dampers to modulate below their minimum ventilation position.

15.2

MIXING DAMPER MINIMUM POSITION

Whenever an air handling unit is operating, a minimum quantity of outside air must be brought into the unit and conditioned in order to meet the ventilation requirements of the people and processes in its zones. The minimum quantity of ventilation air required by a unit is based upon the number of occupants within its zones and the amount of fumes, gasses and odors being generated within the zones. ASHRAE Standard 62 is used as a resource for establishing the minimum ventilation requirements for an application. At no time during the occupied period of operation should the amount of outside air entering the unit fall below this minimum level.

There are periods of time when it is more expensive to condition the outside air entering the AHU than it is to recondition the air returning from the zones. The periods of time when it is more economical to condition the return are:

1. Whenever the outside air temperature exceeds 60 to 70 °F or when the return air enthalpy is greater than the outside air's enthalpy.
2. Whenever the supply air temperature control loop is calling for heat.

When either of these conditions is present, the operating efficiency of the AHU would be increased if the return air damper were commanded 100% open and the outside and exhaust air dampers commanded 100% closed. Unfortunately, allowing the outside air dampers to close completely violates the code requirements that regulate the amount of ventilation air required in occupied facilities. Under these conditions, the mixing dampers are commanded to minimum position. The following subsections describe different methods of maintaining a minimum volume of outside air entering an air handling unit whenever it is operating.

15.2.1 Open Loop Minimum Position Control

Minimum position describes the position of the outside air dampers that maintains the minimum ventilation flowrate requirements of the air handling unit.

Different strategies are used to prevent the outside air dampers from closing below their minimum position. One method commands the dampers to minimum position by sending a fixed control signal to the mixing damper actuators. This signal is sent whenever the unit is calling for heat, the outside air temperature or enthalpy exceeds its high limit set point or when the mixed air temperature control loop tries to modulate the dampers below their minimum ventilation position. The magnitude of the minimum position signal is based upon the percentage of total fan volume that must consist of ventilation air during minimum position operation. For example, if a 3000 cfm air handling unit requires a minimum of 600 cfm of ventilation air during the occupied period, the signal sent to the damper actuators is set to value of the controller's output signal range that represents 20% of its signal span. The value 0.20 was derived by dividing the minimum quantity of air by the maximum capacity of the fan (600 cfm ÷ 3000 cfm).

This is an open loop strategy because it lacks the feedback needed to determine if the correct amount of outside air is actually entering the air handling unit. The nonlinear response of damper assemblies coupled with their leakage, linkage slack and binding blades usually creates a response that cannot be relied upon to maintain minimum ventilation requirements. Consequently, the actual volume flowrate through the damper at minimum position is seldom related to the percentage of stroke of the actuator. In extreme cases, the 20% controller signal may produce an actuator force that is too small to break the dampers free from their closed position. In other cases, the 20% signal may only open the dampers slightly, providing less than required ventilation to the process. Under either of these conditions, the unit does not provide sufficient outside air into its zones and the building is operating in a non-compliance mode with respect to the ventilation code. Conversely, the nonlinear response of the damper may permit excessive quantities of outside air into the system when commanded to minimum position based upon a fixed output signal. Under these circumstances, the operating efficiency of the process is reduced because more outside air than necessary must be conditioned before being sent to the zones. If the open loop minimum position strategy is to be incorporated into a mixed air process, the following formula is used to calculate the actual amount of ventilation air currently entering the mixing plenum:

$$\frac{(rat - mat)}{(rat - oat)} \times 100 = \% \ Outside \ Air \ by \ Volume$$

or

$$(oat \times \%oa) + [rat \times (1 - \%oa)] = mat$$

Percent Outside Air Formula

where: mat = mixed air temperature
        rat   = return air temperature
        oat   = outside air temperature
        %oa = percent of ventilation in the mixed air stream

To establish how much of the mixed air is ventilation air, measure the outside, return and mixed air temperatures and enter their values into the top formula. Solve for the value of %OA.

To initially set up mixing dampers for an open loop minimum position strategy, determine the amount of ventilation air that will be required in the application as a percentage of supply fan's capacity (cfm). Measure the current outside and return air temperatures. Enter the percentage of outside air and temperatures into the bottom formula and solve it for the mixed air temperature. Place a thermometer in the a location where the outside and return air streams are throughly mixed. Vary the control signal to the mixing damper actuators until the mixed air temperature matches the calculated value, indicating the correct volume of ventilation air is present in the mixing plenum. This procedure will compensate for the nonlinear response associated with the dampers but does not overcome the problems associated with response hysteresis, increased friction, wind and accumulations of dirt on the mechanical components.

15.2.2 Closed Loop Minimum Position Control

To improve the accuracy of the minimum position strategy by accounting for the actual response of the damper assembly, the *temperature* method of calculating the volume of ventilation air demonstrated in the previous subsection can be used in DDC applications to control the dampers. In this closed loop strategy, a proportional or PI control loop algorithm is used to modulate the dampers to their minimum position instead of using fixed signal positioning. The percent outside air formula is input into the application program. The mixed air temperature needed to supply the correct percentage of outside air is calculated every 15 minutes using the current outside and return air temperatures. The set point is entered into the controller's transfer function and the dampers are *modulated* to maintain the control point near set point. The modulating controller continues to adjust its output signal until the control point equals the calculated minimum position set point. If the dampers are bound closed, affected by wind, rubbing against the sides of the frame, etc., the control signal continues to increase until they break free, insuring the correct amount of ventilation air is being brought into the AHU.

Example 15.1

> A 20,000 cfm application requires a minimum of 3600 cfm of ventilation air. The outside air dampers must provide 18% ($3600 \div 20000$) of the fan's capacity during all occupied periods. The microprocessor multiplies the outside air temperature by 0.18 and adds the result to the return air temperature times 0.82. Adding these to values together determines the mixed air set point required to insure 18% of the mixed airstream consists of ventilation air. The mixed air temperature set point is input into the controller's transfer function algorithm and the dampers modulate to maintain 18% outside air flow. Recalculating the set point periodically (every 15 minutes) insures the minimum position is maintained as the outside and return air temperatures change.

This strategy is easily performed by the microprocessor in a DDC system and increases the probability that the correct amount of ventilation air is being brought into the system. This closed loop strategy works well as long as there is an adequate temperature difference (5° minimum) between the outside and return air streams. When the temperature difference decreases below 5°, the positioning of the dampers defaults to the open loop method to maintain response stability. Combining these two strategies increases the amount of time the proper quantity of ventilation air is being brought into the system.

A similar minimum position strategy can be performed using carbon dioxide sensors in place of temperature sensors to calculate the quantity of ventilation air in the mixing plenum. The CO_2 level in the return air and the outside air streams replaces the temperature variables in the percent outside air formula and a CO_2 set point is calculated. A modulating controller algorithm positions the dampers based upon the set point and the CO_2 present in the mixed air. This strategy also works well as long as there is an adequate difference between the CO_2 levels in the outside and return air streams. Presently, the increased cost of CO_2 sensors outweighs any advantages of using this strategy over the temperature method.

15.2.3 Separate Minimum Position Dampers

Some air handling units are constructed with two separate damper assemblies in the outside air plenum. In these applications, one of the dampers is a modulating, opposed blade configuration used for mixed air temperature control and the other is a two-position parallel blade device used for minimum position control. Upon start-up of the AHU, a binary signal is sent to the two-position air flow damper commanding it open. The modulating opposed blade dampers remain closed whenever the unit is calling for heat, the outside air temperature or enthalpy exceeds its high limit set point or when the mixed air temperature control loop tries to modulate the dampers below their minimum ventilation position. The mixed air temperature loop modulates the opposed blade dampers to maintain the mixed air temperature set point. This operation differs from other damper systems because in these applications, the modulating dampers are allowed to close completely. It is not necessary to maintain the dampers at a minimum position because the two-position damper supplies the necessary ventilation air into the system whenever the fan is operating.

In these applications, the minimum position parallel blade damper is set up using the open loop positioning procedure presented in section 15.2.1 to insure that excess amounts of air are not being brought into the system. The parallel blade damper may only need to open a small percentage of its actuator stroke before a sufficient quantity of ventilation air into the system. If the damper is allowed to open 100%, excess air is most likely being brought into the building.

15.3

MIXING DAMPER ECONOMIZER STRATEGY

After the mixed airstream leaves the plenum, it enters the supply air temperature process. Whenever the mixed air temperature is warmer than the supply air temperature set point, the airstream must be cooled before it can be discharged to the zones. The excess heat is extracted from the mixed air by the heat transfer coil of a mechanical cooling system (chilled water or direct expansion refrigeration) and released outside of the building by its condensing unit. The amount of energy consumed by the condensing unit can be reduced by minimizing the enthalpy of the mixed airstream.

In earlier mixing damper control strategies, the outside air dampers were commanded to their minimum position as soon as the outside air temperature exceeded the supply air temperature set point. The rational for this decision was that the chiller was operating and by commanding the outside air dampers to their minimum position, the amount of warm air entering the unit was reduced and therefore, saved chiller energy. This logic was flawed because the energy content of the air streams was not considered in the decision to command the dampers to their minimum position.

Outside air is always less expensive to cool when its energy content is lower than the return air's. This remains true no matter how high the outside air dry bulb temperature is. The objective of an *economizer* cycle is to decrease the amount of energy consumed by the mechanical cooling equipment by allowing the outside air dampers to remain 100% open as long as the outside air's enthalpy is lower than the return airstream's enthalpy.

During periods of operation when the outside air's sensible and latent heat content (enthalpy) is *less than* the enthalpy of the return airstream, the outside air is less expensive to cool. This is true even when the outside air's dry bulb temperature is greater than the return air's dry bulb temperature because enthalpy is a measure the total energy content of the air (temperature + moisture), not just the sensible content. The economizer strategy is only operative when:

1. The outside air temperature exceeds the *supply* air temperature set point.
2. The outside air's enthalpy is *lower* than the return air's enthalpy.

During all other periods of operation the mixed air process is maintained using the mixed air temperature control strategy.

15.3.1 Methods of Enabling the Economizer Control Strategy

During the cooling season there are periods of time when the outside air is more expensive to cool than the building's return air. During these intervals, the economizer strategy must be disabled to minimize the cost of mechanical cooling thereby increasing the operating efficiency of the process. There are two methods used to disable the economizer strategy. The first method measures the enthalpy of the outside and return air streams to determine whether the mixing dampers should be allowed to operate in economizer mode. This control strategy is called an *enthalpy based economizer cycle*. The second technique monitors the dry bulb temperature of the outside airstream to measure its sensible heat content. The scheme is called a *dry bulb economizer cycle*. Both strategies permit the airstream with the lower energy content to pass across the cooling coil, minimizing the amount of heat the mechanical system must remove from the airstream. The following subsections outline the response of these two control strategies.

15.3.2 Enthalpy Based Economizer Cycle

An enthalpy based economizer strategy requires four sensors to measure the enthalpy of the two air streams. Each airstream uses a dry bulb temperature sensor (thermistor, RTD, pneumatic bulb) and a sensor that measures the relative humidity or dew point of the airstream. In DDC applications, the microprocessor calculates the enthalpy of both air streams. Whenever the outside air's enthalpy is greater than the return air's enthalpy, the economizer strategy is disabled, commanding the outside and exhaust air dampers to their minimum position and the return air dampers will move to their corresponding open position (100% - minimum position).

Whenever the outside air's enthalpy is less than the return air's enthalpy the outside air dampers are allowed to modulate to 100% open and the return air dampers modulate closed. The outside air dampers will remain open until the enthalpy of the outside air exceeds that of the return air.

Figure 15.2 shows the enthalpy measuring sensors (enclosed with a dashed box) and their position within the air handling system. The outside air temperature

sensor and a humidity sensor are typically located on the exterior of a north wall
to prevent the sun from affecting the temperature sensor's ability to correctly measure the thermal intensity of the outside air. These sensors can also be mounted
inside the outside air plenum of the air handling unit, upstream of any coils. In
DDC applications, the outside air temperature and humidity sensors are typically
configured as a global or shared points. Therefore, only one set is required to measure the outside condition. All other systems requiring outside air data reference
the memory location of these sensors to obtain the shared information. The return
air temperature and humidity sensor is located in the return duct of the air handling system, upstream of the return air dampers. Since each air handling unit
experiences different internal thermal and moisture loads, a set of enthalpy sensors must be installed in each air handling unit that incorporates the enthalpy
based economizer strategy.

15.3.3 Dry Bulb Temperature Economizer Cycle

To obtain most of the savings associated with an enthalpy based economizer strategy without incurring the additional cost of the humidity measuring sensors, the
economizer cycle can be enabled and disabled based upon the difference between
the dry bulb temperatures of the outside and return air streams. In this strategy,
the outside air dampers are allowed to modulate between minimum position and
100% open whenever the outside air temperature is *less than* a predetermined high
limit set point. Once the outside air temperature exceeds the high limit set point,
typically 65 to 75°, the economizer strategy is disabled. The outside and exhaust
air dampers are commanded to their minimum ventilation position and the return

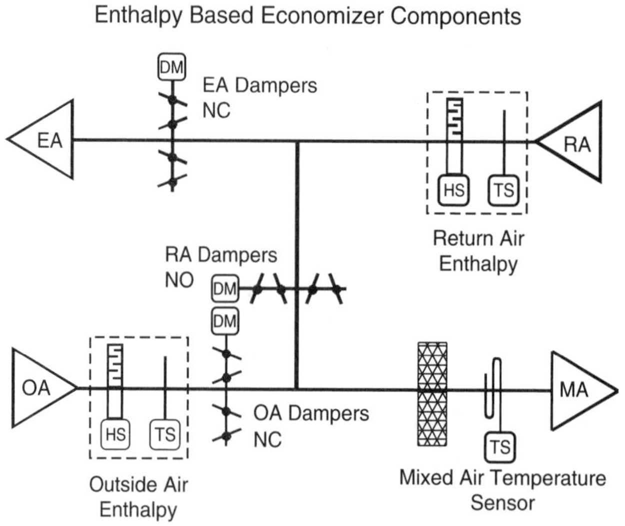

Figure 15.2 Enthalpy Economizer Sensors

air dampers will open proportionately. Figure 15.3 depicts the response of a dry bulb temperature economizer.

A psychrometric chart is used to correctly establish the high limit dry bulb temperature of the outside air that corresponds with the expected return air enthalpy conditions. By comparing the expected return air enthalpy with the average outside air humidity conditions that occur during the summer, a dry bulb temperature can be selected that allows effective economizer operation during most operating conditions. Figure 15.4 shows a schematic portrayal of a section of a psychrometric chart showing typical conditions for the mid-northern latitudes. To determine the temperature at which the economizer cycle should be disabled, the expected return air conditions are plotted on the chart. Figure 15.4 shows the return air conditions at $80°$ and 40% relative humidity, yielding and enthalpy of 29 btu/lb of air (solid lines). The high limit outside air dry bulb temperature is found by drawing a vertical line at the intersection of the typical outside air relative humidity and the return air enthalpy line (dashed line). If the average relative humidity is 70%, the intersection at the 29 btu/lb return air conditions line occurs at $70°$. This is the high limit temperature set point. It is the maximum outside air dry bulb temperature that can be brought into the air handling unit that still has a lower energy content than the return airstream. To increase the reliability of this method, a few degrees are subtracted from the $70°$ to account for higher morning humidity values and sensor drift. Typically, the high limit set point ranges between $60°$ and $70°$ in areas with 60 to 80% summer humidity readings. Figure 15.5 shows the components of a dry bulb economizer strategy.

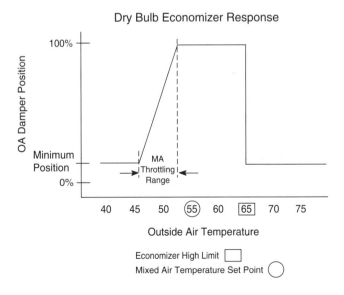

Figure 15.3 Dry Bulb Economizer Response

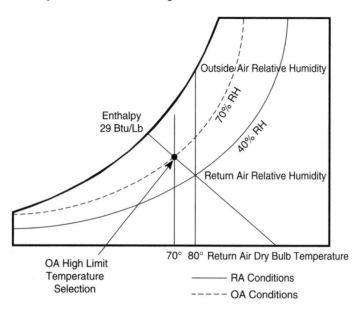

Figure 15.4 The Psychrometric Chart Used to Determine the Economizer High Limit Set Point

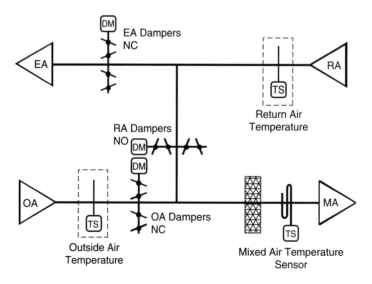

Figure 15.5 Dry Bulb Temperature Based Economizer Components

15.4

COMMANDING DAMPERS TO THEIR NORMAL POSITION

The outside, return and exhaust air dampers are commanded to their *normal position* whenever the air handling unit's fan is commanded off. This prevents cold outside air from migrating into the building during periods of scheduled unoccupancy and emergencies. Commanding the dampers to normal position also limits the development of natural convection currents within the building. These currents remove conditioned building air which is replaced by unconditioned infiltration air.

The mixing dampers remain in their normal position during warm-up strategies when there are no occupants in the building. Under these circumstances, there is no ventilation air required because there are no people expected in the building during the pre-occupied warm up operation. The mixing dampers should be enabled a short time before occupancy to flush the building of fumes and odors that may have accumulated in the air during the unoccupied period.

15.5

ENTHALPY-BASED ECONOMIZER LOOP DESIGN AND OPERATION

15.5.1 Enthalpy-Based Economizer Control Loop Components

The enthalpy-based economizer control strategy requires the hardware components shown in Table 15.1

Table 15.1 ENTHALPY-BASED ECONOMIZER CONTROL LOOP COMPONENTS

Number	Quantity	Description	Signal Type
1	1	OA Temperature Sensor (−20 to 120 °F)	Analog Input
2	1	MA Temperature Sensor (20 to 100 °F)	Analog Input
3	1	RA Temperature Sensor (40 to 100 °F)	Analog Input
4	2	OA and RA Humidity Sensors (0 to 100%)	Analog Input
5	1	DDC Controller Panel w/ DI, BI, DO & BO	All
6	3	OA, RA and EA Damper Actuators (4 to 20 mA)	Analog Output

15.5.2 System Drawing

A System drawing for an enthalpy-based economizer strategy is shown in Figure 15.6.

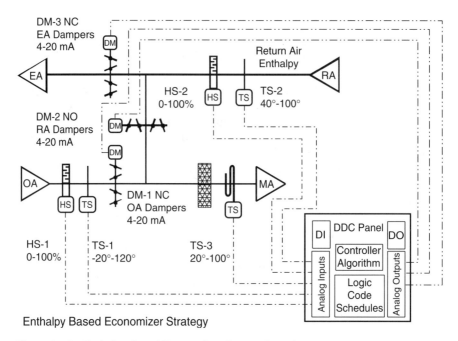

Enthalpy Based Economizer Strategy

Figure 15.6   Enthalpy Based Economizer System Drawing

15.5.3 Sequence of Operation

The sequence of operation for the enthalpy based economizer strategy illustrated in Figure 15.6 is explained in this section.

Overview

The mixing dampers shall modulate to maintain the mixed air temperature at a fixed set point of 56 °F at the sensor's location.

Energy Management Strategies

The mixing dampers economizer logic is enabled based upon the outside and return air enthalpies to minimize the mechanical cooling required to condition the supply airstream.

Pre-Occupied Conditioning

Warm-Up: When the air handling unit is in the morning warm-up mode the mixing dampers (DM-1, DM-2 & DM-3) shall remain in their normal position (4 mA).

Purge: When the air handling unit is in a summer purge mode, the outside and exhaust dampers are commanded 100% open and the return air dampers shall be commanded closed to allow the cooler outside air to purge the building (20 mA).

Occupied Operation

When the air handling unit is commanded into occupied mode, the mixed air temperature control logic is enabled.

Damper Operation During the Heating Mode Operation: When the air handling unit is in heating mode, the mixing dampers shall be modulated at minimum position based upon the mixed air temperature as measured by sensor TS-3 to maintain a 20% ventilation requirement in the zones.

Damper Operation During the Economizer Mode Operation: When the air handling unit is in its cooling mode, the mixing dampers are allowed to modulate beyond their minimum position to maintain the mixed air temperature at set point (56 °) as measured by sensor TS-3.

The direct acting, proportional + integral controller algorithm generates a 4 to 20 mA output signal to modulate the NC outside and exhaust dampers (DM-1, DM-3) along with the NO return air dampers (DM-2) to maintain the mixed air temperature set point of 56 °F.

As the mixed air temperature increases, the signal to the mixing damper actuators also increases, modulating the outside and exhaust dampers toward 100% open and modulating the return air dampers proportionately closed.

Disabling the Economizer Mode Operation: The DDC system calculates the outside enthalpy every 15 minutes based upon sensors TS-1 and HS-1 and the return air enthalpy every 15 minutes based upon the input from sensors TS-2 and HS-2. Whenever the outside air's enthalpy exceeds the return air's enthalpy the mixing dampers are modulated at their minimum position based on the temperature measured by TS-3 to maintain a 20% ventilation requirement in the zones.

Unoccupied Operation

When the air handling unit fan is commanded off, the mixing dampers (DM-1, DM-2 & DM-3) are commanded to their normal positions (4 mA).

Emergency Mode Operation

Whenever the air handling unit's fan is commanded off due to a mixed air low temperature limit trip, fan failure to prove operation, smoke detector trip or other system fault, the mixing dampers shall be commanded to their normal position (4 mA).

15.5.4 Flowchart

Figure 15.7 is a flowchart of the enthalpy based economizer strategy sequence of operation.

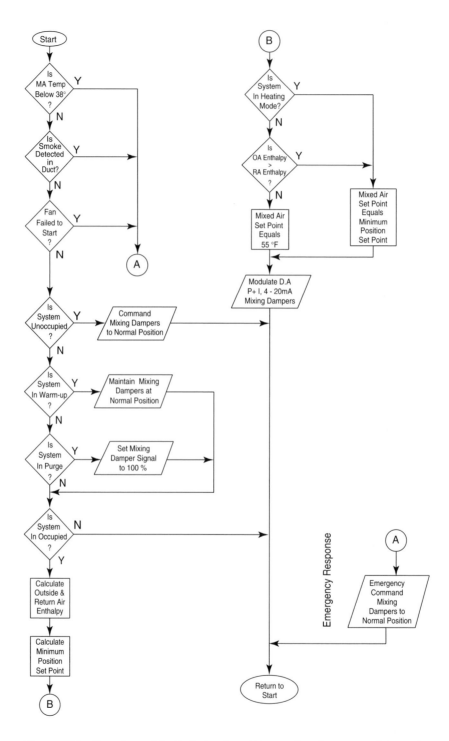

Figure 15.7   Flowchart of the Enthalpy Based Economizer Sequence of Operation

15.5.5 Flowchart Analysis

The *start* oval indicates the beginning of the sequence of operation. The three decision diamonds following the start symbol determine whether any system fault is present that requires the operation of the mixing dampers to be disabled. If the temperature of the air downstream of the heating coil is less than 36° or smoke is detected in the return air duct or the supply fan is commanded on but it fails to prove it is operating, the mixing dampers are commanded to their normal position and an alarm is generated at the work station computer. When all faults have cleared, the next decision diamond instructs the DDC system to determine whether the system is in its unoccupied mode of operation.

If the system is in its unoccupied mode, the dampers are commanded to their normal position and remain there until the system enters its pre-condition mode. Once the pre-condition mode begins, the system fan is commanded on and the DDC system determines whether the building requires a warm-up or purge cycle. If the system is commanded to warm-up, the mixing dampers remain in their normal position. The supply air temperature control loop maintains a higher set point to supply warm air to the zones. If the zones are above their set points and the outside air temperature is below the zone temperature, the system can purge the building by commanding the outside and exhaust dampers fully open and closing the return air dampers. This permits all the warm air to be exhausted from the building and be replaced with cooler ventilation air. Once the warm up or purge cycle is complete, the system enters its occupied mode of operation.

During the occupied mode, the minimum position mixed air temperature set point and the enthalpies of the outside and return air streams are calculated every 15 minutes. If the supply air temperature control loop is calling for heat or the outside air enthalpy is greater than the return air enthalpy, the minimum position mixed air temperature set point is used for the mixed air temperature control loop. Under all other operational conditions the set point remains at 56° in accordance with the sequence of operation. The end of the flowchart shows the mixing dampers are modulated using the correct set point and the returns to start and the sequence of operation is repeated.

15.6

SUMMARY

Mixed air temperature control is commonly called the economizer strategy because it minimizes the cost of cooling air with mechanical equipment. By allowing the lower energy containing airstream to pass across the cooling coil, the quantity of heat that must be removed from the air is reduced along the energy required by the mechanical system.

To insure that the lower energy airstream is being cooled, the economizer cycle is disabled by either a dry bulb temperature high limit set point or a comparison of the outside and return air enthalpies. Once the outside air exceeds the high limit set point, it is less costly to condition the return airstream. When the outside air is more costly to condition, they are commanded to their minimum position. Minimum position can be maintained using a proportional or proportional + integral control algorithm to modulate the mixing dampers to maintain an adjustable mixed air set point. The set point is a function of the amount of ventilation air required to satisfy the process and the temperatures of the outside and return air streams. Modulating the dampers by this method overcomes many of the problems associated with open loop positioning methods that base flowrate on the control signal magnitude. Even if modulating is not possible or feasible, the open loop minimum position is still determined based on temperatures. A positioning relay can be added to the mixing damper actuators to minimize some of the problems associated with friction in the damper assembly when using an open loop minimum position strategy.

15.7

EXERCISES

Determine if the following statements are true or false. If any portion of the statement is false, the entire statement is false. Explain your answers.

1. Mixed air temperature control strategies vary the amount of outside, return and exhaust air entering the mixing plenum to maintain the mixed air temperature.
2. The outside and exhaust air dampers are normally closed devices.
3. When the dampers are commanded to their minimum position, enough outside air must enter the mixing plenum to maintain its temperature set point.
4. When the outside air temperature is below the return air temperature, the enthalpy based economizer cycle is always enabled.
5. Air handling units with two-position minimum outside air dampers can under certain conditions, close the modulating dampers 100% during occupied time.
6. In mixed air temperature applications that do not incorporate an economizer strategy, the outside air damper modulates to minimum position whenever the outside air temperature exceeds the mixed air temperature set point.
7. The dry bulb economizer strategy operates the mixing dampers whenever the outside air temperature is below the high limit temperature set point.
8. When the mixed air temperature strategy is operational and the outside air temperature increases, the output signal from the controller will decrease,

opening the outside and exhaust dampers and proportionately closing the return air dampers.

9. Enthalpy based economizer strategies require temperature and humidity sensors to determine the energy content of an airstream.

10. The mixing dampers are commanded to minimum position whenever the supply air temperature loop is calling for cooling.

Respond to the following statements, questions and problems completely and accurately, using the material found in this chapter.

1. How does an economizer strategy increase the operating efficiency of a system in comparison to a basic mixed air temperature control strategy?

2. Why can a closed loop minimum position strategy maintain minimum position better than an open loop strategy?

3. Why is proper damper sizing important to the modulation of the mixing dampers?

4. Calculate the percentage of outside air when the following temperatures exist:

OA Temperature	RA Temperature	MA Temperature	% Outside Air
68°	78°	68°	
45°	74°	58°	
	75°	56°	35%

5. What procedure is used to calibrate the minimum position signal for mixing dampers using an open loop strategy?

6. If during the calibration procedure for determining the minimum position of an outside air damper the mixed air temperature is too high, based upon the calculated value, must the actuator signal to the mixing dampers be increased or decreased?

7. Draw a flowchart that represents a modulating control loop's response to an error.

8. Under what operating conditions are the return air dampers completely closed.

9. Using the line diagram below and the material presented in the chapter, develop the following documents for the dry bulb economizer process: (1) a parts list; (2) a sequence of operation; (3) a flowchart

10. Using the material presented in the chapter, develop the following documents for an enthalpy based economizer process for an air handling unit that has two damper racks, a two-position minimum position damper and a modulating mixed air temperature damper. Use the system drawing in Figure 15.8 as a reference, develop a:
    1. A parts list.
    2. A sequence of operation.
    3. A flowchart.

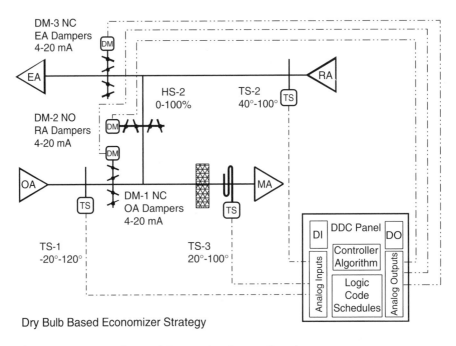

Dry Bulb Based Economizer Strategy

Figure 15.8 Dry Bulb Based Economizer System Drawing

CHAPTER 16

Heating Control Strategies

All heating control strategies vary the amount of sensible heat that is being transferred into a controlled medium to maintain their process requirements. Closed loop applications maintain heating set points in air handling units and zones. Open loop strategies are used to prevent a process temperature from falling below its low limit set point. The following sections detail the operation of the more common HVAC heating processes.

16.1

OVERVIEW OF THE DIFFERENT HEATING CONTROL STRATEGIES

Preheating, heating and reheating are three categories of heating processes found in air handling unit applications. All of these processes generate similar responses to changes in their measured variable but are designed to perform different functions. *Preheat applications* are implemented to warm the outside air as it enters an air handling unit to temperatures in the vicinity of 50°. When the air is preheated to a temperature that serves the cooling requirements of the air handling unit, the air is described as being *tempered*, not heated. The ventilation air is tempered to prevent it from freezing water or condensate inside coils located in the air handling unit. *Heating applications* are control strategies where the air entering the heating coil is already about 50° and additional heat is needed to raise the supply air temperature to a set point above 90°. *Reheat applications* use small coils located in the building's zones to add additional heat to a zone to maintain its temperature set point.

307

The thermal energy used in multiple zone commercial heating processes may be supplied by steam, hot water or electrical current. In these larger systems, a boiler is used to generate the steam or hot water used to transport energy to the heating processes. Steam transfers its *latent* heat as it condenses, raising the temperature of the air passing across the coil. The condensate (condensed steam) is piped back to the boiler where more energy is added to the liquid converting it back into steam. Hot water can be generated with a boiler or through the use of a steam to hot water heat exchanger (converter). Hot water is piped to a process where it transfers *sensible* heat to the air passing across the coil. The control agent in both steam and hot water applications is modulated using a control valve.

Electric current can be pulsed or modulated into a process using contactors or modulated using solid state thyristor switches. Thyritors vary the amount of current entering an electric coil much like a valve modulates steam or hot water. Additional binary safety controls are required in all electric coil installations to monitor air flow and provide a high temperature limit to prevent overheating the area surrounding the coil. Electric coils have an advantage over steam and hot water coils in preheat applications because they never freeze. The disadvantage of specifying electric coils is the higher cost of electrical energy compared to natural gas or oil. The following sections describe the control strategies used to modulate hot water or steam into a heat transfer coil.

16.2

PREHEAT CONTROL STRATEGIES

Preheat coils differ from regular heating coils in their function and location. The three primary differences between the two processes are:

1. Preheat coils raise the temperature of large quantities of outside air being brought into an air handling unit to meet the ventilation requirements of the system.

2. Preheat control strategies temper the ventilation air entering the air handling unit, raising its temperature to approximately 50 °F in contrast to heating coils that raise the air temperature to levels above 90 °F. Tempering describes a heating process where the coil adds sufficient heat to the ventilation air stream to bring its temperature up to 50 to 60 °F. This air is used to offset the cooling load in interior zones.

3. Preheat coils are typically located upstream of the air handling unit's filters, in the outside airstream. Heating coils are typically positioned downstream of the filters in order to heat the mixed airstream.

Preheat coils are used in applications where the temperature of the mixed airstream drops below 40° during normal operation. This condition is likely to occur in applications where a large volume of outside air is being drawn into the unit to satisfy the ventilation requirements of the system. Preheat coils are also required in air handling units that condition 100% outside air. These units have no return air ducts. All of the conditioned air is exhausted from the building and replaced with outside air. In some single zone air handling unit applications, the preheat and heating functions can be performed by one coil incorporating face and bypass dampers. This application is shown in section 16.3.2.

The purpose of a preheat coil control strategy is two-fold. Its first function is to prevent the preheat coil from freezing, the second is to prevent cold outside air from freezing any equipment or components within the air handling unit or duct systems. If the preheat control strategy fails, the entire mechanical enters an emergency mode of operation and must shut down to minimize possible damage.

16.2.1 Two-Position Preheat Control Strategy

A two-position strategy opens the control valve 100% whenever the outside air temperature approaches freezing.

Two-position preheat control is an open loop strategy. The sensor is located upstream of the coil so it is unable to measure the amount of heat actually being added to the airstream by the preheat coil. The sensor measures the outside air temperature before it enters the air handling unit. The controller is calibrated to maintain the ventilation air's temperature above a *low limit* temperature set point. The controller opens the steam or hot water valve whenever the outside air temperature falls below the low limit set point (36 to 40 °F). Once the valve is commanded open, the coil operates at 100% capacity until the outside air temperature rises above the set point plus one-half of the control differential (38 to 42 °F). This simple strategy prevents the preheat coil and other system components from freezing when the outdoor air temperature is below 32 °F.

The two-position response of this strategy creates temperature control problems as the outside temperature rises above 25 °F. A preheat coil must be sized with sufficient capacity to temper all the air entering the unit during winter design temperature conditions. Consequently, the coil only tempers the incoming air when the outside air temperature is equal to or below the winter design temperature. If the design winter temperature is 0° and the coil is sized to produce a 55° rise in the air's temperature. The coil will add too much heat to the air during all other periods of operation. Consequently, when the outside temperature approaches 38°, the preheat discharge air temperature may exceed 90°. That temperature is too warm for zones that still require cooling from the source of tempered air.

An open loop preheat control strategy will prevent coils from freezing but it also consumes excess energy during most of its operation, reducing the operational efficiency of this type of process. The excessively warmed mixed airstream may require mechanical cooling to lower the temperature of the preheated air for zones requiring cooling. These disadvantages outweigh the advantages gained by the loop's simplicity in installation, calibration and operation.

16.2.2 Closed Loop Preheat Control

To retain the freeze protection characteristic inherent in a two-position coil strategy and also improve the efficiency and temperature control of the preheat process, a face and bypass damper assembly can be used in conjunction with the preheat coil. This strategy alters the number of control loops required to maintain the preheat discharge air temperature from one (open loop) to two (closed loop). Face and bypass dampers are installed on the upstream side of the preheat coil to regulate the amount of air that passes through the coil. Figure 16.1 shows a rendering of a face and bypass damper assembly.

The face and bypass damper assembly consists of a set of normally closed dampers stacked above of a set of normally open opposed blade dampers. The normally open dampers are used to vary the flow of air across the coil surface. For this reason the face dampers are also known as the coil dampers. The air that does not pass across the preheat coil is bypassed over the top of the coil through the normally closed bypass dampers. The face and the bypass dampers are linked together to one actuator shaft so one signal controls both sets of dampers. As the signal to the actuator increases, the normally open coil dampers modulate toward their closed position as the normally open bypass dampers open.

A face and bypass damper assembly permits control over the air temperature leaving the coil that is otherwise impossible with a two-position control valve. During operation at design winter temperatures, the face dampers are 100% open and the bypass dampers are 100% closed. All the air passes across the coil, as it would in an open loop control strategy. The coil is sized to permit the proper temperature rise so the control valve could also be 100% open. As the outside air temperature begins to increase above the design winter outside air temperature, the

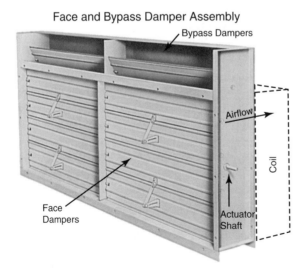

Face and Bypass Damper Assembly

Figure 16.1 Face and Bypass Damper Assembly

face dampers begin to modulate closed and the bypass dampers modulate proportionately open. The air bypassing the coil does not receive any heat from the coil. This cold air mixes with the warm air downstream of the preheat coil where the closed loop temperature sensor is located. The blending of the unheated bypass air with heated air allows the temperature of the mixture to be maintained at its set point. As the volume of bypassed air in the mixture increases, the preheat coil's discharge temperature decreases.

To prevent the preheat coil from freezing, the same open loop control strategy used for a two-position loop is employed. Whenever the outside air temperature falls below the low limit set point (38 to 40 °F), the control valve opens 100% to prevent the coil from freezing. When the valve is commanded wide open, the face and bypass control loop maintains the downstream temperature at set point under all load conditions. When the outside air temperature is above the low limit set point, the face and bypass dampers are commanded to their normal position and the control valve is modulated to maintain the preheat discharge temperature set point.

16.3

ILLUSTRATION OF A STEAM PREHEAT CONTROL STRATEGY

This example illustrates a steam preheat process for a 100% outside air AHU that uses a steam coil with integrated with face and bypass dampers and a modulating steam valve.

16.3.1 Steam Preheat Control Loop Components

The steam preheat control strategy incorporating face and bypass dampers and a modulating steam control valve requires the hardware components shown in Table 16.1.

Table 16.1 STEAM PREHEAT CONTROL LOOP COMPONENTS

Number	Quantity	Description	Signal Type
1	1	OA Temperature Sensor (−20 to 120 °F)	Analog Input
2	1	Preheat Temperature Sensor (20 to 180 °F)	Analog Input
3	1	DDC Controller Panel	All
4	1	Steam Valve and Actuator (0 to 10 volts)	Analog Output
5	1	Parallel Blade Damper & Actuator (0 or 10 volts)	Digital Output
6	1	Face & Bypass Damper w/ Actuator (0 to 10 v)	Analog Output
7	1	Damper End Switch	Digital Input
8	1	Low Limit Temperature Control *	Digital Input

*The binary input from the low temperature limit controller should be used as an indication only. Additional contacts inside the device must be electrically interlocked with the fan's control circuit to disable the fan whenever the temperature of the air downstream of the preheat coil falls below 36 °F.

16.3.2 System Drawing

A system drawing for a steam coil with integrated face and bypass dampers process is shown in Figure 16.2.

16.3.3 Sequence of Operation

The sequence of operation for the steam preheat coil with face and bypass dampers is illustrated in Figure 16.2 is explained in this section.

Process Overview

The preheat steam valve and face and bypass dampers are modulated in sequence to maintain the preheat discharge air temperature at a set point of 55 °F.

Occupied Operation

Whenever the air handling unit's fan is commanded on, the two-position NC outside air damper assembly (DM_1) is commanded 100% open. When the damper end switch (ES_1) closes, indicating the dampers have opened, the fan is commanded on.

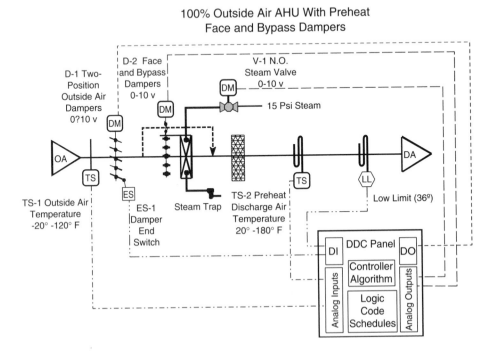

Figure 16.2   Steam Preheat With Face and Bypass Dampers

Outside Air Temperature Less than 40 °F   When the outside air temperature is below 40° as sensed by the outside air sensor (TS_1), the preheat coil steam valve (V_1) is commanded 100% open. A direct acting, proportional + integral control algorithm generates a 0 to 10 volt output signal to modulate the NO face and NC bypass dampers (DM_2) in unison to maintain a preheat discharge air temperature set point of 55° at the location of temperature sensor TS_2.

As the preheat coil's discharge air temperature increases, the signal to the face and bypass damper actuator increases, opening the bypass damper and proportionately closing the face damper. This response lowers the temperature of the mixture downstream of the coils.

Outside Air Temperature Greater Than 40° and Less Than 55 °F   When the outside air temperature is above 40° as sensed by the outside air sensor (TS_1), the face and bypass dampers are commanded to their normal position (0 volts) thereby opening the face damper and closing the bypass damper. The face dampers remain fully closed and all the air entering the unit passes through the bypass damper. A direct acting, proportional + integral control algorithm generates a 0 to 10 volt output signal to modulate the NO steam control valve (V_1) to maintain a preheat discharge air temperature at its 55° set point.

As the preheat coil's discharge air temperature decreases, the signal to the control valve's actuator decreases, modulating the valve open to raise the preheat discharge air temperature.

Outside Air Temperature Greater Than 55 °F   When the outside air temperature is above 55° as sensed by the outside air sensor (TS_1), the face and bypass dampers are commanded to their normal position (0 volts) opening the face damper and closing the bypass damper. The steam control valve (V_1) is also commanded closed (10 v) to disable the preheat process.

Unoccupied Operation

Outside Air Temperature Greater Than 40 °F   When the air handling unit's fan is commanded off, the outside air dampers and the face and bypass dampers (DM_1, DM_2) are commanded to their normal positions (0 volts). If the outside air temperature is above 40°, the steam valve (V_1) is also commanded closed (10 volts).

Outside Air Temperature Less Than 40 °F   When the outside air temperature falls below 40°, the steam valve (V_1) is enabled to prevent the coil from freezing. A direct acting, proportional + integral control algorithm generates a 0 to 10 volt output signal to modulate the NO steam valve (V_1) to maintain an unoccupied air temperature set point of 70° at the location of sensor TS_2.

Emergency Operation

Whenever the air handling unit fan is commanded off due to the fan's failure to prove operation, a smoke detector indicating smoke or a low temperature limit trip as sensed by TS_3, the outside dampers, face and bypass dampers and steam valve are commanded to their normal or failsafe positions.

16.3.4 Flowchart

A flowchart depicting the logic in the sequence of operation for the system shown in Figure 16.2 is shown in Figure 16.3. Section 16.3.5 provides an analysis of this flowchart.

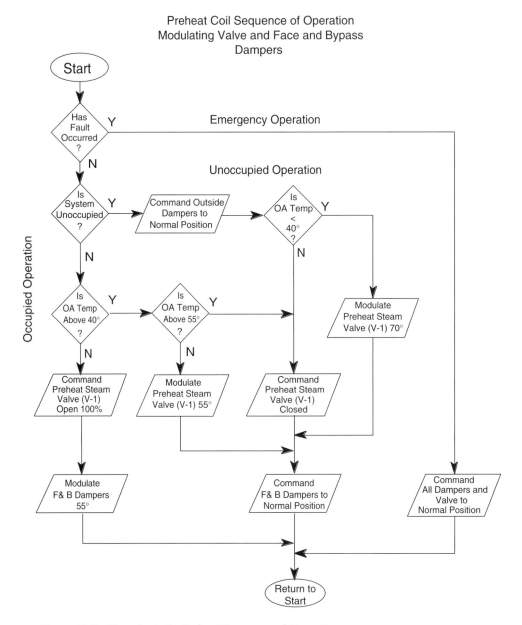

Figure 16.3   Flowchart of a Preheat Sequence of Operation

16.3.5 Flowchart Analysis

The start oval indicates the beginning of the logic of sequence of operation for this preheat process. The first decision diamond determines whether the system has experienced an emergency or fault condition. If it has, the program bypasses all other logic and commands the process control devices to their normal position. The program remains in this loop until the fault condition is cleared.

The second decision diamond determines if the system is in its unoccupied mode of operation. If it is, the outside air dampers are commanded to their normal position and remain there until the system commences its occupied mode. Another decision diamond determines whether the outside air temperature falls below 40° during the unoccupied period. Under these circumstances, infiltration of air through the outside air dampers may pose a freezing problem to the system. To minimize these problems, the steam valve is modulated to maintain a temperature of 70° within the duct.

During the unoccupied period, the control valve is commanded closed whenever the outside air temperature rises above 40°. The face and bypass dampers are commanded to their normal position and the program returns to start.

When occupied mode begins, the outside air dampers are commanded open. Once the end switch makes, the fan is commanded on. A decision diamond determines whether the outside air temperature is below 40°. Under these conditions, the steam control valve is commanded 100% open and the face and bypass dampers are modulated to maintain the preheat discharge temperature at 55°.

When the outside air temperature is above 55°, the steam valve is commanded closed. If the outside air temperature is above 40° and below 55°, the face and bypass dampers are commanded to their normal position and the steam valve is modulated to maintain 55°.

16.3.6 Strategy Summary

Configuring a preheat control loop with a modulating control valve and face and bypass dampers improves the response of the process in comparison to a two-position control valve configuration. To protect the air handling unit's coils from freezing when the outside air temperature drops below 32°F, the preheat valve is commanded to 100% open in similarity with the two-position strategy. If the control valve was allowed to modulate, the condensate in the preheat coil may freeze at lower steam flowrates. By opening the valve and modulating the face and bypass dampers to control the discharge air temperature, the coil is protected and the temperature is maintained at its desired set point.

16.4

ILLUSTRATION OF A HOT WATER PREHEAT CONTROL STRATEGY

This example illustrates a hot water preheat process for an AHU that uses a hot water coil with a three-way modulating valve and coil circulating pump. The preheat process is integrated with the mixing dampers to maintain the mixed air temperature set point at 55°.

Hot water coils can be used in place of steam coils to provide the thermal energy requirements for preheat applications. In these applications, the hot water valve is configured to modulate the flow of hot water into the coil to maintain the discharge air temperature under all load conditions. Therefore, no face and bypass dampers are required. To prevent the preheat coil from freezing, a secondary circulating pump is used to maintain a full load flowrate of variable temperature water in the coil at all times. The pump sustains a fluid velocity in excess of three feet per second, which prevents the water from freezing at temperatures well below 0 °F.

The coil circulating (secondary) pump also maintains a constant temperature across the coil's face. This prevents the formation of stratified levels of water within the coil. If colder layers were allowed to form at the bottom of the coil, they could freeze when exposed to ventilation air temperatures below freezing. The hot water preheat coil control strategy generates the same response as a steam preheat application that uses a modulating valve and face and bypass dampers. The only difference between the strategies is the hot water system replaces the face and bypass dampers with a modulating control valve and secondary circulating pump.

In air handling units that require large volumes of ventilation air but also recirculate return air, the hot water preheat coil's response is integrated with the normal mixing air temperature control scheme. The combined process maintains a mixed air temperature set point instead of a preheat discharge temperature set point to increase the overall operating efficiency of the air handling unit. Integrating these two processes limits the amount of energy added by the preheat coil to the amount required to maintain the mixed air temperature set point. Therefore, when larger amounts of return air are being recirculated, the temperature of the air leaving the preheat coil is allowed to drop below 50° to prohibit the necessity of enabling the mechanical cooling system.

16.4.1 Hot Water Preheat Control Loop Components

The hot water preheat control strategy incorporating a modulating valve and a secondary circulating pump requires the hardware components shown in Table 16.2.

16.4.2 System Drawing

A system drawing for a hot water preheat control process is shown in Figure 16.4.

Table 16.2 **HOT WATER PREHEAT CONTROL LOOP COMPONENTS**

Number	Quantity	Description	Signal Type
1	1	Mixed Air Temperature Sensor (40 to 100 °F)	Analog Input
2	1	DDC Controller Panel	All
3	1	Outside Air Temperature Sensor (−20 to 120 °F)	Analog Input
4	1	Modulating 3 Way Hot Water Valve (4 to 20 mA)	Analog Output
5	1	Circulating Pump Start/Stop Contactor	Binary Output
6	1	NO Return Air Dampers (4 to 20 mA)	Analog Output
7	1	NC Outside Air Dampers (4 to 20 mA)	Analog Output
8	1	Low Temperature Limit Controller (36°)	Binary Input

Hot Water Preheat Coil Integrated With Mixing Dampers

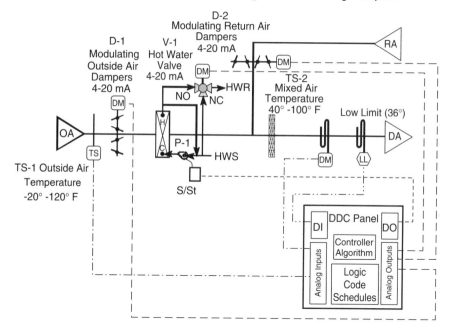

Figure 16.4   Hot Water Coil Preheat Control Strategy

16.4.3 Sequence of Operation

The sequence of operation for the hot water preheat coil illustrated in Figure 16.4 is explained in this section.

Process Overview

The preheat coil's hot water valve is modulated in sequence with the mixing dampers to maintain the mixed air temperature set point of 55 °F.

Occupied Operation

Whenever the system is in occupied mode, the mixing dampers are enabled to permit the necessary flow of ventilation air into the unit. Whenever the outside air temperature is below the mixed air temperature set point *and* the mixing dampers are commanded to minimum position, the secondary circulating pump (P_1) is commanded on.

The hot water valve is sequenced with the mixing dampers so it cannot open until the mixing dampers are at their minimum position. This prohibits the heating of excess ventilation air when the dampers are opening beyond minimum position to reduce the temperature of the mixed air.

The direct acting, proportional + integral preheat control valve and mixing damper control algorithm generates a 4 to 20 mA output signal that modulates the NO hot water valve (V_1) and the mixing dampers (DM_1, DM_2) to maintain the mixed air temperature set point (55°) at the location of sensor TS_2.

As the mixed air temperature increases, the output signal to the hot water valve and mixing dampers also increases, modulating the valve toward its closed position. If the mixed air temperature continues to rise, the increasing output signal closes the valve and opens the outside air dampers beyond their minimum position.

When the outside air temperature is above 45° and the dampers have been modulating beyond their minimum position for one hour, the circulating pump (P_1) is commanded off.

Unoccupied Operation

When the air handling unit is in its unoccupied mode, the mixing dampers are commanded to their normal position. When the outside air temperature is above 40°, the circulating pump (P_1) is commanded off and the preheat coil valve (V_1) is commanded closed (20 mA). When the outside air temperature is below 40°, the secondary pump (P_1) is commanded on and the preheat hot water valve (V_1) is modulated to maintain the mixed air plenum temperature at 70°, as measured by temperature sensor TS_2.

Emergency Operation

Whenever the air handling unit fan is commanded off due to the fan's failure to prove operation, a smoke detector indicating smoke or a low temperature limit trip as sensed by LL_1, the mixing dampers (D_1, D_2) and the preheat valve (V_1) are commanded to their normal position. Secondary pump (P_1) is commanded on if the outside air temperature is below 36°.

16.4.4 Strategy Summary

The hot water preheat control strategy can fulfill its responsibilities of preventing freezing air from entering the mixing plenum and preventing the its coil from freezing. The preheat valve is sequenced with the mixing dampers to maximize

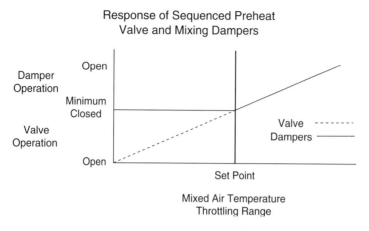

Response of Sequenced Preheat
Valve and Mixing Dampers

Figure 16.5   Preheat and Mixing Damper Response

the efficiency of the process by prohibiting heating whenever the mixing dampers are opening beyond their minimum position. Figure 16.5 shows the response of the sequenced valve and dampers. Note: the control valve operates between 100% open and closed whenever the control point is lower than the set point. During these periods the mixing dampers are at their minimum position. As the control point exceeds the set point, the valve is closed and the outside air dampers modulate open beyond their minimum position and the return air dampers close proportionately.

16.5

ILLUSTRATION OF A HOT WATER HEATING CONTROL STRATEGY

This example illustrates a hot water heating process for an AHU. The heating process is integrated with the mixing dampers to maintain the mixed air temperature set point at 55°.

Heating processes differ from preheat processes in their purpose, air temperatures and coil location. Heating processes are used to raise the temperature of the mixed airstream to set points within a range of 90 to 150 °F. Therefore, a heating coil is located downstream of the mixed air plenum. Most HVAC applications use hot water as the control agent in heating applications although steam and electric sources are also used.

Secondary circulating pumps are not required on the coil if the air entering the coil has already been tempered by a preheat coil or as a result of mixing the outside and return air streams. When a circulating pump is not used, a two-way control valve is used to modulate the flow of the hot water. In these applications, the control agent entering the coil displaces some of the cooler water out of the top of the coil. During light loads, this can create a warmer surface at the bottom of the coil than at the top, causing the air downstream of the coil to become stratified.

In applications where the air entering the heating coil may approach freezing or where the heating coil has a large surface area, a secondary circulating pump and three-way valve are utilized. These components prevent the coil from freezing and maintain a consistent temperature across its surface to reduce stratification of the air leaving the coil. Figure 16.6 shows the different piping configurations for two and three-way valve applications. In both piping schemes, the control valves fail open to permit some heat transfer during in the event of a loss of control signal. Heating valves are typically installed on the leaving side of the coil reducing the temperature requirements of the valve body, maintain the system pressure on the coil at all times and keep the coil full of water.

16.5.1 Hot Water Preheat Control Loop Components

The hot water heating control strategy incorporating a modulating two-way control valve requires the hardware components shown in Table 16.3.

Table16.3 HOT WATER HEATING CONTROL LOOP COMPONENTS

Number	Quantity	Description	Signal Type
1	1	Outside Air Temp Sensor (-20 to $120\ °F$)	Analog Input
2	1	Mixed Air Temperature Sensor (0 to 100 °F)	Analog Input
3	1	Heating Coil Air Temp Sensor (40 to 160 °F)	Analog Input
4	1	Return Air Temperature Sensor (40 to 120 °F)	Analog Input
5	1	Two Way Hot Water Valve (3 to 15 psi)	Analog Output
6	1	Mixing Damper Actuators (3 to 15 psi)	Analog Output
7	1	Low Limit Controller (36°)	Binary Input
8	1	DDC Controller Panel	All

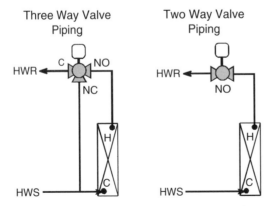

Figure 16.6   Two- and Three-Way Piping Configurations for Hot Water Heating Applications

16.5.2 System Drawing

A system drawing for a hot water heating control process is shown in Figure 16.7

16.5.3 Sequence of Operation

The sequence of operation for the hot water heating coil process illustrated in Figure 16.7 is explained in this section.

System Overview

The two-way hot water control valve modulates to maintain the heating coil discharge set point. The heating valve is interlocked through software with the mixing dampers to prevent the addition of heat into the airstream whenever the outside air damper is open beyond its minimum position.

Pre-Occupied Conditioning

When the system is commanded to pre-occupied warm up, the mixing dampers (DM_1, DM_2, DM_3) remain in their normal position (3 psi) and the hot water valve (V_1) is modulated to maintain a 120° heating coil discharge air temperature as sensed by TS_3.

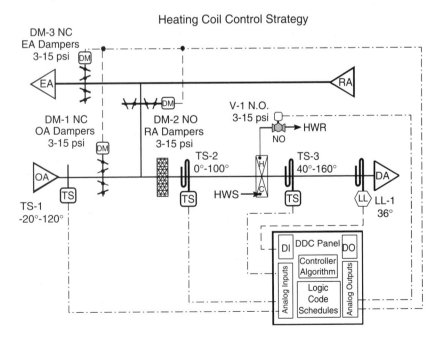

Figure 16.7 Hot Water Coil Heating Application

Whenever the system is commanded to pre-occupied purge, the hot water valve (V_1) is commanded closed (15 psi) and the mixing dampers are modulated to maintain a 55° mixed air temperature set point.

Occupied Operation

Heating Mode   When the air handling unit is in heating mode, the mixing dampers are commanded to their minimum position. The heating discharge air temperature control loop maintains the process at its 125° set point as measured by sensor TS_3.

The direct acting, proportional + integral hot water valve controller algorithm generates a 3 to 15 psi output signal to modulate the two-way, normally open hot water valve (V_1).

As the air leaving the heating coil increases in temperature, the output signal to the hot water valve increases, modulating the valve toward its closed position.

Economizer Mode   Whenever the return air temperature increases above 73°, the heating process is disabled because the zones no longer need additional heat and the economizer mode begins. The heating control valve is commanded closed (15 psi) and the mixing dampers are allowed to modulate beyond their minimum position to maintain a mixed air temperature set point of 55° as measured by TS_2. Any additional heat required in a zone is supplied by its perimeter heating system.

Unoccupied Operation

When the air handling unit fan is commanded off, the outside, exhaust and return air dampers (DM_1, DM_2 , DM_3) are commanded to their normal position (3 psi). When the outside air temperature is above 40°, the hot water heating valve is commanded closed. If the outside air temperature falls below 40°, the heating coil valve (V_1) is modulated to maintain the coil discharge temperature at 70°, as measured by temperature sensor TS_3.

Emergency Operation

Whenever the air handling unit fan is commanded off due to the fan's failure to prove operation, a smoke detector indicating smoke or a low temperature limit trip as sensed by LL_1, the mixing dampers (D_1, D_2, D_3) and the heating control valve (V_1) are commanded to their normal position.

16.5.4 Strategy Summary

The heating coil control strategy is relatively simple. The normally open valve modulates in a closed loop configuration to maintain the discharge air temperature set point. The heating valve is typically sequenced with the chilled water valve to maintain the discharge air temperature of the air handling unit using one control loop.

16.6

ZONE HEATING APPLICATIONS

Reheat coils, fan coil units, unit ventilators and perimeter fintube convectors are all heating applications that use heat transfer coils to add sensible heat to individual zones within a building. Each of these applications uses a normally open valve that is modulated by a zone thermostat to maintain its set point. The control strategy used in all these applications is similar to an air handling unit's heating process. The primary difference lies in the use of a thermostat as the sensing/controller device in zone applications. The use of a thermostat allows the occupants to locally adjust the set point. The air handling unit's set points can only be adjusted by the building operator or control technician.

16.7

EXERCISES

Determine if the following statements are true or false. If any portion of the statement is false, the entire statement is false. Explain your answers.

1. Heating strategies add latent heat to a process to offset its thermal losses.
2. Heating valves are normally open to supply heat to the coil upon loss of their control signal.
3. Preheat coils controlled with an open loop strategy are located downstream of the air handling unit filters.
4. Preheat coils are needed to prevent large quantities of cold air from freezing components in a system or building.
5. As the temperature increases downstream of the face and bypass dampers of a preheat coil, the signal to the damper actuator decreases to open the bypass damper.
6. Heating processes with normally open valves have direct acting controller transfer functions.
7. Secondary circulating pumps are used to protect steam coils from freezing and to maintain a uniform coil temperature across their surface.
8. If the sensor of a preheat process is located upstream of the coil, the loop requires a two-position control valve.
9. Tempered air has a temperature within the range of 70 to 90 °F.
10. Reheat applications use small heating coils that operate with the same strategy as air handling unit heating coils.

Respond to the following statements, questions and problems completely and accurately, using the material found in this chapter.

1. Why is the open loop control of a preheat coil less efficient than a closed loop strategy?

2. Why does air that is too hot for present conditions introduce instability into other control loops?

3. What are the differences between a preheat coil application and a heating application?

4. Why must the outside air dampers be commanded to their minimum position before a preheat valve opens?

5. Draw a flowchart for the preheat sequence of operation described in Section 16.4.

6. Develop a flowchart analysis for the flowchart developed for Problem 5. Use Section 16.3.5 as a reference source.

7. What is the response of a preheat coil with a steam valve and face and bypass dampers if the outside air temperature increases from 32° to 42°?

8. Draw a flowchart for the hot water heating process described in Section 16.5.0.

9. Develop flowchart analysis for the flowchart developed for Problem 8. Use Section 16.3.5 as a reference source.

10. Develop a list of materials, system drawing, sequence of operation and a flowchart for a reheat coil that serves a zone. The temperature is controlled by a zone thermostat and a normally open reheat coil valve.

CHAPTER 17

Cooling Control Strategies

17.1

OVERVIEW OF THE DIFFERENT COOLING STRATEGIES

Air handling unit cooling processes use chilled water or direct expansion (DX) coils to remove sensible and latent heat from an airstream. The absorbed heat is transferred outside of the facility using air cooled condensers, cooling towers or evaporative condensers. Cold water is produced by a chiller in chilled water applications. In chilled water systems, the water passes through a shell and tube evaporator (chiller bundle) where the heat that was absorbed from the airstream is transferred to the refrigerant. The chilled water leaves the bundle and is pumped to the coil locations throughout the building. The cooling coil's capacity is varied by modulating the flow of water through the coil using a two-way or three-way control valve. Most chilled water processes incorporate a closed loop control strategy to maintain the coil's discharge air temperature at its set point.

There are some chilled water coils that have been installed *without* a control valve. Consequently, these coils received 100% of their chilled water flowrate whenever the chiller and its pumping equipment are operating. These processes are characterized as *running wild*. In these applications, the downstream air temperature is a function of the temperature of the chilled water arriving at the coil. There is no way to maintain the temperature of the air exiting the coil at a desired set point. This piping configuration wastes energy and requires reheat systems to

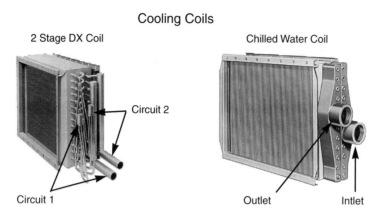

Figure 17.1 Cooling Coils

maintain the temperature in the zones. These systems are not installed in new buildings due to their inherent inefficiencies. To improve the controllability of existing processes, three-way valves are installed in the coil's piping circuit to allow control of the air temperature using a closed loop strategy.

Direct expansion cooling systems do not use water to transport the energy between the chiller bundle and the air handling unit's coil. These systems transport the refrigerant directly to the air handling unit's coil. The liquid refrigerant flows into the heat exchanger coil and absorbs heat directly from the airstream passing across the coil's surfaces. The capacity of a DX coil is a function of the flowrate of refrigerant entering the coil and the suction pressure maintained in the coil. These parameters are based upon the superheat setting of the coil's thermostatic expansion metering valve and are not typically adjusted after the system has been commissioned.

Various strategies are used to vary the cooling capacity of processes using DX coils. In single coil applications, solenoids can be used to unload some of the cylinders in a compressor to reduce its pumping volume thereby increasing the suction pressure in the coil and reducing its capacity. In larger applications, multiple evaporator coils can be installed side by side in the air handling unit. As the load increases, the DX coils are enabled one by one to increase the total cooling capacity of the system. Figure 17.1 shows a two stage direct expansion coil and a chilled water coil.

17.1.1 Interfacing the Building Control System With Condensing Units

Chiller and direct expansion cooling systems are purchased with their own packaged control systems designed to correctly start, stage and shut down their compressors and other related equipment. Various time delays and safety strategies are incorporated in these controls to prevent the system from operating under conditions that would be detrimental to the equipment. A facility's DDC control system is not used to perform the operating strategies of the condensing units. Instead, they are used to enable the condensing unit and to monitor its operation. A binary output point from the building automation system enables the condensing unit whenever the outside air temperature exceeds a predetermined set point (55°) and the building is occupied. All compressor cycling and safety strategies are performed by the equipment's own electronic control system. Binary input points from the building automation system can be used to monitor the operating status of compressors, pumps and condenser fans to permit the units to be monitored from remote locations. Analog inputs can also be used to monitor the supply and return temperatures of the chilled water, condenser water or other process variables.

When the condensing unit is manufactured by the same company as the building control system, all of the operating parameters of the equipment are mapped from the onboard electronic controls to the operator's work station without the need of installing additional hardware points.

17.2

ILLUSTRATION OF A MODULATING CHILLED WATER CONTROL STRATEGY

This example illustrates a chilled water cooling process that uses a three-way modulating valve.

Modulating closed loop control strategies are used with chilled water coil applications. These processes are very similar in design and response to the modulating heating coil application described in Chapter 16. Both of these control strategies use direct acting controllers along with a two or three-way control valve in a closed loop configuration. The difference between these strategies is the normal position of their control valves. A heating process uses a normally open valve while a chilled water process requires a normally closed valve.

Unlike hot water applications, chilled water coils do not require a secondary circulating pump to prevent the coil from freezing. Secondary chilled water pumps are only used in large coil designs to maintain a uniform temperature across the surfaces of the coil.

17.2.1 Chilled Water Cooling Coil Control Loop Components

The chilled water control strategy incorporating a three-way modulating control valve requires the hardware components shown in Table 17.1.

Table 17.1 CHILLED WATER CONTROL LOOP COMPONENTS

Number	Quantity	Description	Signal Type
1	1	Supply Air Temperature Sensor (40 to 100 °F)	Analog Input
2	1	Return Air Temperature Sensor (40 to 100 °F)	Analog Input
3	1	Outside Air Temp Sensor (−20 to 120 °F)	Analog Input
4	1	3 Way Chilled Water Valve (5 to 10 v)	Analog Output
5	1	DDC Controller Panel	All

17.2.2 System Drawing

A system drawing for a cooling process incorporating a chilled water coil is shown in Figure 17.2.

17.2.3 Sequence of Operation

The sequence of operation for the cooling process incorporating a chilled water coil and three-way valve that is illustrated in Figure 17.2 is explained in this section.

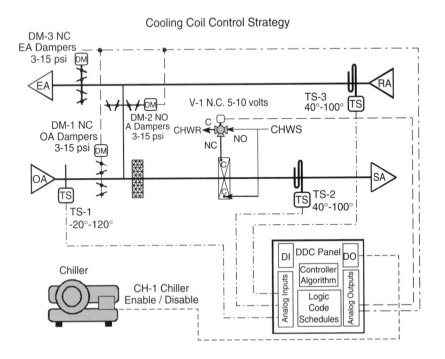

Figure 17.2   Chilled Water Valve Control Strategy

System Overview

The system's chiller is enabled by the building automation system during occupied periods whenever the outside air temperature is greater than the supply air temperature set point (55°). The chilled water valve is modulated to maintain the supply air temperature at its set point.

Pre-Occupied Conditioning

The mechanical cooling strategy is only enabled during pre-conditioning periods if the mixing damper's purge cycle cannot reduce the supply air temperature below 55°.

   If the outside air is too warm to purge the building, the chiller is enabled and the chilled water valve is modulated to maintain a 55 °F supply air temperature measured by the supply air temperature sensor TS_2. When the outside air temperature is below the economizer high limit set point, the outside and exhaust air dampers (DM_1, DM_3) are commanded 100% open and the return air damper (DM_2) will close. When the outside air temperature is greater than the economizer high limit set point (68°) the mixing dampers remain in their normal position and the chilled water coil conditions 100% return air.

Occupied Operation

The operation of the chilled water coil is based upon the relationship between the outside air temperature and the supply air set point.

Outside Air Temperature Less Than the Supply Air Temperature Set Point
When the outside air temperature is less than the supply air's temperature set point (55°) as sensed by the outside air sensor TS_1, the chiller and its associated equipment are disabled. Below this temperature all cooling requirements are met by the economizer strategy of the mixing dampers.

Outside Air Temperature Greater Than the Supply Air Temperature Set Point
When the outside air temperature as sensed by the outside air sensor TS_1 rises above the supply air temperature set point (55°), the chiller system and its associated equipment are enabled (CH_1). The chiller's packaged control system maintains the temperature of the supply water leaving the chiller bundle at 42 °F.

   The supply air temperature control loop maintains a 55° supply air temperature set point as measured by supply air temperature sensor TS_2. The direct acting, proportional + integral supply air temperature control algorithm generates a 5 to 10 volt output signal to modulate the NC chilled water valve (V_1).

   As the temperature of the air leaving the cooling coil increases, the output signal to the chilled water valve also increases, modulating the valve toward its wide open position.

Unoccupied Operation

When the air handling unit's fan is commanded off, the chilled water valve is commanded to its normal position (5 volts) and the chiller is disabled (CH_1).

Safety Operation

Whenever the air handling unit's fan is commanded off due to the fan's failure to prove operation, a smoke detector indicating smoke or a low temperature limit trip, the mixing dampers (D_1, D_2, D_3) and the cooling control valve (V_1) are commanded to their normal position.

17.2.4 Flowchart

A flowchart depicting the logic in the sequence of operation for the system shown in Figure 17.2 is shown in Figure 17.3. Section 17.2.5 provides an analysis of this flowchart.

17.2.5 Flowchart Analysis

The start oval denotes the beginning of the logic for the chilled water coil's control strategy. The first decision diamond determines whether the system is in its unoccupied mode of operation. When the system is in the unoccupied mode, the chiller system is disabled and the chilled water valve is commanded to its normal position (closed). These points remain in this condition until the system enters its pre-occupied mode.

Once the AHU enters its pre-condition mode, a decision diamond determines whether the zones should be purged with cool outside air. If the AHU is commanded into purge mode another decision determines if the outside air temperature is greater than 55°, the set point for pre-conditioning using the mechanical cooling system. When the outside air temperature is below 55°, the chiller remains off and the chilled water valve stays in its normal position. When the temperature is above 55° the chiller system is enabled and the chilled water valve is modulated to maintain a supply air temperature set point of 55°.

During occupied mode operation another decision determines if the outside air temperature is below 55°, the occupied mode supply air temperature set point. When the temperature is below 55°, the chiller system remains disabled and the chilled water valve is commanded to its normal position. When the outside air temperature is above 55° the chiller system is enabled and the valve modulated to maintain a supply air temperature of 55°.

17.2.6 Strategy Summary

The strategy for cooling coils incorporating modulating chilled water valves is straight forward and easy to program. If a primary, secondary or condenser water

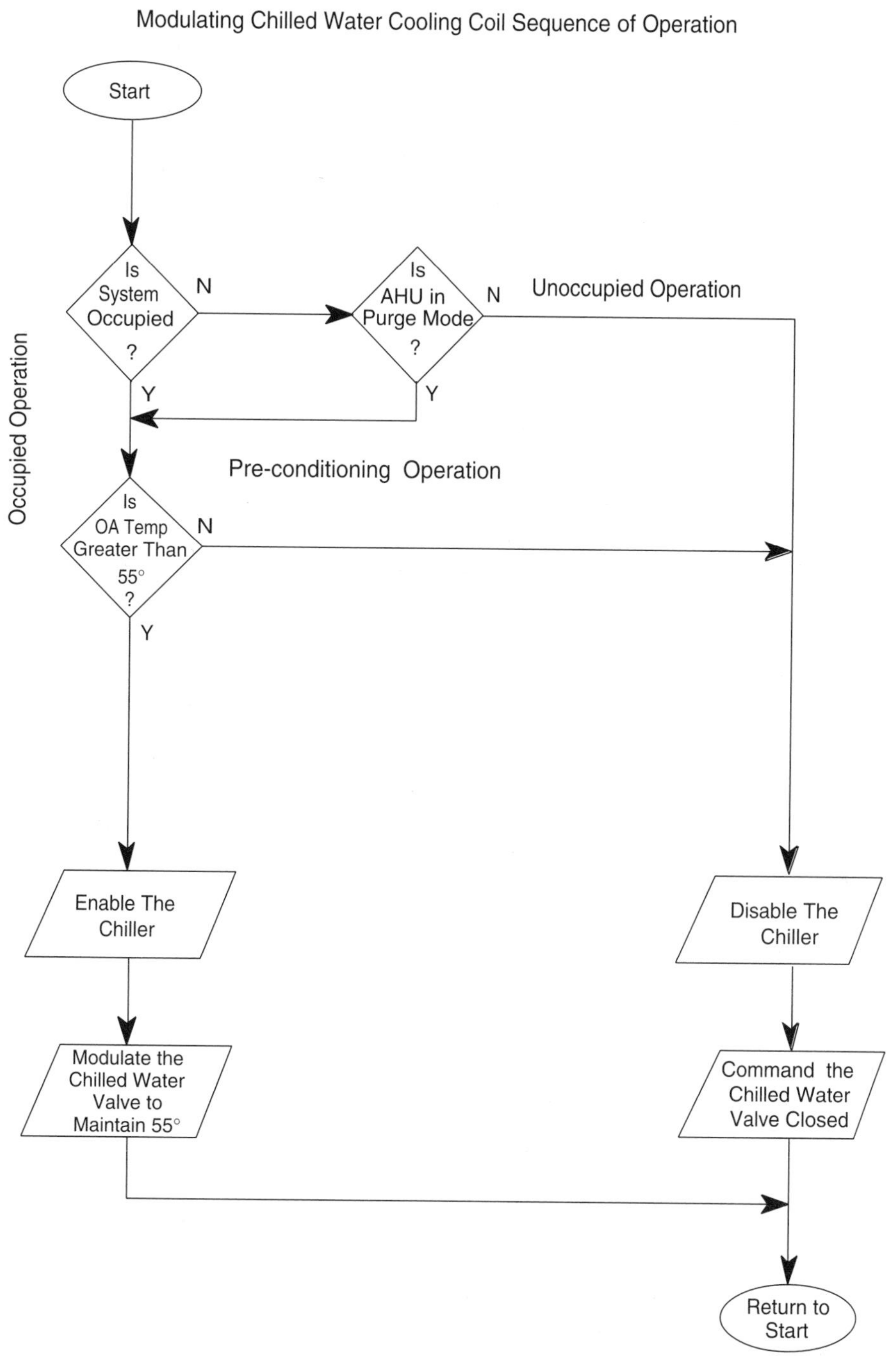

Modulating Chilled Water Cooling Coil Sequence of Operation

Figure 17.3 Chilled Water Coil Flowchart

circulating pump is not under the control of the chiller's packaged control system, they must be commanded on whenever the chiller is enabled.

In air handling units that have their hot and chilled water coils located in the same duct, the hot and chilled water valves are often sequenced using different actuator spring ranges or DDC programming logic. This strategy prohibits both valves from being open at the same time which would reduce the operating efficiency of the temperature control process.

17.3

ILLUSTRATION OF A DISCHARGE AIR TEMPERATURE CONTROL STRATEGY INCORPORATING SEQUENCED HOT AND CHILLED WATER VALVE ACTUATOR SPRINGS

This example illustrates the response of a temperature control loop that modulates hot and chilled water valves with the same controller output signal.

17.3.1 Sequenced Chilled and Hot Water Coil Control Loop Components

The control strategy incorporating chilled and hot water control valves sequenced by their actuator springs and operated by a single controller requires the hardware components shown in Table 17.2.

Table 17.2 SEQUENCED CHILLED AND HOT WATER COIL CONTROL LOOP COMPONENTS

Number	Quantity	Description	Signal Type
1	1	Supply Air Temperature Sensor (40 to 140 °F)	Analog Input
2	1	Return Air Temperature Sensor (40 to 100 °F)	Analog Input
3	1	Outside Air Temp Sensor (−20 to 120 °F)	Analog Input
4	1	NO Hot Water Valve (3 to 8 psi spring)	Analog Output
5	1	NC Chilled Water Valve (9 to 15 psi spring)	Analog Output
6	1	Low Limit Temperature Controller (36°)	Digital Input
7	1	DDC Controller Panel	All

17.3.2 System Drawing

A system drawing for a sequenced chilled and hot water coil control process is shown in Figure 17.4.

17.3.3 Sequence of Operation

The sequence of operation for the sequenced chilled and hot water coil control process shown in Figure 17.4 is explained in this section.

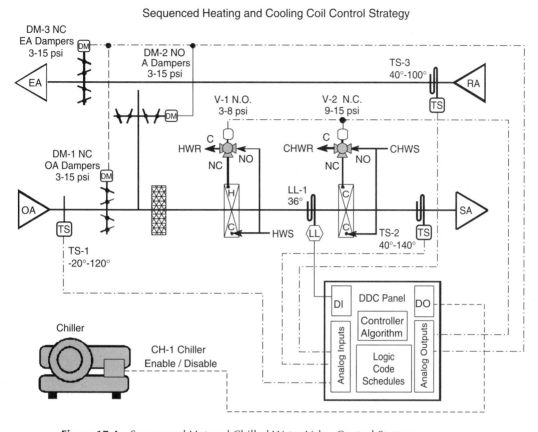

Figure 17.4   Sequenced Hot and Chilled Water Valve Control Strategy

System Overview

The supply air temperature control loop modulates the hot and chilled water valves in sequence (by spring range) to maintain the process set point. (Both of these coils can be programmed to operate during the warm up or purge pre-occupied conditioning cycle as described in section 17.2.3.)

Occupied Operation

The supply air temperature control loop is programmed to maintain a 60° set point as measured by supply air temperature sensor TS_2. The direct acting, proportional + integral supply air temperature control algorithm generates a 3 to 15 psi output signal to modulate the 3 to 8 psi NO hot water valve (V_1) and the 9 to 15 psi NC chilled water valve (V_2).

As the supply air temperature increases, the output signal to the hot and chilled water valves also increases, modulating the NO hot water valve toward its closed position. As the temperature continues to rise, the hot water valve closes completely and the chilled water valve begins to modulate open to maintain the set point.

Unoccupied Operation

When the air handling unit's fan is commanded off, the hot and chilled water valves are commanded to their normal position (hot water open and chilled water closed).

Emergency Operation

Whenever the air handling unit's fan is commanded off due to the fan's failure to prove operation, a smoke detector indicating smoke or a low temperature limit trip, the mixing dampers (D_1, D_2, D_3) and the hot and chilled water control valves (V_1, V_2) are commanded to their normal position.

17.3.4 Strategy Summary

Modulating the hot and chilled water valves in sequence insures that simultaneous heating and cooling cannot occur during normal operation. One sensor and controller are used to control both final controlled devices using a single output signal. A small deadband is usually included between the spring ranges to be sure both valves are completely closed when the controlled variable is equal to the set point. This scheme also allows both valves to be controlled using one analog output point thereby reducing the cost of the project's hardware.

17.4

ZONE COOLING APPLICATIONS OVERVIEW

Fan coil units and unit ventilators are applications that use smaller cooling coils located in individual building zones. Both of these applications use a cooling coil whose capacity is controlled using a normally closed valve. The valve modulates to maintain the zone at its set point. Figure 17.5 shows a schematic diagram of a unit ventilator application. These applications are described in more detail in Chapter 20.

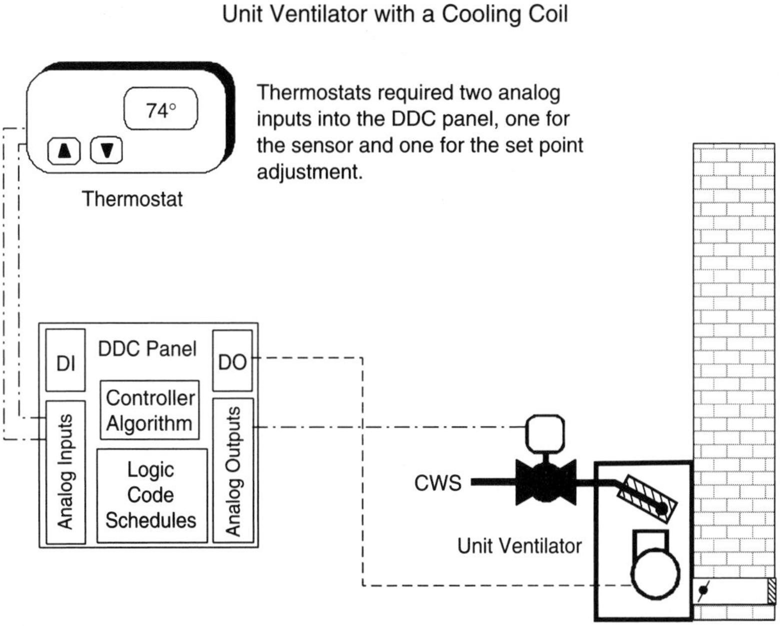

Unit Ventilator with a Cooling Coil

Figure 17.5   Unit Ventilator With Chilled Water Coil

17.5

EXERCISES

Determine if the following statements are true or false. If any portion of the statement is false, the entire statement is false. Explain your answers.

1. Cooling strategies remove latent heat from a process to reduce its moisture content.
2. Cooling valves are normally open to supply cooling to the process upon loss of its control signal.
3. Direct expansion cooling coils need circulating pumps to maintain a uniform surface temperature.
4. Liquid line solenoids are used to modulate the capacity of multiple DX coil installations.
5. As the temperature downstream of a cooling coil increases, the signal to the chilled water valve decreases to maintain the air temperature set point.
6. Multiple DX evaporators and compressor staging allows a two-position controlled process to approach a modulating response.

7. Three-way chilled water valves maintain a constant flowrate through the cooling coil while varying its temperature.
8. The temperature sensor for a binary DX coil installation is located upstream of the coil.
9. When the signal to the chilled water valve is set to 0%, the discharge air temperature increases.
10. Chillers have their own control package that maintains the chilled water supply temperature at a desired set point.

Respond to the following statements, questions and problems completely and accurately, using the material found in this chapter.

1. What are some of the disadvantages of permitting chilled water coils to run wild?
2. What are some of the advantages of using chilled water coils in place of DX coils?
3. What is the response of a chilled water control loop when the outside air temperature exceeds the dry bulb economizer limit?
4. The discharge air temperature set point for the chilled water coil is 60° and the dry bulb economizer limit is 68°. What is the position of the chilled water valve when the outside air temperature is 62° and the throttling range of the cooling loop is 3°?
5. Why does the chilled water valve's position remain the same after the outside air's enthalpy exceeds the return air's enthalpy causing the economizer cycle to be disabled, commanding the mixing dampers to minimum position?
6. What is the purpose of the binary output point of a DDC panel that terminates inside the chiller's control panel?
7. Draw a flowchart for the sequence of operation described in Section 17.3.
8. Develop a sequence of operation that combines the enthalpy based economizer mixed air temperature strategy from Chapter 15 and the chilled water coil's temperature control strategy.
9. Develop the flowchart for the sequence developed in Item 8.
10. Develop a DDC system drawing for Item 8.

CHAPTER 18

Humidity Control Strategies

18.1

OVERVIEW OF HUMIDIFICATION STRATEGIES

Air handling unit *humidification* processes inject steam directly into the supply airstream to increase the amount of moisture it delivers to the building zones. The moisture laden air mixes with the dryer zone air, raising the room's relative humidity. Most humidifiers for comfort applications add sufficient moisture to the supply air to raise the relative humidity within its zones to a value between 30% and 50%. Humidification processes transfer mass (water) and energy (heat) into the airstream.

Humidification is required in dry climates and during cold weather months when drier air is brought into the building to satisfy its ventilation needs. When this dry air mixes with the zone air, the moisture level of the zone drops. Low moisture levels increase the number of health complaints and the spread of colds and viruses within the building. When the relative humidity is elevated above 30%, the growth of bacteria and the spread of illnesses are reduced. Higher levels of moisture in the air also reduce the generation of static electricity which can destroy sensitive electronic parts and erase information stored computer related magnetic media.

Relative humidity higher than 40% promotes excessive condensation and frost on window surfaces when the outside air temperature is below freezing. Window condensation is an indication that the mass and energy being expended to increase the humidity level of the air are being transferred to the window surfaces, decreasing the overall process efficiency. Window condensation also inhibits visibility and produces puddles of moisture on window sills and floors that annoy building occupants. It also promotes the development of molds and mildew which are precursors to poor indoor air quality.

Humidification processes modulate the amount of steam injected into the supply airstream. All steam humidity applications use a tube or grid that injects the steam into the airstream. The steam is injected through holes in the tubes that face against the flow of the airstream. This configuration generates the turbulence necessary for the proper absorption of the moisture droplets as depicted in Figure 18.1.

18.1.1 Humidification Control Strategies

The flowrate of the steam transferred into the supply air is varied as a function of the moisture content of the air in the zones. Generally, the humidity sensor is located in the return air duct of the air handling unit. In this position, it measures the average moisture level of all the zones connected to the air handling unit. Since moisture quickly migrates throughout an area to equalize its vapor pressure, this strategy is adequate for most comfort HVAC processes. In more critical applications, humidity sensors can be installed in each critical zone and control logic used to determine the zone with the greatest demand for humidity. That zone will modulate the humidifier until its load is satisfied or until another zone has the higher demand.

Humidifying strategies for occupant comfort applications are typically configured to maintain a minimum level of humidity within a zone during cold weather operation. Therefore, these strategies are designed to prevent the conditioned

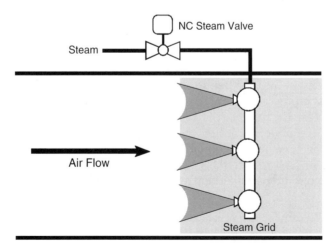

Figure 18.1   Steam Humidifier Grid

space for exceeding a *low limit*. These strategies have no means of limiting the maximum humidity level that may occur in the zones. They do not maintain a modulating humidity set point within the zone but allow the humidity to float within a range above the low limit set point.

All humidifying control loops require an additional sensor installed in the supply duct, a few feet downstream of the humidifier grid. This sensor monitors the relative humidity in the supply air after moisture has been added by the humidifier. Its controller commands the steam valve toward its closed position whenever the relative humidity exceeds a high limit set point of 80%. If the supply air's humidity is allowed to exceed this level, condensation is likely to begin to form on the cooler duct surfaces. Under normal operating conditions, the supply air's relative humidity should remain below this high limit when the zone set point is kept between 30% to 40% and the humidifier and its control valve are sized correctly.

18.2

ILLUSTRATION OF A HUMIDIFYING CONTROL STRATEGY

This example illustrates a steam humidifying process that uses a two-way modulating control valve.

The steam humidifier is located downstream of the heating coil to improve the absorption of the moisture in the warmer airstream. A two-way normally closed control valve is used to modulate the steam into the process. The use of a normally closed valve requires a reverse acting controller action. Reverse acting control loops reduce the controller's output signal to the valve's actuator as the relative humidity in the airstream increases. This modulates the valve toward its closed position, reducing the flow of steam into the supply air.

Humidifier steam valves must fail closed upon loss of the control signal to prevent uncontrolled steam flow into the duct. Under these conditions, the steam would condense as the supply air's humidity approached 100%. The resulting liquid would saturate the duct's insulation rendering it useless and providing an environment for the growth of mildew, mold and bacteria.

18.2.1 Steam Humidifier Control Loop Components

The steam humidification control strategy requires the hardware components shown in Table 18.1.

Table 18.1 STEAM HUMIDIFIER CONTROL LOOP COMPONENTS

Number	Quantity	Description	Signal Type
1	2	Duct Humidity Sensors (0% to 100%)	Analog Input
2	1	NC Steam Valve & Actuator (3 to 15 psi)	Analog Output
3	1	DDC Controller Panel	All

18.2.2 System Drawing

A system drawing for a steam humidifying process incorporating a modulating steam valve is shown in Figure 18.2.

18.2.3 Sequence of Operation

The sequence of operation for the steam humidifying process incorporating a modulating two-way valve that is illustrated in Figure 18.2 is explained in this section.

System Overview

The humidifier's steam valve is modulated to maintain a return air relative humidity set point of 35%.

Pre-Occupied Conditioning

The humidity control strategy is enabled whenever the air handling unit's fan starts for the pre-occupied or occupied modes of operation.

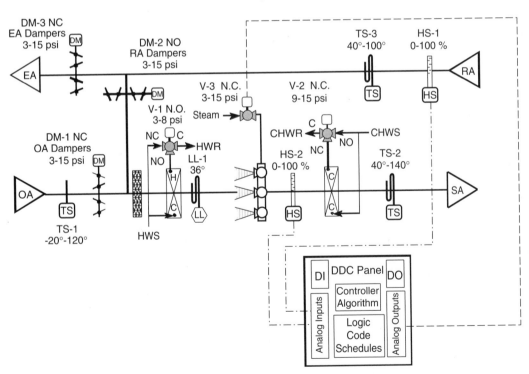

Figure 18.2   Steam Humidification Process

Occupied Operation

The relative humidity control loop is programmed to maintain a 35% return air humidity set point as measured by return air humidity sensor HS_1. The reverse acting proportional + integral humidity control algorithm generates a 3 to 15 psi output signal to modulate the NC humidifier steam valve (V_3).

As the relative humidity of the return air increases, the output signal to the steam valve decreases, modulating the valve toward its closed position.

If the humidity level of the supply air measured by HS_2 exceeds 80%, the reverse acting, proportional + integral return air humidity control algorithm is instructed to use the supply air's relative humidity sensor HS_2 to maintain a high limit relative humidity set point of 80%. The 3 to 15 psi output signal modulates the NC steam valve (V_3) to maintain 80% relative humidity in the supply duct until the return air humidity measured by HS_1 exceeds 36%. Once the return air has exceeded its set point, the loop is instructed to use sensor HS_1 and the set point is changed back to 35%.

Unoccupied Operation

When the air handling unit's fan is commanded to unoccupied mode, the humidifier steam valve (V_3) is commanded closed.

Emergency Operation

Whenever the air handling unit fan is commanded off due to the fan's failure to prove operation, a smoke detector indicating smoke or a low temperature limit trip, the humidifier steam valve (V_3) is commanded closed.

18.2.4 Flowchart

A flowchart depicting the logic in the sequence of operation for the system shown in Figure 18.2 is shown in Figure 18.3. Section 18.2.5 provides an analysis of this flowchart.

18.2.5 Flowchart Analysis

The start oval indicates the beginning of the humidifier control logic. The first decision diamond determines whether the system is in its unoccupied mode of operation. When the system is in the unoccupied mode, the humidifier's steam valve is commanded closed and remains closed until the pre-occupied conditioning mode begins.

During the pre-occupied and occupied modes, a decision diamond determines whether an alarm condition exists in the system that requires the humidifier to be disabled. Another decision diamond determines if the supply air's relative humidity is above its high limit set point of 80%. If it is, the humidifier steam valve is modulated to maintain a maximum supply air relative humidity of 80% using

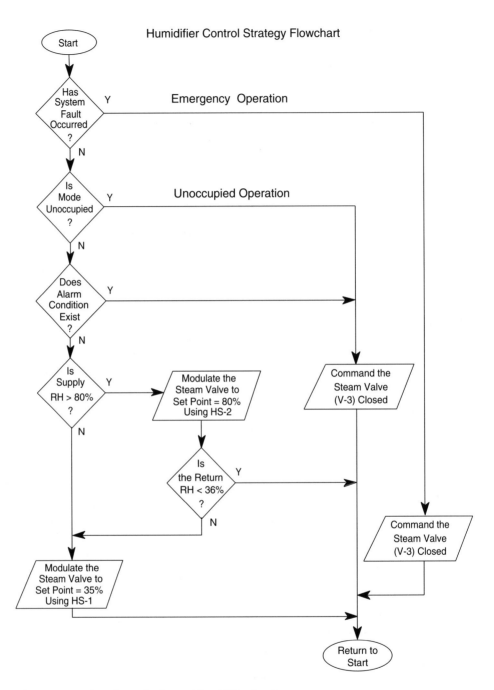

Figure 18.3   Flowchart of a Steam Humidification Process

the *supply air* humidity sensor. The system operates in this mode until the relative humidity of the return air exceeds its set point. Once this occurs, the humidifier steam valve is modulated using the *return air* humidity sensor to maintain a set point of 35%.

18.2.6 Strategy Summary

Changes in the steam pressure or environmental conditions that call for a wide open valve may drive the supply air relative humidity above 80%. These conditions are more likely to occur when the return air relative humidity is below the lower limit of its throttling range. Operating the control loop to maintain 80% supply air relative humidity shortens the time required to bring the zone humidity levels back to their design requirements. Once the return air sensor indicates the average humidity in the zones has returned to their proper level, control reverts back to the return air humidity sensor. This logic is easy to perform using the programming capabilities of a DDC system. Additional safety logic can be added to the program that commands the steam valve closed and generates an alarm if the supply air relative humidity exceeds 90% for a short period of time. This indicates a mechanical failure may exist that needs attention.

18.3

OVERVIEW OF THE DEHUMIDIFICATION PROCESSES

Dehumidification is required during the warm weather months when the ventilation air contains a high quantities of water vapor. Moist air adversely affects occupant comfort, paper handling processes and encourages the growth of molds and mildew on cool surfaces. Comfort dehumidification processes typically use chilled water or direct expansion coils to remove the necessary moisture out of the airstream. When the relative humidity of the supply airstream is above its set point, the cooling coil's primary function is the removal of latent heat (moisture) from the air. Its secondary function is to sensibly cool the air. To dehumidify the air, the DX or chilled water coil is modulated to maintain a surface temperature below the dewpoint of the airstream. This causes water vapor to condense out of the air on to the coil fins. When the coil is operating in a dehumidifying mode, the air leaving the coil will be colder than its *temperature* set point. A reheat coil is used to raise the discharge air temperature to its dry bulb set point.

The air handling unit's heating coil is located downstream of the cooling coil in applications where a humidity high limit set point must be maintained. This position allows the heating coil to be used as the reheat coil during summer operation. Its normally open control valve is modulated to raise the supply air temperature up to its set point whenever the cooling coil is operating in a dehumidifying mode. Another option is to leave the AHU heating coil upstream of the cooling coil and use zone reheat coils to temper the airstream before it enters the zone.

18.4

ILLUSTRATION OF A DEHUMIDIFYING CONTROL STRATEGY

This example illustrates a dehumidifying process that uses a chilled water cooling coil and a hot water reheat coil.

A dehumidification strategy requires multiple control loops to permit the cooling coil to perform its two functions, maintaining a sensible heat set point (temperature) or a latent heat set point (humidity). The cooling process control loop operates cooling coil as previously described in Chapter 17. The dehumidification process control strategy is similar to the cooling strategy except that it uses a dewpoint or relative humidity sensor in place of a discharge air temperature sensor. During the dehumidification operation, the supply air temperature is maintained by modulating a hot water valve as described in Chapter 16. Note: this control strategy limits the maximum humidity level allowed in the zone during high humidity operation. It does not maintain a modulating humidity set point.

18.4.1 Dehumidifier Control Loop Components

The dehumidification control strategy requires the hardware components shown in Table 18.2.

Table 18.2   DEHUMIDIFIER CONTROL LOOP COMPONENTS

Number	Quantity	Description	Signal Type
1	2	SA + RA Temperature Sensors (40° to 140 °F)	Analog Input
2	1	Supply Air Humidity Sensor (0% to 100%)	Analog Input
3	1	OA Temperature Sensor (−20 to 120 °F)	Analog Input
4	1	NO Two Way Valve with Actuator (3 to 15 psi)	Analog Output
5	1	NC Two Way Valve with Actuator (3 to 15 psi)	Analog Output
6	1	DDC Controller Panel	All

18.4.2 System Drawing

A system drawing for a dehumidifying process incorporating a chilled water coil is shown in Figure 18.4.

18.4.3 Sequence of Operation

The sequence of operation for the dehumidifying process incorporating a chilled water coil that is illustrated in Figure 18.4 is explained in this section.

System Overview

The chilled water coil operates to provide thermal cooling or dehumidification of the mixed airstream. The chilled water valve modulates in response to one of two

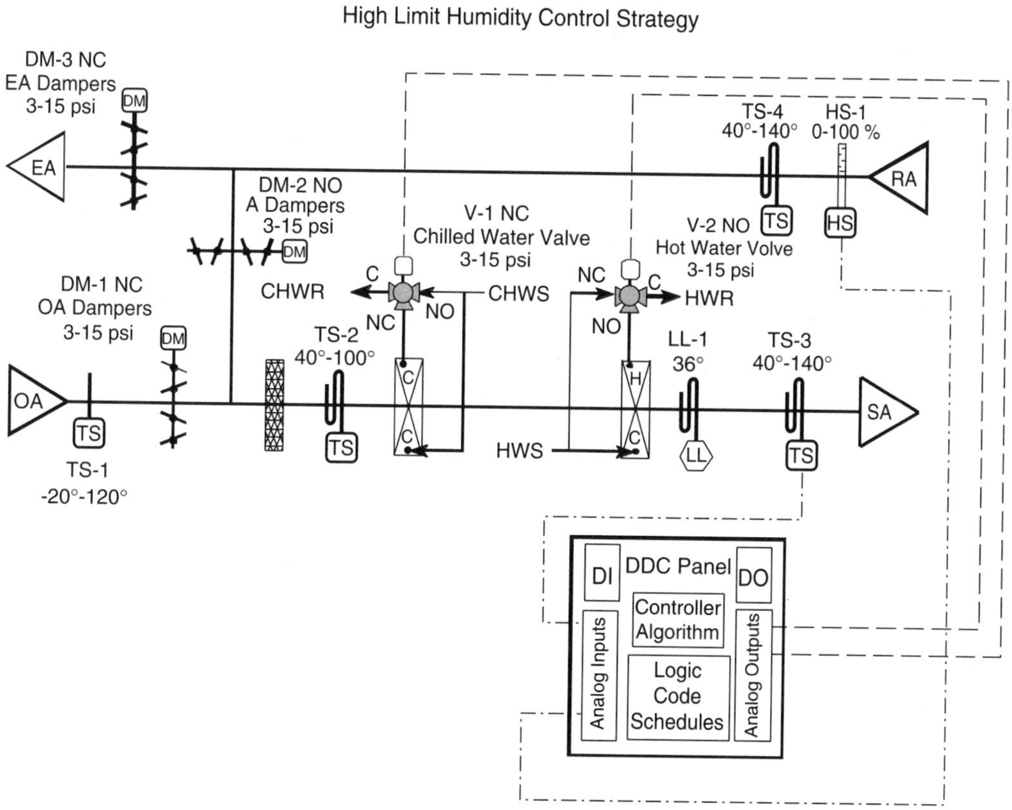

High Limit Humidity Control Strategy

Figure 18.4   Chilled Water Coil Dehumidification With Reheat

conditions, the process error between the humidity set point (60%) and the relative humidity of the return air or the process error between the temperature of the supply air and the supply air temperature set point (55°). The relative humidity requirement (dehumidification) takes precedent over the thermal requirement of the system.

Pre-Occupied Conditioning

The chilled water valve is modulated whenever the relative humidity of the return air exceeds 60% or if the mixing damper purge cycle is inadequate in bringing the zone temperature to set point before occupancy time.

When the return air's relative humidity is greater than the set point (60%), the normally closed chilled water valve is commanded 100% open to dehumidify the supply airstream. If the supply air temperature drops below its set point, the heating valve is modulated to raise the supply air temperature to 55° as described in the Heating Strategy section.

When the return air relative humidity is below 60% and the outside air is above the supply air temperature set point, the chilled water valve is modulated to maintain the supply air set point and the heating valve remains closed.

When the unit is in warm-up mode, the chilled water valve remains closed and the hot water valve is modulated to maintain a supply air temperature set point of 120°.

Occupied Operation

The operation of the chilled water coil is based upon the relationship between the outside air temperature, return air humidity and the supply air temperature set points.

Outside Air Temperature Is Less Than the Supply Air Temperature Set Point

When the outside air temperature as sensed by the outside air sensor TS_1, is below the supply air temperature set point (55°) and the return air relative humidity (HS_1) is below its set point of 60%, the chilled water valve is commanded closed. The mixed air control loop modulates the mixing dampers to maintain the mixed air temperature set point (55°) using the mixed air temperature sensor TS_2.

Heating/Reheating Control   When the supply air temperature falls below its set point, heating is required. The supply air temperature sensor TS_3 sends a signal to the controller to maintain the supply air temperature set point of 55°. The direct acting proportional + integral controller algorithm generates a 3 to 15 psi output signal to modulate the normally open hot water valve (V_2).

As the supply air temperature increases, the signal to the hot water valve also increases, modulating the hot water control valve toward its closed position.

Cooling   When the mixed air temperature control loop cannot maintain its set point (55°) and the return air relative humidity is below its high limit set point (60%) the chilled water coil is used to maintain the supply air temperature set point. The hot water valve (V_2) is commanded closed to prevent simultaneous heating and cooling from occurring.

TS_3 is used in the modulation of the chilled water valve as long as the return air relative humidity remains below the set point (60%). The direct acting proportional + integral controller algorithm generates a 3 to 15 psi output signal to modulate the normally closed chilled water valve V_1.

As the supply air temperature increases, the signal to the chilled water valve also increases, modulating the valve toward its wide open position.

Dehumidification Control   When the return air relative humidity (HS_1) rises above the humidity set point (60%), the chilled water valve (V_1) is commanded 100% open to reduce the moisture content of the mixed air passing across the coil.

The reheat coil valve (V_2) is modulated to maintain the supply air temperature at set point (55°) as measured by TS_3.

Unoccupied Operation

When the air handling unit fan is commanded to unoccupied mode, the chilled water valve is commanded closed. When the outside air temperature is below 36°, the hot water valve is modulated to maintain a 75° temperature in the vicinity of the coils using sensor TS_3.

Emergency Operation

Whenever the air handling unit's fan is commanded off due to the fan's failure to prove operation, a smoke detector indicating smoke or a low temperature limit trip, the chilled water valve (V_1) is commanded closed. If the outside air temperature is below 36°, the hot water valve is commanded open, otherwise it is commanded closed.

18.4.4 Flowchart

A flowchart depicting the logic in the sequence of operation for the system shown in Figure 18.4 is shown in Figure 18.5.

18.4.5 Strategy Summary

The dehumidifying strategy illustrated in this section is used to prohibit the zone's relative humidity from rising above a high limit set point. It is not intended to maintain the zone humidity at a desired set point. Whenever two modulating water valves are used to heat, cool and dehumidify, separate control output signals must be sent to each of the valve actuators to permit simultaneous heating and cooling during periods of dehumidification. The valves cannot be sequenced with valve springs or reheating would not be possible. During periods of normal heating or cooling, one of the valves is commanded closed to prohibit simultaneous heating and cooling.

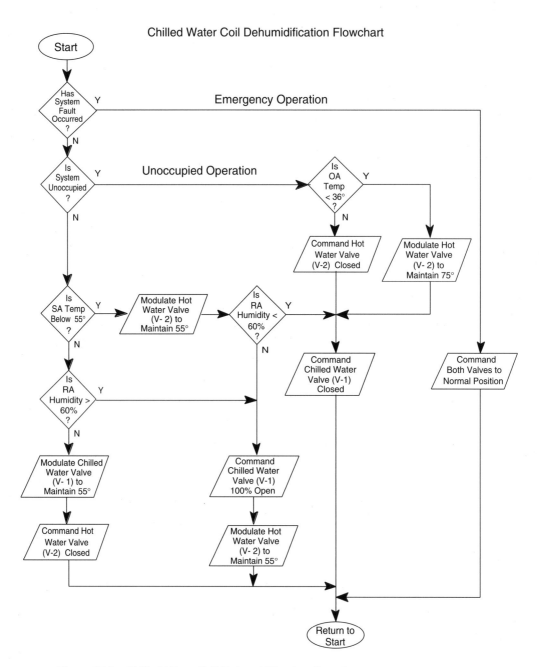

Figure 18.5 Chilled Water Coil Dehumidification Flowchart

18.5

MAINTAINING A CONSTANT HUMIDITY SET POINT

In the previous humidity applications, the controls and mechanical equipment were configured to prevent the relative humidity from exceeding either a high or low limit set point. To maintain the humidity control point at a fixed set point throughout the year, both the humidification and dehumidification equipment and their associated control hardware must be installed on the air handling unit.

The control strategy for constant humidity set point applications incorporate the details described in the previous applications. Figure 18.6 depicts the system diagram for this application. The operational strategy for this configuration prevents the simultaneous humidifying and dehumidifying of the airstream although simultaneous heating and cooling is still permitted. If at any time, humidification is required to increase the moisture content of the air in the zones, the steam grid is allowed to operate. This type of system is commonly used in laboratory, hospital and manufacturing applications where the process requirements must be maintained at a constant temperature and relative humidity.

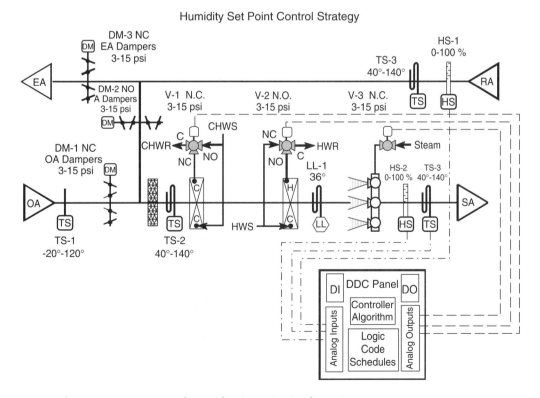

Figure 18.6 Year Round Humidity Set Point Configuration

18.6

CHEMICAL DEHUMIDIFICATION

Processes that require extremely low levels of moisture in the air must operate their dehumidifying coils at temperatures that cause the condensed moisture to freeze on the coil surfaces. This ice must be periodically removed by a defrosting process to maintain the heat transfer rate of the coil and to prevent the coil from becoming completely blocked by ice. The defrosting process periodically interrupts the dehumidification function, producing oscillations in the temperature and humidity levels within the zone.

Chemical dehumidification processes use a large, rotating wheel of desiccant material to *adsorb* moisture from the airstream and expel it from the building. The wheel is positioned between the mixed and regenerative or exhaust air ducts as shown in Figure 18.7. The desiccant, usually silica gel or lithium bromide compounds, attracts and chemically bonds with the water vapor molecules. The desiccant wheel adsorbs moisture from the mixed airstream without condensing it on its surface. The saturated desiccant is rotated into another chamber and exposed to a high temperature airstream that releases the moisture thereby reactivating the desiccant so it can be reused in the process.

As the mixed air (return plus ventilation air) passes through the desiccant material, it gives up its moisture in an exothermic reaction. Exothermic reactions give off heat as a by-product of the chemical reaction. The warm, dry air exits the wheel and passes through a chilled water coil where it is re-cooled to the supply air temperature set point.

At any interval in time, a portion of the wheel is located in the mixed air duct while the remaining portion of the desiccant wheel is passing through the regenerative air duct. An internal baffle has seals that prevent the regeneration air from passing into the dehumidified airstream. The regenerative air is heated by a hot water coil or gaseous fuel burner before it passes through the moisture laden desiccant wheel. The hot air draws the moisture off of the desiccant's surface, drying (regenerating) the desiccant so it is capable of adsorbing moisture as it re-enters the mixed airstream. The hot moist air exits the building through dampers and a louver similar to an exhaust air duct. These systems come with packaged controls that operate the dehumidifying equipment. The wheel's speed of rotation is kept constant while the regeneration air is maintained by a modulating control loop. The entire unit is cycled on and off with a binary control loop using a humidistat.

Chemical dehumidifiers can be used in place of chilled water, brine or DX coils in applications requiring low dewpoint temperatures. They can also used in applications where it is cost effective to reduce the size of the cooling system by transferring the coil's latent heat load to the chemical dehumidifier. The use of a desiccant dehumidifier reduces the chiller size, saving electrical energy as it increases the consumption of natural gas or other fuel needed to regenerate the desiccant in the dehumidifier.

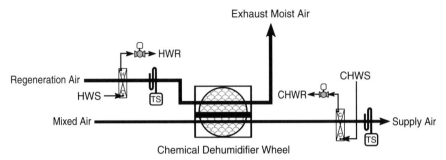

Chemical Dehumidifier

Figure 18.7 Desiccant Dehumidification Configuration

EXERCISES

Determine if the following statements are true or false. If any portion of the statement is false, the entire statement is false. Explain your answers.

1. Humidification processes add moisture to the supply air during cold weather operation.
2. Humidifiers are used to prevent the zone moisture level from exceeding a high limit.
3. The hot water coil in an air handling unit is located downstream of the chilled water coil in humidifying applications.
4. In a dehumidifying application, as the relative humidity control point of the return airstream exceeds its high limit set point, the cooling coil is controlled using a humidity sensor.
5. In a dehumidifying application, the heating coil only operates when the high limit humidity set point has been exceeded.
6. Desiccant wheels absorb moisture without condensing it.
7. When a chilled water coil is used for cooling and dehumidifying, the actuators for the chilled and hot water valves have sequenced spring ranges.
8. Humidifiers are located upstream of the heating coil to induce better absorption of the moisture by the air stream.
9. Humidity controllers are direct acting to modulate a normally closed valve.
10. A dehumidifying control loop is direct acting, using a normally closed chilled water valve.

Respond to the following statements, questions and problems completely and accurately, using the material found in this chapter.

1. Where would the humidity high limit sensor be located in Figure 18.1?
2. Why does the steam flow of a grid type of humidifier discharge against the direction of the air flow?
3. Why are humidifier steam valves normally closed devices?
4. Why can the humidification strategy described in Section 18.2 only maintain a low limit?
5. Why is the heating coil located downstream of the cooling coil in dehumidification processes?
6. Add the necessary control devices and logic to the low humidity limit system depicted in Figure 18.2 so the mixing dampers operate with a dry bulb economizer strategy. Develop a new list of materials and add the necessary logic to the sequence of operation and flowchart.
7. Why do the hot and chilled water valves shown in Figure 18.4 have the same actuator spring range instead of being sequenced?
8. Develop the sequence of operation for the humidity set point process depicted in Figure 18.6.
9. Develop the flowchart for the sequence of operation developed for Item 8.
10. Develop a sequence of operation for Figure 18.7.

CHAPTER 19

Duct Static Pressure Control

FAN PRESSURE

Duct static pressure control loops are used in air handling units that vary the capacity of their fans based upon the thermal load in their zones. By varying the static pressure in the supply duct, the volume of air leaving the fan also changes in a manner that balances the fan's capacity with the total thermal load of the zones. Air handling unit supply fans generate air pressure within the distribution ducts to create air movement. Conditioned air is propelled from the fan's outlet (high pressure) to the zone diffusers (lower pressure) by the pressure difference generated by the fan. The force needed to move the air comes from the energy drawn by the fan's motor. Electrical energy is converted into rotational power that rotates the fan blades at their design speed. The rotation of the fan blades transfers most of the energy drawn from the power source to the airstream, increasing its *total pressure.* Total pressure consists of two components, static pressure and velocity pressure. *Static pressure* is a force that presses against the inside surface of the ducts. It increases as the fan's speed increases, compressing more air into the duct.

Velocity pressure is a force that is applied in the direction of air movement in the duct. Velocity pressure generates the movement of air through the duct.

Static pressure in a duct acts as a reservoir of fan energy that can be used at a later time to generate air movement. It is used to overcome the friction created when air moves through the duct. Static pressure is converted into velocity pressure to maintain the flow of air to the end of the duct.

Whenever the air distribution system calls for an increase in the amount of air flowing through the duct system, by opening a control damper, a portion of the static pressure (energy) is converted into velocity pressure to produce the necessary increase in the flowrate of the air. The fan replaces the static pressure as it is used by the system. Conversely, if the system calls for a reduction in air flow by closing a damper, some of the velocity pressure is converted into static pressure, raising the duct pressure. Under these conditions, a larger proportion of the total pressure generated by the fan is converted into static pressure due to the decrease in the system's air flowrate. Throughout a fan's operation, the total pressure transferred to the airstream remains relatively constant while the proportions of static and velocity pressure vary with the dynamic changes of the dampers in the air distribution system.

19.1.1 Variable Volume Fan Systems

Variable air volume (VAV) air handling units *vary* the amount of energy the fan transfers to the airstream in response to changes in the *cooling load* within its zones. VAV duct distribution systems incorporate special control dampers located in each zone to alter the quantity of air entering the zone in response to a signal from the zone's thermostat. When the cooling load in a zone decreases, the signal from the thermostat commands the damper to throttle toward its minimum position, reducing the amount of cool air entering the zone. When the cooling load increases, the damper is commanded open to allow more conditioned air to enter the zone, offsetting the increase in its heat gain. Figure 19.1 shows a variable air volume zone box and its modulating damper which is repositioned by a signal from the thermostat.

A reduction in the volume of air flowing into a zone reduces the distribution system's need for velocity pressure because less air must be moved through the ductwork. To decrease the velocity pressure of the airstream, some of the fan's energy is converted back into static pressure. As more zone dampers throttle toward their minimum position, the duct's static pressure continues to increase because the fan is transferring more energy into the airstream than is needed to satisfy the system's load. If the amount of energy transferred by the fan could be automatically reduced to the level required by the zone loads, the process efficiency would be increased and the air distribution system would still have sufficient pressure to maintain the necessary air flow to supply the zones.

There are two methods commonly used to vary the amount of total energy transferred into the airstream by the fan. The first method incorporates a variable frequency drive (VFD). This electronic device modulates the amount of energy

VAV Zone Terminal Box

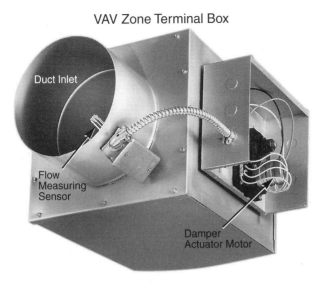

Duct Inlet

Flow
Measuring
Sensor

Damper
Actuator Motor

Figure 19.1   VAV Zone Terminal Box

transferred to the fan motor based upon the static pressure in the duct. The second method uses dampers installed on the inlet housing of the fan to vary the amount of air flowing into the fan. Both of these methods are described in more detail in the following subsections.

19.1.2 Variable Frequency Drives

Variable frequency drives are special electronic motor controllers that vary the speed of three phase induction motors in response to changes in a control signal. These devices were developed to capitalize on a physical relationship that exists between the rotational speed of an induction motor and the sinusoidal frequency of its AC power source. The speed of an AC induction motor is proportional to the frequency of its power source as described by the formula:

$$rpm = \frac{(120 \times source\ frequency)}{(number\ of\ poles\ in\ the\ motor\ stator)}$$

There are only two values in this formula that determine a motor's speed of rotation, the number of magnetic poles used in its construction and the frequency of its power source. The term *120* is a mathematical constant that converts between the time elements of frequency (seconds) and the motor's rotational speed (minutes) along with a multiple of two that accounts for the effect of both the north and south poles in a motor's pole set. Since the number of magnetic poles is a function of the motor's construction and cannot be changed, the frequency of the motor's power source is the only variable that can be altered to produce changes in a motor's rotational speed. Variable frequency drives alter a motor's speed by varying the frequency of the power it supplies to the motor. As the frequency generated by the

VFD decreases, the motor's speed also decreases. VFDs are also called variable speed drives (VSD) because they alter the speed of the motor.

19.1.3 Inlet Vane Dampers

Inlet vane dampers are also used as a method to vary the flowrate of air through a duct distribution system based upon the thermal load in the zones. Inlet vane dampers normally have closed blades mounted on the inlet opening of the fan housing. They alter the volume flowrate through the fan by changing the amount of energy that can be transferred between the rotating fan blades and the airstream. Figure 19.2 shows the inlet vane dampers mounted to the side of the fan's inlet housing.

Inlet vane dampers reduce the amount of kinetic energy that is transferred to the air by *pre-spinning* the entering airstream in the same direction as the fan blade's rotation. As the air is spun in the direction of rotation of the fan blade, the relative velocity between the airstream and the blades decreases. The amount of reduction is related to the amount of spin induced into the airstream. The amount of spin is a function of the position of the inlet damper blades. As the blades modulate toward their closed position, they impart more spin to the entering airstream, reducing the difference in velocity between the air and the fan blades. As the amount of energy transferred to the airstream decreases, the total pressure transferred to the air leaving the fan housing also decreases. Any reduction in

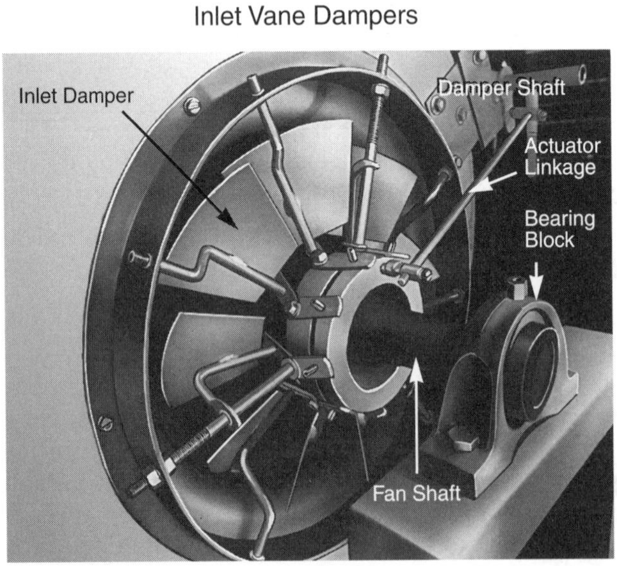

Inlet Vane Dampers

Figure 19.2 Inlet Vane Dampers

total pressure produces a corresponding decrease in the static and velocity pressure of the air in the duct, thereby reducing the flowrate through the distribution system. The inlet vane dampers also change the fan's pressure-volume performance curve. As the vanes throttle from fully open to closed, changes in the performance curve generate reductions in the energy drawn by the motor. Reductions in air flowrate also decrease the noise produced by the air moving through the duct system.

The reduction in horsepower experienced by inlet vane damper systems will be less than that achieved using a variable frequency drive in the same application. The reason for the reduction in efficiency is based upon the fan laws. When the static pressure of the air in the duct changes in response to changes in the position of the inlet vane dampers, the change in the amount of power drawn by the fan motor is a function of the *square* of the pressure change (ΔP^2). When the static pressure changes in response to changes in the frequency of the motor's power source, the change in the amount of power consumed by the fan motor is a function of the cube of the change in the motor speed (Δrpm^3). When selecting the best method to modulate static pressure, an economic analysis is performed to determine the life cycle cost comparison between the higher capital cost of a variable frequency drive and the simpler, less costly inlet vane dampers.

19.1.4 Static Pressure Control Loop Response

The duct static pressure control loop maintains the static pressure set point with respect to the pressure at a reference location. The set point value is selected so the static pressure at the sensor's location is high enough to insure that all zones located downstream of that location have sufficient pressure available to provide their design load air flowrate. Since the static pressure upstream of the sensor will always be greater than the pressure at the sensor location, the zone control dampers located upstream of the sensor will always have sufficient static pressure to provide their design load flowrate.

The position of the inlet vane dampers or the output frequency of the VFD is varied in response to changes in the duct static pressure. As the thermal load in the zones decrease, their dampers begin to close and the static pressure in the duct increases. The static pressure sensor measures this change in duct pressure and sends a corresponding signal to the controller. The controller responds by reducing the signal to the inlet vane dampers or VFD, reducing the capacity of the fan. As the fan capacity is reduced, the static pressure in the duct decreases back toward its set point. The control loop configuration for a duct static pressure process is the same whether a variable speed drive or inlet vane dampers are used as the final controlled device. The sensor and its location in the duct along with the controller's mode and action are the same regardless of the type of final controlled device. The output signal from the controller either terminates at an inlet vane damper actuator or on the input terminals of the variable frequency drive.

19.2

STATIC PRESSURE SENSORS

Static pressure control strategies use a differential static pressure sensor to measure the control point inside the air duct. The sensor's probe is a simple pitot tube or straight piece of $\frac{1}{4}$" diameter copper pipe inserted into the duct at the proper location within the distribution system. The pitot or copper tube is inserted perpendicular to the duct wall as shown in Figure 19.3. It extends into the duct, $\frac{1}{2}$ to 1 inch beyond the duct wall. The orientation of the sensor is critical to the proper operation of the control loop. If the tube is tipped slightly into the airstream it will measure the static pressure plus some additional velocity pressure. Consequently, the sensor will generate an inaccurate signal of a higher magnitude than the actual static pressure. If the tube is tipped away from the direction of flow of the airstream, it records a lower than actual pressure. The sensor may also induce pulsations in the output signal due to air turbulence that forms on the downstream side of the sensor.

The pitot tube measures the pressure that the air exerts on the duct wall. A capacitor or mechanical sensor compares the pressure in the duct to a reference pressure and generates an output signal that is proportional to the difference in

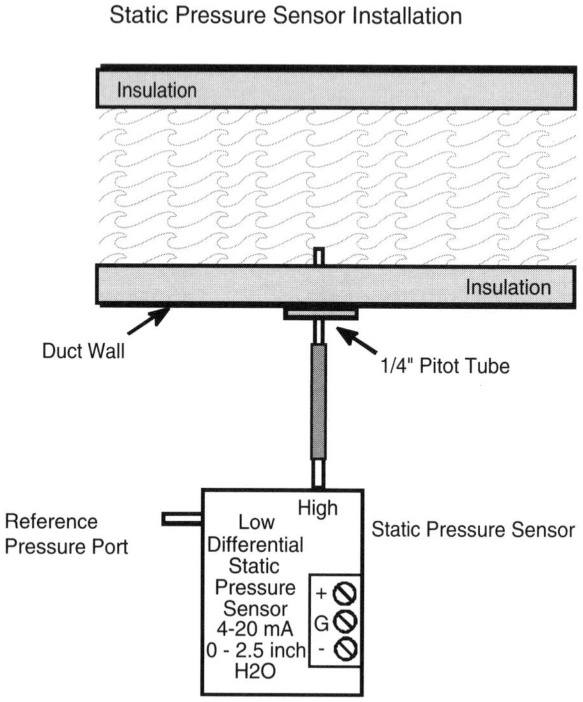

Static Pressure Sensor Installation

Figure 19.3   Static Pressure Sensor Installation Detail

static pressure. Each sensor housing has two ports that are connected to the sensing element's signal tubing. One of the ports is labeled as the high pressure port and the other as the reference or low pressure port. In supply duct static pressure applications, the high pressure port is connected to the pitot tube that is inserted through the duct wall. The reference port is either left open to the ambient surrounding the sensor housing or piped to another location in the building. If the reference port is located any distance away from the sensor housing, a large diameter tube must be used to minimize the response time of the sensor. If the reference port is located in a turbulent pressure zone, near exit doors, elevator shafts or open windows, the loop may become unstable due to the fluctuations of the reference pressure coupled with the transmission lag of the sensor element's signal.

Static pressure sensors are typically located about two-thirds of the distance downstream of the fan in the longest or critical duct run. This general location is satisfactory for most installations. The static pressure set point can be modified slightly during system commissioning so the last VAV box has sufficient flow during design load operation. The sensor's output signal changes whenever the static pressure in the duct or the reference pressure changes. When the reference pressure decreases, the inlet vane damper or VFD reduces the air flow through the fan to reduce the differential pressure measured by the sensor. The same response occurs when the reference pressure remains constant but the pressure in the supply duct increases.

19.3

ILLUSTRATION OF A SUPPLY AIR STATIC PRESSURE CONTROL STRATEGY

This example illustrates a supply air static pressure process incorporating inlet vane dampers on the fan's housing.

A reverse acting static pressure control loop modulates the fan's normally closed inlet dampers by varying the control signal to the pneumatic actuator in response to changes in the static pressure of the air in the supply duct. As the static pressure in the supply duct decreases, the reverse acting controller's output signal increases, opening the inlet vanes. This action decreases the amount of spin imparted to the air entering the fan thereby increasing the energy transferred to the airstream and its flowrate. This increase in flowrate raises the static pressure in the duct back to its set point.

19.3.1 System Drawing

A system drawing for a supply duct static pressure control strategy incorporating inlet vane dampers is shown in Figure 19.4.

Static Pressure Control Strategy Using Inlet Vane Dampers

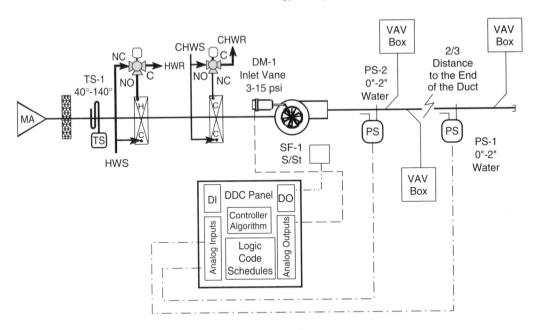

Figure 19.4   Supply Fan Static Pressure Control Strategy

19.3.2 Sequence of Operation

The sequence of operation for the supply air static pressure process incorporating inlet vane dampers that is illustrated in Figure 19.4 is explained in this section.

System Overview

The supply air static pressure control loop modulates the inlet vane dampers on the supply fan to maintain the static pressure of the supply air duct at 1.3 inches of water. The static pressure control strategy operates whenever the air handling unit's fan is commanded on.

Pre-Occupied Conditioning

The system fan is commanded on at the beginning of the pre-occupied conditioning period. The static pressure control loop modulates the supply fan inlet vane dampers as outlined in the Occupied Operation section.

Occupied Operation

The static pressure sensor (PS_1) sends a 4 to 20 mA signal to the DDC panel proportional to the static pressure measured in the duct. The reverse acting, proportional + integral control algorithm generates a 3 to 15 psi output signal to modulate the normally closed inlet vane damper actuator (DM_1) to maintain the loop set point of 1.3 inches of water.

When the supply duct static pressure increases in response to the VAV zone dampers throttling toward their minimum position, the output signal to the damper (DM_1) decreases, modulating the inlet vane dampers toward their closed position thereby reducing fan capacity and motor power consumption.

Unoccupied Operation

When the system enters its unoccupied period, the air handling unit's supply fan is commanded off and the inlet vane dampers are commanded to their normal (closed) position.

Emergency Operation

Whenever the air handling unit's fan is commanded off due to the fan's failure to prove operation, a smoke detector indicating smoke, a low temperature limit trip or high static pressure at the fan outlet, the supply fan is commanded off and the inlet vane dampers are commanded to their normal (closed) position.

If the static pressure in the duct rises above 1.8 inches of water column, as sensed by P_2, the supply fan will be commanded off and the inlet vane dampers (DM_1) commanded to normal position. An alarm will be initiated by the system database notifying the building operator that a problem exists and the system will have to be manually reset after the problem has been repaired.

19.3.3 Flowchart

A flowchart depicting the logic in the sequence of operation for the system shown in Figure 19.4 is shown in Figure 19.5. Section 19.3.4 provides an analysis of this flowchart.

19.3.4 Flowchart Analysis

The start oval indicates the beginning of the supply fan's static pressure control logic. The first decision diamond determines whether a system fault has occurred. If a fault occurred, the fan is commanded off and the inlet vane dampers commanded to their normal position. The next decision determines if the unoccupied period is over. If it is, the fan is started and the inlet vane dampers modulate to maintain the static pressure set point. If the system is in its unoccupied mode of operation, the fan is commanded off and the inlet vane dampers commanded to their normal position.

19.3.5 Strategy Summary

Variable air volume systems are the most common type of air handling unit currently being installed in larger commercial buildings. The ability of this configuration to handle a multiple thermal loads simultaneously with a minimum of reheat

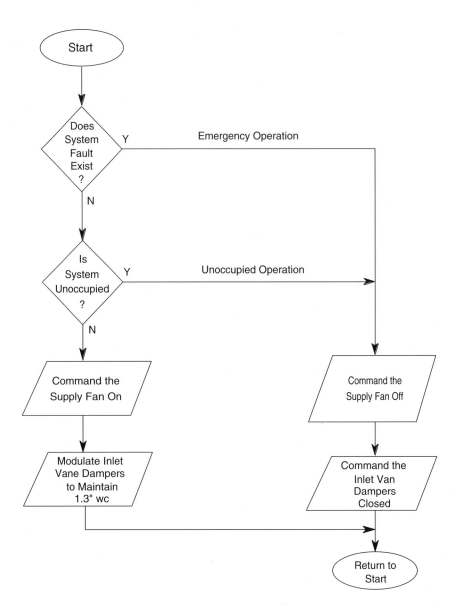

Figure 19.5   Supply Air Static Pressure Control with Inlet Vane Dampers Flowchart

results in an efficiently operating system. This type of air handling unit only retains its energy efficient status if the equipment was installed correctly and the control loops are kept properly tuned.

19.4

ILLUSTRATION OF A DUAL FAN STATIC PRESSURE CONTROL STRATEGY

This example illustrates a dual fan static pressure process incorporating VFDs serving the supply and return fan motors.

When a variable air volume air handling unit has a supply and a return fan, they must be modulated correctly to prevent building pressurization problems from occurring. These problems decrease the efficiency of the process and can cause the static pressure loops to become unstable. A return fan is sized to move less air than the system's supply fan. The amount of return air volume is equal to the supply fan volume minus the fixed exhaust requirements (lavatory, mechanical room, closet etc. fans) of the zones. Reducing the return volume slightly beyond the fixed exhaust quantity positively pressurizes the building, reducing the infiltration of unconditioned air into the space. When the supply fan's capacity is being modulated, the return fan must also be modulated to maintain the fixed relationship between the two airflow rates.

Some installations modulate the final controlled devices of both fans using the same control signal. Consequently, as the supply fan increases (or decreases) its capacity in response to changes in the static pressure of the supply duct, the return fan's response follows (tracks). This strategy requires the return fan's inlet vane damper linkage or its VFD characteristics to be properly calibrated to maintain the correct relationship between the supply, return and fixed exhaust air volumes. If the fans are not balanced correctly, building doors will whistle as excess air escaped through the cracks or the doors become difficult to open because the building was under a slight negative pressure (vacuum).

Another method of synchronizing dual fan operation is through the use of air flow measuring stations installed in the supply, return and outside air ducts. Flow measuring stations use pitot tubes or vortex shedding sensors to measure the velocity of the air flowing through the duct. The velocity is calculated from the velocity pressure measured by the flow sensor. The pitot or vortex sensors are assembled into a grid that is mounted in the duct, perpendicular to the airflow. The size of the grid is based upon the area of the duct so it can measure the entire flow path. Each sensor on the grid generates a signal that is proportional to the velocity of the air at that location in the duct. The signals from the individual sensing units are mathematically averaged together to generate a signal that represents the average velocity pressure of the air passing through the air flow measuring station. When the velocity is multiplied by the duct's area, the volume flowrate is calculated (cfm).

Indoor air quality issues in some buildings have driven the HVAC industry toward the use of air flow measuring stations located in the supply, return and outside air ducts to obtain a physical measurement representing the air volumes in these ducts. These stations are used with DDC systems that can use their mathematical and custom programming capabilities to insure that the correct amount of ventilation air is brought into the building during all modulating conditions. When the amount of ventilation air entering a VAV system must be verified and monitored, air flow measuring stations are the only strategy that allows the operator to read and trend the flowrate.

Pitot tubes can be used in applications where the air's velocity remains above 700 feet per minute. When the air's velocity drops below this threshold, the pitot tube sensor's accuracy changes to $+/- 40\%$. Vortex shedding sensors are a flow measuring device that is more sensitive at lower flow rates than most pitot tubes. These sensors place a trapezoidal shaped obstruction in the flow path of a tube. As air passes around the obstruction, eddy currents are produced in direct proportion to the velocity of the air passing around the trapezoid's edges. Each eddy current produces a pressure pulse in the airstream that is counted by an electronic circuit connected to the sensor. Figure 19.6 shows a schematic of the vortex shedding sensor.

The sequence of operation for dual fan systems uses a differential static pressure sensor to maintain the *static pressure set point* in the supply duct to modulate the capacity of the supply fan as described in Section 19.2. The return air fan's capacity is modulated to maintain a volume *flowrate set point* that is equal to the measured supply air volume minus the fixed exhaust requirements of the system. The fixed exhaust requirements may be reduced by a small volume to permit the establishment of a slight positive pressure within the building.

19.4.1 Dual Fan Static Pressure Control Loop Components

The dual fan static pressure control strategy requires the hardware components shown in Table 19.1.

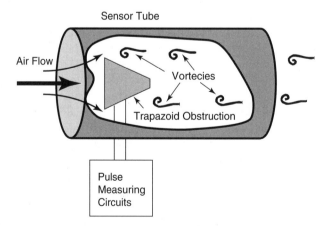

Figure 19.6   Vortex Shedding Sensor

Table 19.1 DUAL FAN STATIC PRESSURE CONTROL LOOP COMPONENTS

Number	Quantity	Description	Signal Type
1	1	Static Pressure Sensor (0 to 2.0″ water)	Analog Input
2	2	Variable Frequency Drives (4 to 20 mA)	Analog Outputs
3	3	Air Flow Measuring Stations (4 to 20 mA)	Analog Inputs
4	2	Isolation Control Relays Start/Stop Drives	Digital Outputs
5	2	Static Pressure Limit Switches	Digital Inputs
6	1	DDC Controller Panel	All

19.4.2 System Drawing

A system drawing for a dual fan static pressure control process incorporating variable frequency drives is shown in Figure 19.7.

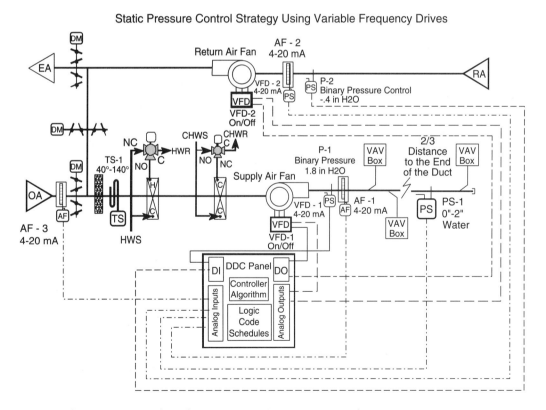

Figure 19.7   Supply and Return Air Static Pressure Control

19.4.3 Sequence of Operation

The sequence of operation for the dual fan static pressure process incorporating variable frequency drives that is illustrated in Figure 19.7 is explained in this section.

System Overview

The supply air static pressure control loop modulates the supply fan's variable frequency drive (VFD_1) to maintain the supply duct static pressure at 1.5 inches of water. Flow measuring stations are installed to measure the supply air flowrate and calculate the return air's flowrate set point.

The return fan's drive is modulated to maintain the difference in flowrate between the supply and return air quantities equal to the fixed exhaust requirements (1500 cfm) of the building.

The outside air flow measuring station (AF_3) is used to modulate the mixing dampers to maintain minimum position airflow. These strategies are enabled whenever the fans are commanded on based upon their programmed operating schedule. Both analog and binary signals are required to operate a variable frequency drive. The fans use a binary output signal from the DDC panel to cycle the isolation control relays that start and stop the drives. The analog signal is used to modulate the drive's output frequency.

Pre-Occupied Conditioning:

The fans are commanded on at the start of the pre-conditioning period. The static pressure control loop modulates the supply fan to maintain 1.5 inches of water pressure at the supply static pressure sensor's (PS_1) location. The return air flow measuring station (AF_2) is used to modulate the return air fan's capacity as outlined in the Occupied Operation section.

Occupied Operation

Supply Fan Control A static pressure sensor (PS_1) transmits a 4 to 20 mA signal to the DDC panel indicating the pressure in the supply duct at the sensor's location. The reverse acting, proportional + integral control algorithm generates a 4 to 20 mA output signal to modulate the supply fan's variable frequency drive (VFD_1) to maintain the loop set point of 1.5 inches of water.

As the static pressure in the supply duct increases in response to the VAV zone dampers throttling toward their minimum position, the output signal to the VFD decreases, modulating the drive's output to a lower frequency thereby reducing the supply fan's capacity and power consumption.

Return Fan Control As the static pressure in the supply duct modulates VFD _1 to a different frequency, the air flow volume measured by AF_1 also changes. An algorithm in the DDC panel subtracts the building's fixed exhaust requirement

(1500 CFM) from the control point of the supply air measuring station (AF_1). The result is the volume flowrate set point for the return fan.

The air flow measuring station AF_2 transmits its signal to the DDC panel that indicates the current return air flowrate. The reverse acting, proportional + integral control algorithm generates a 4 to 20 mA output signal that is sent to the return fan's variable frequency drive (VFD_2) to modulate the output frequency of the return fan motor to maintain the new volume flowrate set point.

As the static pressure in the supply duct decreases in response to the VAV zone dampers throttling toward their wide open position, the signal from the DDC panel to VFD_1 increases, modulating the drive to a higher frequency thereby increasing the supply fan's volume flowrate. A new set point for the return fan's capacity is calculated. The signal to the return fan's VFD increases to increase the speed of the fan to maintain the higher return air flowrate.

Mixed Air Minimum Position Damper Control   The mixing dampers modulate to maintain a mixed air temperature set point. Whenever the system is commanded to minimum position, the mixing dampers modulate based upon the signal generated by air flow measuring station AF_3. The set point for the mixing damper ventilation air control loop is calculated at 20% of the current supply air volume being measured by station AF_1.

Unoccupied Operation

When the air handling unit's fan is scheduled off, the binary signals to both VFD isolation control relays are commanded off and the analog signals to both VFDs are commanded to 0% (4 mA).

Emergency Operation

Whenever either air handling unit fan is commanded off due to the presence of a system fault, the supply and return fans are commanded off and the analog signals to both VFDs are commanded to 0%.

If the static pressure in the supply duct rises above 2.8 inches of water, as sensed by P_1, or falls below -2.4 inches of water as sensed by P_2, the supply and return fans are commanded off through their isolation control relays and both analog signals for the VFDs are commanded to 0%. The DDC database generates an alarm that must be acknowledged by the building operator before the fans can be commanded back on.

19.4.4 Strategy Summary

Variable frequency drives are more energy efficient in terms of power consumption, power factor and starting current than inlet vane dampers. Variable frequency drive technology has developed to a point where the unit cost of a device has fallen to a level where they are cost competitive with inlet vane dampers. Combining a VFD with air flow measuring stations allows for a control strategy that can maintain correct building pressures and minimum ventilation air quantities while maximizing operational energy efficiency.

19.5

EXERCISES

Determine if the following statements are true or false. If any portion of the statement is false, the entire statement is false. Explain your answers.

1. Fans are used to transfer kinetic energy from their motor to the airstream.
2. As the static pressure in the duct increases, the power drawn by the fan's motor also increases.
3. Static pressure stores energy in the duct that is converted into kinetic energy to maintain air movement through the duct.
4. The faster a fan blade rotates, the higher the total pressure it creates in the duct.
5. As the zone dampers in the system modulate toward their closed position, the velocity pressure in the supply duct decreases and the static pressure increases.
6. Static pressure control loops have direct acting actions.
7. Inlet vane dampers are normally closed devices.
8. Static pressure sensors measure the differential pressure between the static and velocity pressures in a duct.
9. Variable speed drives alter the frequency of the fan motor's power source to alter the flowrate through the fan.
10. Inlet vane dampers pre-spin the air to reduce the transfer of kinetic energy to the airstream.

Respond to the following statements, questions and problems completely and accurately, using the material found in this chapter.

1. What advantage do normally closed inlet vane dampers have over normally open inlet vane dampers with respect to starting a fan?
2. What is the speed of a 4 pole motor operating at 45 hertz?
3. What is the response of the variable frequency drive if the present error of a static pressure control loop is zero and the reference pressure suddenly increases?
4. Explain how inlet vane dampers and VFDs alter the flow of air through a duct system.
5. Why are air flow measuring stations used in dual fan static pressure control applications?

6. Develop the flowchart for Figure 19.7.
7. Using the system depicted in Figure 19.4 as a reference, what is the response of the static pressure control loop when the outside air temperature drops from 90º to 80º in one hour? (Hint: the thermal loads within the zones decrease)
8. Why is a differential static pressure sensor and a flow measuring station required in the supply duct of a dual fan system? Could the same strategy be performed with the static pressure sensor removed? Explain your answer.
9. Using the system depicted in Figure 19.7 as a reference, what is the response of the return air volume flowrate control loop when the zone air temperatures drop? Explain your answer.
10. Using the system depicted in Figure 19.7 as a reference, what is the response of the supply air static pressure control loop when the fixed exhaust fans in the building are inadvertently turned off? Explain your answer.

CHAPTER 20

Zone Terminal Device Control

Zone terminal devices are components located in the vicinity of a zone that are used to maintain its temperature set point. These devices may be auxiliary parts of the building's air handling unit or stand-alone equipment that provides 100% of the thermal conditioning within a zone. Variable air volume boxes, constant volume boxes, fan coil units and unit ventilators are a few of the commonly used zone terminal devices. The following sections outline the operational and control characteristics of a sample of these terminal devices.

20.1

SINGLE DUCT VARIABLE AIR VOLUME BOXES

Variable air volume boxes are used to modulate the flow of conditioned air entering a zone in response to its thermal load. The zone's thermostat modulates a control damper located inside the VAV box in response to changes in temperature. As the temperature in the zone increases, the VAV damper opens more to increase the volume of cooler air entering the zone. Conversely, as the temperature in the zone decreases, the VAV damper modulates toward its minimum flow position to decrease the volume of cooler air entering the zone. This prevents the zone temperature from dropping below its set point.

The conditioned air that is delivered to the VAV box from the air handling unit is called *primary* air. VAV air handling systems are primarily cooling systems. The temperature of the primary air is maintained at a set point within the range of 50 to 65 °F by the air handling unit's supply air temperature control loop. The primary air temperature must always be cool enough to offset the cooling load in the warmest zone when its VAV damper is 100% open. If a zone requires additional heat because the primary air is too cool for its thermal load, the VAV box can be configured with a reheat coil. The reheat coil is sized with sufficient capacity to add the necessary heat to the primary airstream before it enters the zone. Baseboard fin tube convectors can also be installed around the perimeter of the zone instead of using the reheat coil. The operation of any heating devices within the zone is integrated with the VAV damper control system to maximize the operating efficiency of the process by prohibiting the addition of heat into the zone before the damper has been modulated to its minimum position.

20.1.1 VAV Box Construction

VAV boxes are constructed with a sheet metal enclosure that houses a modulating damper assembly along with the necessary control devices as depicted in Figure 20.1. The damper varies the amount of primary air that enters the zone. Pressure *dependent* VAV boxes position the damper using the room thermostat's output signal. Pressure *independent* boxes reset the flowrate set point of a volume controller with the zone thermostat's output signal to position the damper. Both of these applications are described in the following sections.

Internal Components of a VAV Box

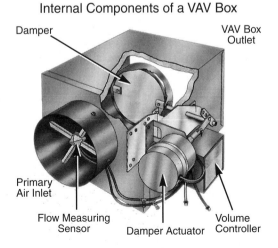

Figure 20.1   VAV Box Pictorial Diagram

20.1.2 Pressure Dependent VAV Boxes

The volume of air flowing through any type of damper or valve is related to the pressure difference that exists across the device's flow opening. As the pressure drop across the VAV air damper increases, the volume of air passing through the box also increases.

In an operating system, the static pressure in the supply duct varies in response to changes in the flowrates throughout the entire duct system. As these changes occur, the pressure drop across the VAV box dampers also varies. These changes alter the flowrate of primary air into the zones, even through those boxes whose dampers are not currently modulating. VAV boxes that respond to changes in duct's static pressure are called *pressure dependent* boxes. These boxes have no mechanical or control system that compensates for the normal pressure fluctuations that occur in the supply duct.

Pressure dependent boxes use their thermostat's output signal to reposition the VAV damper. As the zone's temperature decreases, the output signal from the thermostat modulates the damper toward its minimum or closed position. The closure of the damper causes the supply duct's static pressure to momentarily increase before the supply air static pressure control loop responds. During this period of transition that results from time lags in the system, the flowrate of primary air increases through all the connected zones utilizing pressure dependent boxes. The primary air's flowrate returns to normal after the supply fan's static pressure control loop responds to the increases in duct pressure.

Pressure dependent boxes are less expensive to build than boxes that are designed to be pressure independent. They do not require any additional, more sophisticated, control devices to compensate for changes in the primary air duct's static pressure. Pressure dependent boxes can be used in applications with larger capacitances that can absorb the momentary fluctuations in air flowrate caused by changes in the supply duct's static pressure. The inability of pressure dependent boxes to maintain tight control over their volume flowrate makes their processes less efficient than those incorporating pressure independent boxes.

20.1.3 Pressure Independent VAV Boxes

Pressure independent boxes were developed to overcome the problems and inefficiencies associated with pressure dependent VAV boxes. Pressure independent boxes utilize a primary and a secondary control loop to maintain the correct flowrate through the VAV damper whenever variations occur in the duct's static pressure. The primary control loop maintains a volume flowrate set point by modulating the box damper in response to changes in the supply duct's static pressure. The secondary control loop resets the flowrate set point of the primary loop based upon the zone's current temperature. This control strategy results in a VAV box response that is immune to the normal fluctuations that occur in the supply duct's static pressure.

Pressure independent boxes require an additional sensor and controller to maintain the volume set point (cfm). A pitot tube, vortex shedding sensor, flow ring, hot wire anemometer or other sensor is used to measure the velocity pressure of primary air entering the VAV box. The velocity pressure is related to the velocity of the airstream, which when multiplied by the area of the pri   air duct yields the volume flowrate through the box. Whenever the static pressure in the primary air duct changes, the closed loop volume control strategy utilizes the sensor's signal to reposition the damper in a manner that maintains the correct volume flowrate through the damper. These variations in damper position occur without any change in the input signal from the zone's thermostat.

The set point of the volume flowrate control loop is based upon the zone's temperature and the minimum and maximum flowrate limits (throttling range) of the VAV box controller. These minimum and maximum limits define the range of volume flowrates for a VAV box application and are calculated during the design stage of the system. The minimum flowrate set point is based upon the ventilation and air change requirements of the zone. The maximum set point is based upon the design day cooling load requirements and the temperature of the primary air. These set points are input into each VAV box controller when the building is commissioned.

The zone's thermostat is used to maintain the desired temperature set point. The output signal from the thermostat resets the volume flowrate set point of the VAV box based upon the current error between the zone's set point and current temperature. Whenever the zone calls for cooling, the signal from the thermostat resets the volume flowrate set point to a higher value, increasing the flowrate of primary air entering the zone. The volume control loop still maintains the box's pressure independence at its new set point. When the cooling load of a zone decreases, the thermostat's output signal reduces the volume flowrate set point, reducing the quantity of air entering the zone through the box damper. The graph in Figure 20.2 depicts the relationship between the thermostat signal (zone temperature) and the VAV box volume set point. Pressure independent boxes allow for better zone temperature control through the addition of the box volume

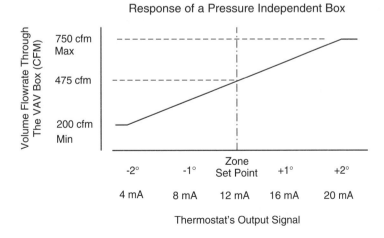

Figure 20.2   Pressure Independent VAV Box Response

control loop. These loop components add additional cost to the box which is recovered through greater process operating efficiencies.

20.1.4 VAV Box Configurations

VAV boxes are available in a multitude of different configurations and sizes to meet the requirements of most systems and zones. They can be specified with pressure dependent or independent flow characteristics, normally open or normally closed primary air dampers, secondary fans, pneumatic or electronic controls, pneumatic or electric actuators or reheat coils. The relationship between damper modulation, fan control and reheat valve operation can be specified before shipping or altered in the field to fit the characteristics of the process. Consequently, the same general VAV box type can be configured to operate with one of several different sequences of operation. Although there are a variety of box sequences commonly used, they all share the same fundamental control responses, similar to the examples shown in the following sections.

20.2

ILLUSTRATION OF A COOLING ONLY VAV TERMINAL UNIT

This example illustrates a zone temperature control process incorporating a cooling only VAV box terminal unit.

A cooling only variable air volume box has no reheat coil and modulates the flow of cool primary air into a zone to maintain its temperature set point. They are used in applications where a zone only experiences net heat gains throughout the year. Interior zones that are surrounded by conditioned spaces generally have occupancy and equipment loads that produce an excess amount of heat year-round. These zones are prime candidates for cooling only VAV boxes.

Cooling only boxes can be used in exterior zones if an auxiliary means of heat transfer into the zone is provided. Baseboard convectors are typically installed on exterior walls to add sensible heat to the zone and reduce the heat transfer by radiation to the cold wall surfaces. If the convector system is sized to serve the winter design load, cooling only VAV boxes can be used to supply ventilation air during the winter and cooling in the summer. Reheat coils are also available that are mounted on the outlet side of the VAV box. These coils transfer additional heat to the zone during cold weather or warm-up operations. All heating strategies must be integrated with the box's cooling functions to prevent the VAV box from cooling the zone when the heat source is in the process of raising its temperature.

20.2.1 Sequence of Operation

The sequence of operation for the zone temperature control process incorporating a cooling only VAV box that is illustrated in Figure 20.3 is explained in this section.

Cooling Only - Pressure Independent VAV Box

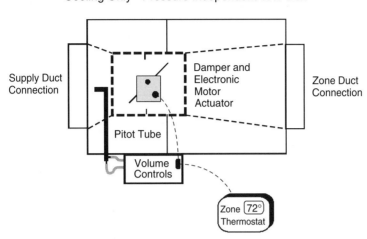

Figure 20.3   Cooling Only VAV Box

System Overview

The zone's thermostat resets the flowrate set point of the cooling only VAV box to vary the volume of primary air flow entering the zone to maintain its temperature set point.

Occupied Operation

The DDC electronic thermostat has a programmed cold weather set point of 72° and a warm weather set point of 75°. The thermostat also has a limited adjustment range equal to the set point $+/-$ 2° to provide occupant control over the zone's temperature.

The thermostat generates a 4 to 20 mA signal that is transmitted to the VAV box volume controller. The direct acting controller algorithm modulates the normally closed VAV box damper between its minimum and maximum flowrate positions based upon the zone's temperature offset.

As the zone's temperature increases, the thermostat signal to the VAV box volume controller also increases, resetting the volume flowrate set point to a higher value. The VAV box controller generates an increasing signal to its primary air damper actuator, modulating the damper toward its maximum flow position. The Figure 20.4 shows a graph of the response of the damper throughout the throttling range of the thermostat.

Unoccupied Operation

During the air handling unit's unoccupied period, the primary air flow to the VAV box ceases. The primary air damper is commanded closed to reduce natural convection currents from forming between the zone and the air handling unit.

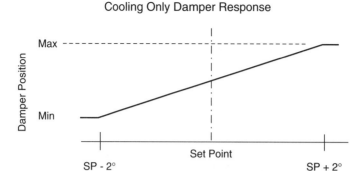

Figure 20.4   Cooling Only VAV Box Damper Response

20.3

ILLUSTRATION OF A VAV TERMINAL UNIT WITH A REHEAT COIL

This example illustrates a zone temperature control process incorporating a VAV box terminal unit with an integrated reheat coil.

A cooling VAV box with an integrated reheat coil sequences the modulation of primary air damper with the heating coil's control valve to maintain the zone temperature set point. They are used in applications where the zone experiences both heat gains and losses throughout the year.

The reheat coil warms the primary air as it leaves the VAV box. Reheat coils are only permitted to operate after the zone's thermostat signal has reduced the primary air flowrate to its minimum flow limit. This strategy minimizing the mass of air being reheated. Reheat VAV boxes are connected to diffusers that discharge air over the exterior walls to reduce the radiation effects of cold surfaces. Figure 20.5 shows photographs of hot water and electric reheat coils used in VAV box applications.

20.3.1 Sequence of Operation

The sequence of operation for the zone temperature control process incorporating a VAV box with an integrated reheat coil that is illustrated in Figure 20.6 is explained in this section.

System Overview

The zone's thermostat modulates the flow of primary air and hot water to maintain the zone's temperature set point. The normally closed primary air damper is sequenced through software with the normally open reheat coil control valve. The reheat valve can only begin to modulate open after the primary air damper has modulated to its minimum flowrate setting.

VAV Box Coils

Hot Water Coil

Electric Reheat Coil

Figure 20.5   Reheat Coils for VAV Boxes

VAV Box With Reheat Coil

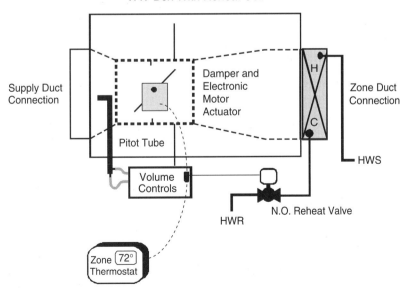

Figure 20.6   Cooling VAV Box With Reheat

Pre-Occupied Conditioning

When the air handling unit is commanded on and the zone temperature is below set point, the VAV box damper and the reheat valve are commanded 100% open to quickly bring the zone temperature to its occupied set point. Once the zone reaches set point, the VAV box modulates as described in the occupied operation section.

Occupied Operation

The DDC electronic thermostat has a programmed cold weather set point of 72° and a warm weather set point of 75°. The thermostat also has a limited adjustment range of set point $+/- 2°$ to provide occupant control over the zone temperature.

The thermostat sends a 4 to 20 mA signal to the volume controller located on the VAV box. The direct acting controller algorithm modulates the sequenced, normally closed primary air damper with the normally open reheat valve as a function of the room temperature.

As the zone temperature decreases, a reduction in the signal to the volume controller reduces the volume flowrate set point. The volume controller reduces its output signal to the box damper actuator motor, modulating the damper toward its minimum flow position. If the zone's temperature continues to drop, the normally open reheat valve begins to modulate open. The graph in Figure 20.7 shows the response of the damper and reheat valve throughout the throttling range of the thermostat.

Unoccupied Operation

During the air handling unit's unoccupied period, the primary air flow to the VAV box ceases. The primary air damper is commanded closed to reduce natural convection currents from forming between the zone and the air handling unit. The reheat valve is also commanded closed.

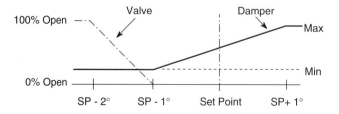

Figure 20.7   VAV Box with Reheat Damper and Valve Response

20.4

FAN POWERED VAV TERMINAL UNITS

Fan powered VAV boxes have a fractional horsepower squirrel cage fan located inside its enclosure to increase the volume of air discharged from the box under certain operational conditions. Fan powered VAV boxes use the space above the ceiling as a return air plenum. The box recycles some of the warmer air in this plenum to reduce the amount of energy that would otherwise be added by the reheat coil. Fan powered boxes have a primary air damper similar to those found in other VAV box designs. An second intake opening exists next to the primary air access that provides the passage between the return air plenum and the VAV box fan. This opening typically has a throw-away filter mounted over its entrance. The construction of the VAV box fan is dependent upon whether the box fan operates continuously or intermittently. Figure 20.8 shows a schematic of both types of fan powered boxes.

20.4.1 Constant Volume Fan Powered VAV Boxes

Constant fan powered boxes have their fan installed in *series* with the primary air duct and damper. This fan only operates when the air handling unit's fan is operating. These boxes produce a constant flowrate of air into the zone. The fan draws air from the primary air duct and from the return air (ceiling) plenum, mixes them together as they pass through the fan and discharges the mixture into the zone. The temperature of the discharged air is varied by regulating the amount of primary air in the mixture.

Constant fan VAV boxes produce a constant volume, variable temperature response. The constant volume maintains the design air distribution patterns within the zone because the flowrate through the diffusers remains the same. This response is less noticeable to the zone occupants than systems that vary flowrates. Varying flowrates generate changes in the air movement and diffuser noise levels. It also changes the air distribution and velocity patterns within the zone as the dampers throttle between their minimum and maximum flowrates. Constant fan VAV boxes are used to reduce the size of the air handling unit's fan by supplying a portion of the fan power requirements at each box location.

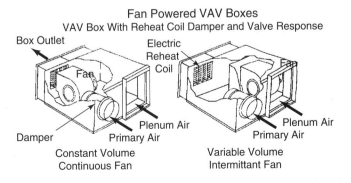

Figure 20.8   Fan Powered VAV Boxes

20.4.2 Variable Volume Fan Powered VAV Boxes

Intermittent fan powered VAV boxes have the fan installed in *parallel* with the primary airstream. When the box fan is commanded on, it draws all of its air from the return air plenum. It discharges this airstream into the box, mixing it with the air passing through the primary air damper. The mixed air is discharged into the zone. The box fan can operate whether the air handling unit's fan is on or off.

Intermittent fan powered boxes are used in applications where a typical VAV box operating at its minimum position flowrate, is not capable of moving sufficient air into the zone to provide adequate air circulation. In these applications, the box fan is cycled on when the primary air damper modulates to its minimum position, drawing warmer air from the ceiling plenum and mixing it with the cooler primary air. The resulting increase in the volume of air being discharged into the zone is at a temperature that maintains thermal comfort and air movement within the zone. In both continuous and intermittent VAV box fan powered applications, the box fan and reheat valve can be configured to cycle on during unoccupied periods to maintain a low limit temperature set point in the zone.

20.5

ILLUSTRATION OF AN INTERMITTENT FAN POWERED VAV TERMINAL UNIT WITH A REHEAT COIL

This example illustrates a zone temperature control process incorporating an intermittent fan powered VAV box terminal unit with an integrated reheat coil.

An intermittent fan powered VAV box with a reheat coil sequences the operation of the primary air damper, box fan and reheat coil to maintain the zone's temperature set point. The reheat coil is used to supply additional heat to the zone during periods when the plenum air is not sufficiently warm to satisfy the heating load of the zone. Using a strategy similar to that used for VAV boxes with reheat coils, the heating coil and intermittent fan are only activated after the zone thermostat has reduced the flowrate of primary air to its minimum value.

20.5.1 Sequence of Operation

The sequence of operation for the zone temperature control process incorporating an intermittent fan powered VAV box with an integrated reheat coil that is illustrated in Figure 20.9 is explained in this section.

System Overview

The zone's thermostat modulates the primary air flow into the zone to maintain its temperature set point. The normally closed primary air damper is sequenced

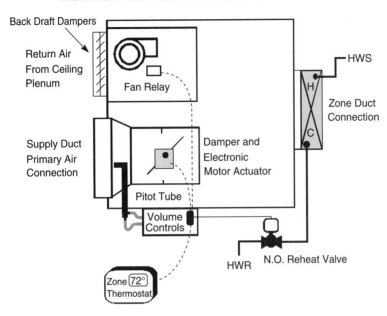

Figure 20.9   Intermittent Fan Powered VAV Box With Reheat

through software with the normally open reheat coil control valve and the box fan to provide energy efficient comfort conditioning.

The box fan is commanded on whenever the primary air damper has modulated to its minimum position. When the fan is operating, the return air backdraft damper opens fully due to the static pressure differential created by the fan, permitting the warmer plenum air to mix with the primary airstream before exiting the box. The reheat valve modulates open to maintain the zone temperature set point whenever the primary air damper is at its minimum flow position and the zone temperature falls below its set point

Pre-Occupied Conditioning

When the air handling unit is commanded on and the zone temperature is below set point, the box fan is also commanded on. The plenum return air damper opens and the reheat valve is commanded 100% open to quickly bring the zone up to set point. Once the zone reaches set point, the VAV box modulates as described in the occupied operation section.

Occupied Operation

The DDC electronic thermostat has a programmed cold weather set point of 72° and a warm weather set point of 75°. The thermostat also has a limited adjustment range of set point $+/-2°$ to provide occupant control over the zone temperature.

Intermittent Fan Powered VAV Box Damper,
Fan and Valve Response

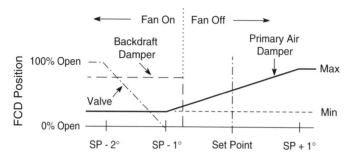

Figure 20.10   Valve and Damper Response for an Intermittent Fan Powered VAV Box

The thermostat sensor sends a 4 to 20 mA signal to the primary air volume controller located on the VAV box. The direct acting controller algorithm modulates the normally closed primary air damper to maintain the zone temperature set point. When the primary air damper is modulated to minimum position in response to a drop in the zone temperature, the box fan is commanded on. If the temperature continues to decrease, the normally open reheat control valve is modulated to maintain the zone temperature set point.

Figure 20.10 shows the damper and valve response for this application. Note, the maximum volume of air that can be drawn by the box fan is less than the maximum volume supplied by the primary air supply. There is also a temperature deadband between the time the box fan is commanded on and the hot water valve begins to modulate open to determine whether the plenum air is sufficiently warm to offset the initial heat load.

Unoccupied Operation

During the air handling unit's unoccupied period, the primary air flow to the VAV box ceases. The primary air damper is commanded closed to reduce natural convection currents from forming between the zone and the air handling unit. If the zone temperature falls below its low limit set point, the box fan is cycled on and the reheat coil valve commanded 100% open to maintain the zone's low limit set point.

20.6

DUAL DUCT BOXES

Dual duct air handling units simultaneously condition two separate airstreams, one hot and one cold, which are transported to each zone through two parallel ducts. A dual duct terminal device called a *mixing box* is located in each zone to blend the two airstreams together to create a discharge air temperature that satisfies the current load in the zone. Supplying a heated and cooled airstream to each zone allows the air handling unit to satisfy the different loads of all the zones simultaneously.

There is no need for reheat coils or perimeter heating systems in most dual duct installations because these systems supply warm air and cold air to each zone throughout the year. This configuration produces a very comfortable environment at a cost of reduced process efficiency.

Figure 20.11 shows photographs of two dual duct mixing boxes. These boxes are available in different configurations for both constant volume and variable volume air handling unit applications. In constant volume box applications, pneumatically controlled boxes incorporate a spring driven volume regulator to compensate for changes in the static pressure of the hot and cold supply ducts. Boxes employing DDC based control systems employ flow measuring devices similar to those in pressure independent VAV boxes to compensate for varying duct static pressures.

There are two modulating dampers in each dual duct mixing box. One damper for each of the inlet air ports. The dampers are configured to operate

Dual Duct Mixing Box

Dual Duct VAV Box with Reheat

Figure 20.11 Dual Duct Mixing Boxes

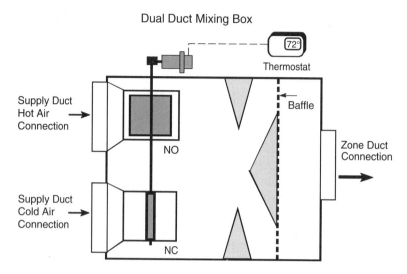

Figure 20.12   Constant Volume Mixing Box

with complementary responses to changes in their control signals. As the signal to the actuator increases, the normally open heated air damper modulates toward its closed position as the cooled air damper modulates opens. Both dampers can be linked together on a common shaft so they operate with one actuator. A sheet metal baffle promotes the mixing of the two air streams before they are discharged from the box. The baffle also attenuates the noise created as air moves past the closing dampers. Figure 20.12 shows a schematic representation of the inside of a constant volume mixing box.

The zone's thermostat connects directly to the damper actuators of the mixing box. At all times other than during design load operation, the mixing box is blending the heated and cooled air streams to satisfy the zone's load. This is an energy intensive process. DDC control systems allow more complex control strategies to be utilized to minimize the mixing of the cold and warm air streams, improving the efficiency of the process. These strategies incorporate individual damper actuators along with two volume flowrate sensors and controllers. The added cost of this control system is justified by the cost avoidance of the energy needed to condition the zone.

20.7

VARIABLE VOLUME DUAL DUCT TERMINAL UNIT

Dual duct terminal units modulate the flow of heated and cooled air into a zone to maintain its temperature set point. A variable volume dual duct box is a pressure independent terminal unit that modulates the flow of air in response to the signal from the zone's thermostat. Two independently operating actuators are required to permit individual modulation of the heated and cooled air thereby limiting the amount of mixing that occurs and improving the process efficiency.

Dual duct VAV mixing boxes have two volume controllers and flow measuring devices to provide pressure independent volume control of the box's discharge air volume. One sensor and controller are used to position the damper in the cold air connection. This control loop is configured to regulate the maximum cold duct air flow needed for design cooling load operation. The other controller and flow measuring device are used to regulate the box's heated air damper to maintain a minimum air flowrate into the zone along with the maximum heated air volume flowrate. The zone thermostat's output signal resets the discharge volume controller flowrate as shown in the graph in Figure 20.14. The shaded box shows the only interval where mixing of heated and cooled air is allowed by the sequence of operation. The shaded box shows the flow of heated air gradually diminishes as the zone's temperature approaches its set point. Under these conditions, the flow of cooled air is increased, mixing with the minimum flow of hot air to maintain a minimum flowrate into the zone to satisfy its ventilation requirements. In the response shown in Figure 20.14 the maximum volume flowrate of heated air is less than the maximum volume flowrate of cooled air due to the higher energy content of the heated airstream.

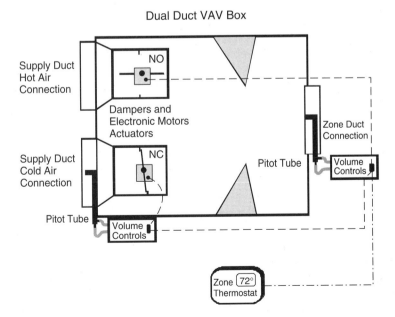

Figure 20.13   Dual Duct VAV Box

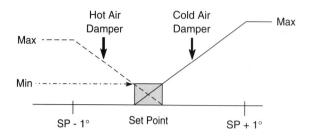

Figure 20.14   Dual Duct VAV Damper Response

20.8

CONSTANT VOLUME TERMINAL UNITS

Constant volume terminal boxes are single duct devices used in applications where the primary air is supplied by a VAV air handling unit but the flowrate into the zone must be maintained at a constant value. Constant volume boxes incorporate volume measuring circuits as described in pressure independent VAV box applications to maintain the volume flowrate regardless of changes in the supply duct static pressure. Once installed, the volume controller is tuned to the correct volume flowrate. The flowrate set point is never reset by another control device. These boxes can be specified with reheat coils for applications where the zone may require supplemental heating. In reheat applications, the thermostat is connected directly to the reheat valve instead of the volume flowrate controller. As the temperature in the zone decreases, the valve is modulated open to reheat the air being discharged from the box. Figure 20.15 shows the schematic diagram for a constant volume box.

20.9

FAN COIL UNITS

Fan coil units are small, distributed air handling units that are located in the conditioned space. They are used to locally heat or cool the zone. Air is drawn from the zone and passed across heating and cooling coils to maintain the zone's temperature set point. Fan coil units are manufactured in a variety of sizes and configurations to satisfy most single zone applications. If sufficient capacity is unavailable in a single unit, multiple units are installed in the same zone to satisfy the thermal load. The fan coil's name is derived from the construction of the device. Each unit

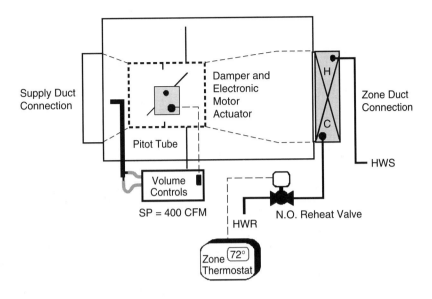

Figure 20.15   Constant Volume Box With a Reheat Coil

has a *fan* to circulate the zone's air and a heat exchanging *coil* or coils to alter the temperature of the air as it passes through the *unit,* hence the name *fan coil unit.* Figure 20.16 shows a photograph of a fan coil cabinet.

Some advantages of using this type of conditioning unit is that the heating and cooling equipment is distributed throughout the building so if one unit malfunctions, the temperature control in the remaining zones is unaffected. They also reduce the amount of mechanical room and floor area required in a building because there are no large air handling units or distribution ducts required in these installations.

The primary disadvantage of using a large number of fan coil units in a building is their inability to bring ventilation air into the zone. A fan coil unit cannot be the only equipment used to meet the thermal requirements of the zone. Since they have no provisions to bring in outside air, the ventilation requirements of the occupants cannot be met. Therefore, another means of supplying the zone with outside air is required to meet local building codes. In these cases, a small 100% outside air handling unit is used to condition and transport the necessary ventilation air to each zone. Another disadvantage of these applications is the increase in manpower required to maintain all the units throughout the building instead of a few air handling units.

These mini air handling systems have control devices and operating strategies identical to those found in air handling units. A wall mounted thermostat is used

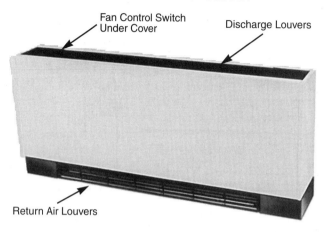

Fan Coil or Unit Ventilator Cabinet

Fan Control Switch
Under Cover

Discharge Louvers

Return Air Louvers

Figure 20.16   Fan Coil Cabinet

to regulate the output capacity of the heating and cooling sections of the unit in response to changes in the zone's temperature. In configurations employing hot and chilled water coils, the control system modulates the valves in sequence to prohibit mixing of energy flows. In units that use direct expansion cooling coils or electric heating elements, the control system energizes the applicable control relays in stages to better balance the capacity of the system with the zone's load.

Low limit zone temperature control strategies are also available in fan coil applications. This strategy allows the fan coil units to cycle their fan on and command the heating valve 100% open whenever the zone's temperature decreases below the its programmed low limit set point. The building automation system can also be integrated with the fan coil unit's control panel to enable and disable the units based upon a time of day schedule to reduce the cost of operating the building mechanical systems. Locally mounted fan controls are typically installed to allow the fan's speed to be controlled by the room occupants. An OFF position should not be specified since it inhibits proper control of the zone temperature and ventilation. Figure 20.17 shows a side view of the internals of a fan coil unit.

Figure 20.18 shows the modulating response of a fan coil unit throughout the throttling range of the process. When the zone's temperature begins to decrease, a reduction in the output signal to the valve actuators modulates the chilled water valve toward its closed position. If the temperature in the zone continues to decline, the hot water valve modulates open. Note, a deadband exists between the signal where the chilled water valve closes and the hot water valve begins to open. This prevents the valves from oscillating when the zone temperature is near its set point.

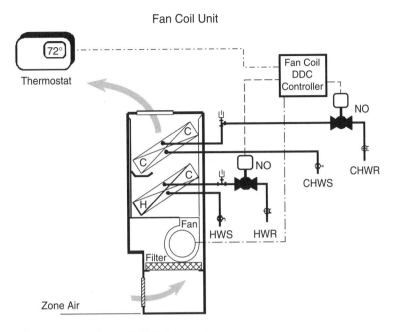

Figure 20.17 Fan Coil Unit Schematic

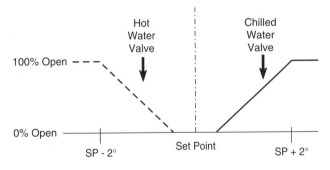

Figure 20.18 Valve Sequence for a Fan Coil Unit

20.10

UNIT VENTILATORS

Unit ventilator terminal units modulate the flow of heating, cooling and ventilation air into a zone to maintain its temperature set point. They have construction and operational characteristics that are similar to fan coil units. The primary difference between these two types of units is the ability of a unit ventilator to supply conditioned outside air to satisfy occupant ventilation requirements. To facilitate a connection to the outside air, unit ventilators must be installed against an exterior wall surface. This permits a short duct to be placed through the wall to outside air louvers recessed into the exterior surface of the building.

The inclusion of a duct to the outside of the building introduces the requirement for additional control strategies that are not needed in fan coil unit applications. Unit ventilators have a set of mixing dampers that can be positioned (binary) or modulated (analog) by their control system. Control strategies must be implemented that command the dampers closed whenever the unit's fan shuts off and to protect the coils from freezing when cold outdoor air is being drawn into the unit. There are three unit ventilator control strategies commonly used for these applications. They were developed by ASHRAE and are described below.

20.11

ILLUSTRATION OF AN ASHRAE CYCLE 1 UNIT VENTILATOR CONTROL STRATEGY

This example illustrates a zone temperature control process incorporating a unit ventilator configured to operate using an ASHRAE Cycle 1 control strategy.

Cycle 1 control applies to unit ventilators that are used in applications that have large quantities of fixed exhaust requirements in the zone that must be made up with conditioned outside air. Whenever the unit ventilator is operating, its outside air damper is commanded 100% open, allowing the maximum design ventilation flowrate into the zone.

20.11.1 Sequence of Operation

The sequence of operation for the zone temperature control process incorporating a unit ventilator configured to operate using an ASHRAE Cycle 1 control strategy that is illustrated in Figure 20.19 is explained in this section.

System Overview

The unit ventilator's fan is commanded on based upon a user defined occupancy schedule. Whenever the unit fan is operating and the zone temperature is above its set point, the outside air damper is commanded 100% open. The heating and cooling valve actuators are sequenced through software to maintain the zone's temperature set point without simultaneously heating and cooling the air. If the zone's temperature falls below its set point, the outside air dampers are commanded closed.

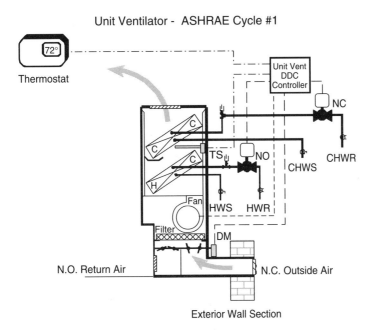

Figure 20.19   ASHRAE Cycle 1 Unit Ventilator Controls

Pre-Occupied Conditioning

A signal from the building automation system commands the building into occupied mode operation and the unit ventilator's fan is commanded on.

Warm-Up   If the zone temperature is below set point, the outside and return air dampers remain in their normal position and the hot water valve is commanded 100% open. When the zone temperature reaches its set point, the outside air damper is commanded 100% open and the control system operates using the occupied sequence of operation.

Cool-Down   If the zone's temperature is above set point, the outside and return air dampers remain in their normal position and the chilled water valve is commanded 100% open. When the zone temperature reaches its set point, the outside air damper is commanded 100% open and the control system operates using the occupied sequence of operation.

Occupied Operation

The DDC electronic thermostat has a programmed cold weather set point of 72° and a warm weather set point of 75°. The thermostat also has a limited adjustment range of set point $+/-2°$ to provide occupant control over the zone temperature. Whenever the zone temperature is greater than or equal to its set point, the outside air damper is commanded open by the unit ventilator's DDC control panel.

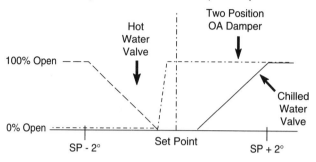

Figure 20.20   Cycle 1 Valve and Damper Response

The thermostat sends a control signal to the unit ventilator's DDC panel which generates output signals to modulate the normally open heating and the normally closed chilled water valves in response to changes in the zone temperature.

As the zone's temperature begins to decrease, a reduction in the output signal from the thermostat to the DDC panel modulates the chilled water valve toward its closed position. If the temperature in the zone continues to decrease below set point, the outside air dampers are commanded closed and hot water valve begins to modulate open. The valve and damper response is shown in Figure 20.20.

Unoccupied Operation

During the building's unoccupied period, the unit ventilator's fan is commanded off. The valves and mixing dampers are commanded to their normal position. If the zone temperature decreases below its programmed low limit set point, the fan is commanded on and the hot water valve remains 100% open. The system continues to cycle the unit using this strategy to maintain the zone's low limit set point.

Emergency Operation

Whenever the temperature downstream of the heating coil drops below the low limit set point, the heating coil discharge temperature sensor overrides the signal from the room thermostat and modulates the hot water valve toward its open position.

20.12

ILLUSTRATION OF AN ASHRAE CYCLE 2 UNIT VENTILATOR CONTROL STRATEGY

This example illustrates a zone temperature control process incorporating a unit ventilator configured to operate using an ASHRAE Cycle 2 control strategy.

An ASHRAE Cycle 2 unit ventilator operates with minimum outside air during heating and cooling operations. An economizer strategy is used to modulate the dampers for free cooling whenever the outside air temperature is below a programmed high limit set point. The hot water valve and the outside air damper are sequenced to allow the outside air damper to modulate beyond minimum position whenever the zone's temperature increases above its set point. Whenever the outside air temperature exceeds the high limit set point, analogous to an economizer dry bulb temperature limit, the outside air damper is commanded to its minimum position. If a cooling coil is present in the unit, it will operate whenever the outside air is too warm to maintain the zone temperature at set point. Cycle 2 is the most commonly used control strategy for unit ventilators because the ventilation requirements are maintained throughout its operation.

20.12.1 Sequence of Operation

The sequence of operation for the zone temperature control process incorporating a unit ventilator configured to operate using an ASHRAE Cycle 2 control strategy that is illustrated in Figure 20.21 is explained in this section.

System Overview

The unit ventilator's fan is commanded on based upon a user defined occupancy schedule. Whenever the fan is operating, the outside air damper is commanded to

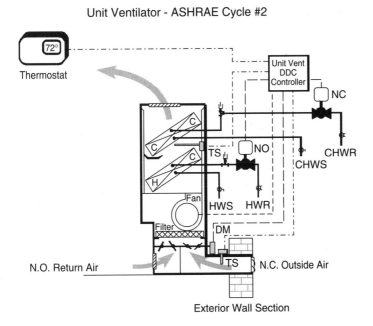

Figure 20.21 Cycle 2 Unit Ventilator Controls for Dual Coil Units

its minimum position. The mixing damper and hot water valve actuators are sequenced with software to maintain the zone's temperature set point without heating more than the minimum amount of outside air.

Pre-Occupied Conditioning

A signal from the building automation system commands the building into conditioning mode and the unit ventilator's fan is started.

Heating   If the zone's temperature is below set point, the outside and return air dampers remain in their normal position and the hot water valve is commanded 100% open. When the zone temperature reaches its set point, the control system operates using the occupied sequence of operation.

Ventilating (Free Cooling)   If the zone's temperature is above its set point and the outside air temperature is below the economizer high limit set point, the outside air dampers are commanded 100% open through the DDC control panel to provide free cooling to the zone.

Cooling   If the zone's temperature is above set point and the ventilation air is too warm to maintain the set point, the outside and return air dampers remain in their normal position and the chilled water valve is commanded 100% open. When the zone temperature reaches its set point the control system operates using the occupied sequence of operation.

Occupied Operation

The DDC electronic thermostat has a programmed cold weather set point of 72° and a warm weather set point of 75°. The thermostat also has a limited adjustment range of set point $+/-2°$ to provide occupant control over the zone temperature. The thermostat sends a control signal to the unit ventilator's DDC control panel which generates output signals to modulate the mixing dampers, normally open hot water and the normally closed chilled water valves in response to changes in the zone temperature.

Whenever the zone is calling for heating, the outside air damper remains at its minimum position and the hot water valve is modulated to maintain the zone's temperature set point.

When the zone is calling for cooling and the outside air temperature is below the high limit set point, the hot water valve modulates closed and the mixing dampers are modulated beyond their minimum position to maintain the zone's temperature set point.

When the zone is calling for cooling and the outside air temperature is above the high limit set point, the mixing dampers are commanded to their minimum position. The chilled water valve is modulated to maintain the zone temperature set point. This valve and damper response is shown in Figure 20.22.

Unoccupied Operation   This response is the same as that of a Cycle 1 unit ventilator. See Section 20.11.1

Emergency Operation   This response is the same as that of a Cycle 1 unit ventilator. See Section 20.11.1

ASHRAE Cycle #2 Valve and Damper Response

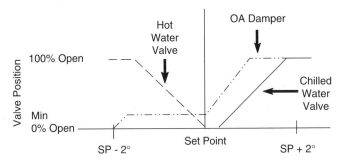

Figure 20.22 Cycle 2 Valve and Damper Response

20.13

ASHRAE CYCLE 3 UNIT VENTILATOR CONTROL STRATEGY

ASHRAE Cycle 3 unit ventilator control uses hardware and software configurations similar to the Cycle 2 unit ventilator strategy. The major difference between Cycle 2 and Cycle 3 lies in how the mixed air dampers are controlled. Under Cycle 3 control, the mixing dampers are modulated to maintain a fixed mixed air temperature set point during the heating, ventilating and mechanical cooling modes of operation. They modulate to maintain 55 to 60° air entering the heating coil at all times using a mixed air temperature sensor instead of the zone's temperature. The sensor located on the downstream side of the heating coil in Cycle 2 is repositioned upstream of the coil to perform this Cycle 3 control strategy. This offers better freeze protection because the cabinet temperature sensor has been relocated to a position upstream of the heating coil. The coils never see air below 55° temperature so they are less susceptible to freezing. Unit ventilators operating under Cycle 3 control will not maintain a minimum flowrate of ventilation air during colder weather. As the outside air temperature decreases, the percentage of ventilation air entering the unit decreases.

20.14

EXERCISES

Determine if the following statements are true or false. If any portion of the statement is false, the entire statement is false. Explain your answers.

1. VAV boxes vary the amount of primary air entering a zone based upon the zone's temperature.
2. VAV boxes are only available with normally open primary air dampers.
3. VAV boxes with reheat coils are primarily used for interior zone applications.

4. Intermittent fan powered terminal boxes are constant volume, variable temperature devices.
5. Dual duct mixing boxes have two primary air dampers.
6. Pressure independent boxes use an additional control loop to maintain a constant volume flowrate set point.
7. ASHRAE has three control configurations commonly used for fan coil unit control strategies.
8. A constant fan powered VAV box has its fan located in series with the primary air stream.
9. Constant volume VAV boxes have a primary air damper controlled by a zone thermostat.
10. The primary air damper and reheat coil of a VAV box are sequenced to prevent simultaneous heating and cooling from occurring.

Respond to the following statements, questions and problems completely and accurately, using the material found in this chapter.

1. Develop a sequence of operation for a pressure independent VAV box.
2. Develop a sequence of operation for a cooling only VAV box located in an exterior zone that is integrated with perimeter heat convectors installed on the exterior walls.
3. A VAV box with an integrated reheat coil is presently heating the zone when a number of high heat generating equipment is turned on. What is the response of the pressure independent VAV box control loop? Explain your answer.
4. Why do VAV boxes have high and low flowrate limits?
5. An intermittent fan powered box is presently cooling a zone with its primary damper open slightly beyond its minimum position. The occupants in the zone leave for lunch, turning off the lights and computers. What is the response of the temperature control system? Explain your answer.
6. What are differences between the control responses of the ASHRAE Cycle 1 and Cycle 3 control strategies?
7. A fan coil unit is operating in its heating mode and a cold front moves through the area dropping the outside air temperature from 36 to 16 °F in 15 minutes. What will be the likely response of the temperature control loop to the change in outside air temperature? Explain your answer.
8. A Cycle 2 unit ventilator is presently mechanically cooling a zone. The outside air temperature drops below the high limit set point. What is a likely response of the temperature control loop? Explain your answer.
9. A Cycle 3 unit ventilator is presently mechanically cooling a zone. The outside air temperature drops four degrees. What is a likely response of the temperature control loop? Explain your answer.
10. Develop a system drawing and description of operation for a unit ventilator operating with the ASHRAE Cycle 3 strategy.

PART 4

Application Response
and Analysis

Single Zone Air Handling Unit Configurations

The remaining chapters of this text analyze the *response* and *interactions* that occur between individual control loops that have been joined together into a control system. Each chapter focuses upon a specific type of air handling unit and the typical processes performed in the unit. The design and operating characteristics of each process are described at the beginning of each chapter. A brief sequence of operation is included along with a system drawing. The remaining sections of each chapter describe the *response* of the system to a change in one if its process variables. This information is presented to develop the skills needed to analyze a dynamic system in the field. Procedural steps that can be used to analyze a specific process problem are also presented.

To get the most out of these chapters, concentrate on the *relationships* and *interactions* that exist between the loops that make up the control system. This method will help to develop a sense of how changes in one part of the system cause responses in other process control loops. These are the skills that are required to become successful in the field related aspects of the control industry.

21.1

SINGLE ZONE, CONSTANT VOLUME AIR HANDLING UNITS

Single zone constant volume air handling units are used in applications that maintain the thermal comfort requirements for a *single building zone*. The zone may be made up of an entire small building, several rooms on one side of a larger building or part the floor area of a large store or warehouse. The identifying feature of any zone is the presence of a *single* thermostat located within the space. The thermostat's output signal controls the capacity of the heating and cooling coils to meet the comfort needs of the zone. If more than one thermostat is installed within the rooms served by an air handling unit, the system is not a single zone air handling unit application.

21.1.1 Constant Volume—Variable Temperature Air Handling Units

Single zone air handling units can be designed to deliver either a constant or variable quantity of air through its supply duct. Constant volume systems have a supply fan that delivers a fixed volume of conditioned air to the zone. The quantity of air delivered by the fan is a function of the zone's design heating or cooling load characteristics. In this type of system, the *temperature* of the supply air is varied to establish a balance between the thermal capacity of the air handling unit and the load in the zone.

21.1.2 Single Fan Units

Single zone air handling systems can be configured with or without a return fan installed in the return air duct. In single fan installations, one supply fan is used to circulate the air between the air handling unit and its zone. In these applications, the low pressure produced at the supply fan's inlet draws a sufficient quantity of air back from the zone via the return air duct. The return air duct remains under a negative pressure whenever the supply fan blade is rotating.

A single fan system lacks the means necessary for generating a positive pressure in the return duct. Consequently, a single fan configuration is incapable of exhausting any of the return air out a set of exhaust dampers located near the air handling unit. Any outside air that is brought into the zone to fulfill ventilation requirements must be relieved using an separate exhaust fan or static pressure relief dampers. The inability to exhaust from the air handling unit limits the implementation of an economizer strategy in the sequence of operation. Single fan systems typically provide a two-position outside air damper that opens when the unit is operating to draw ventilation air into the building. The amount of air brought into the unit is a function of the amount of air that can be relieved from the zone by exfiltration, an exhaust fan or by a static pressure relief damper.

21.1.3 Dual Fan, Single Zone Systems

Return fans are used in single zone applications where the return duct's resistance to flow is greater than the supply fan can overcome. Dual fan systems are designed to modulate their mixing dampers using an economizer cycle. The return fan produces a positive pressure upstream of the return and exhaust air dampers that forces the return air out of the building when the exhaust dampers modulate open.

The operation of a dual fan air handling unit is very similar to that of a single fan installation. The return fan is interlocked with the supply fan's motor starter control circuit so that it starts whenever the supply fan is commanded on. The interlocking control circuit also insures that if either fan trips off due to a system fault, or is manually turned off, the other fan also stops.

21.1.4 Heat Exchangers

The heating and cooling coils of single zone air handling units are installed in *series* with each other with respect to the air flow. The entire mixed airstream must pass through both coils before being transported to the zone. In HVAC terminology, this air handling unit configuration is called a *single path* system. In most single path applications, the heating coil is positioned upstream of the cooling coil, allowing it to be used in both comfort heating and freeze protection strategies. In applications where the cooling coil is used for a dehumidification process, the heating coil can be located downstream of the cooling coil to permit reheating of the colder, dehumidified airstream.

21.1.5 Temperature Controls

The temperature control strategies used with single path air handling units must avoid the simultaneous heating and cooling of the mixed airstream in all but dehumidification strategies. The heating control strategy is disabled before the economizer or cooling strategies are allowed to operate. In units having a hot and chilled water coil, the valve actuators are sequenced using different actuator spring ranges or through DDC software strategies to prohibit simultaneous heating and cooling of the airstream. In electric heating or direct expansion cooling applications, control relays and pressure/electric (PE) switches are used to prohibit the simultaneous operation of both coils. The following sections describe the response of a single zone, dual fan unit that has temperature and humidification processes.

21.2

ILLUSTRATION OF A SINGLE ZONE CONSTANT VOLUME DUAL FAN AHU WITH HUMIDIFICATION

This example illustrates the response of a dual fan, single zone, constant volume air handling unit with an economizer strategy and steam humidification. The system has the following control loops in its control system:

1. Supply and return fan start/stop.
2. Zone air temperature control.
3. Zone humidification control.
4. Low temperature limit protection.
5. Temperature based minimum position control.

21.2.1 System Drawing

A system drawing for a single zone is shown in Figure 21.1.

21.2.2 Sequence of Operation

The sequence of operation for the single zone air handling unit that is illustrated in Figure 21.1 is explained in this section.

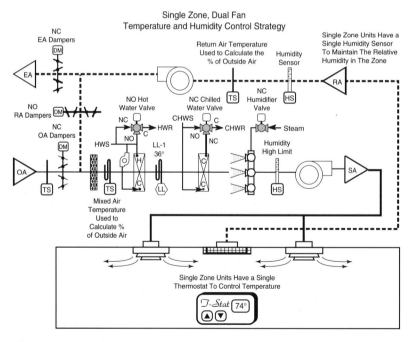

Figure 21.1   Single Zone AHU With Humidity Control Loops

Fan Start/Stop

The supply fan is commanded on based upon a user defined occupancy schedule. The return fan is electrically interlocked with the supply fan so they operate in unison. Whenever the fans are off, the mixing dampers along with the hot, chilled water and humidifier valves are commanded to their normal position.

If either fan trips off due to the presence of a fault condition, the other fan's control circuit is also opened, disabling both fans. The mixing dampers along with the hot, chilled water and humidifier valves are commanded to their normal position.

Low Temperature Limit

If the low limit temperature device that is located downstream of the hot water coil measures a temperature below 36 °F, the supply fan is commanded off, disabling the return fan. The mixing dampers along with the hot, chilled water and humidifier valves are commanded to their normal position.

Zone Temperature Control

The normally open hot water valve is modulated in sequence with the mixing dampers and the normally closed chilled water valve by a direct acting controller to maintain the zone's temperature set point. As the zone's temperature increases, the control signal to the final controlled devices also increases.

Heating   The heating valve operates whenever the zone temperature is less than its set point. The valve will be 100% open when the zone temperature is greater than 2° below set point.

If the zone is in a heating mode and its temperature begins to increase, the hot water valve is modulated toward its closed position. If the temperature continues to rise above the zone's set point, the hot water valve will modulate completely closed and the mixing dampers will begin to modulate beyond their minimum position.

Economizer   The normally closed outside and exhaust air dampers along with the normally open return air dampers modulate in sequence with the hot and chilled water valves to maintain the zone's temperature set point. The economizer cycle begins as soon as the outside air dampers modulate beyond their minimum position.

During periods when the system is heating, the mixing dampers are modulated to maintain their minimum position based upon the temperatures of the outside, return and mixed air streams. A new mixed air set point is calculated every 5 minutes to compensate for changes in the outside or return air temperatures.

As the zone's temperature increases above its set point, the direct acting thermostat increases its output signal to the damper actuators, opening the outside and exhaust dampers beyond their minimum position while proportionately closing the return air dampers.

Once the outside air temperature exceeds the economizer high limit set point, the economizer strategy is disabled and the mixing dampers are modulated to maintain their minimum position outside air flowrate based upon temperature.

Minimum Position Whenever the temperature difference between the outside and return air streams becomes less than 5°, the modulating method of maintaining minimum position decreases in accuracy. This may cause an incorrect quantity of ventilation air to be brought into the zone. Excessive outside air increases conditioning costs while insufficient quantities may pose indoor air quality problems. During these periods, the dampers are positioned based upon the percentage of stroke method to insure adequate ventilation air is entering the system.

Whenever the difference between the outside and return air temperatures is less than 5°, the dampers are commanded to their minimum position by sending a 20% output signal to their actuators.

Cooling The chilled water valve begins to modulate open after the mixed air dampers have reached 100% open and the zone is still calling for additional cooling. The chilled water valve is also modulated whenever the zone temperature is above set point and the economizer strategy has been disabled. The modulating sequence of the valves and dampers is shown in Figure 21.2.

Steam Humidifier

The normally closed humidifier steam valve modulates to maintain the zone's relative humidity at set point. As the relative humidity in the zone decreases, an increasing signal to the steam valve modulates the normally closed valve toward

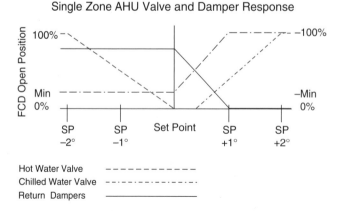

Figure 21.2 Single Zone AHU Valve and Damper Response

its wide open position. If the supply air relative humidity exceeds the supply duct humidity high limit set point of 80%, the valve modulates using the supply air relative humidity sensor until the humidity control point is within the process throttling range. When the fans are commanded off, the steam valve is commanded to its normal position.

21.2.3 Steady State Process Conditions for the Single Zone Unit

Table 21.1 lists the current process conditions for the single zone unit shown in Figure 21.1.

Note the following:

1. The table indicates that the mixed air is made up of 20% outside air by volume [(0.20 x 27°) + (0.80 x 72) = 63°]. The nonlinear installed characteristics of the mixing dampers only required outside air dampers to open 16% of its control signal span before the 20% ventilation requirement was reached.

2. The temperature difference between the outside air and the zone causes heat to transfer out of the building. This energy must be replaced in order to maintain the zone's set point. When the hot water valve is positioned 10% open, the coil transfers the same amount of energy into the zone as is leaving through the walls, ceiling and exhaust system.

3. Exhausting 18% of the zone's air (2% exfiltration) removes moisture that must be replaced. The humidity valve is opened 30% to achieve a balance between the mass transfer of water vapor entering the zone and that being removed by the return fan and exfiltration.

Table 21.1   STEADY STATE PROCESS CONDITIONS

Process	Set Point	Control Point	FCD	Position
Outside Air Conditions		27° DB 65% RH		
Mixed Air Temperature	20% based on Temperature	63°	OA & EA RA Damper	16% Open 84% Open
Zone Temperature	72° DB	72°	Hot Water Valve	10% Open
Zone Humidity	35% RH	35%	Steam Valve	30% Open

21.3

PROCESS RESPONSES TO LOAD CHANGES

The following passages list the probable responses of the air handling unit's control loops to the described change in a process load.

21.3.1 Load Change 1

The zone's temperature rises in response to an increase in its internal heat gains.

Zone Temperature Process Control Response

1. The thermostat measures the increase in the zone's temperature and generates a direct acting, proportional increase in its output signal.
2. The control signal to the valve actuators increases. The hot water valve modulates from its previous steady state position of 10% open, to its new position, 0% open (closed).
3. The increasing output signal of the thermostat commands the outside and exhaust air dampers to modulate beyond their steady state minimum position (16%) toward 30% open while proportionately closing the return air dampers.
4. The changes in valve and damper positions cause the supply air temperature exiting the air handling unit to decrease. The cooler air entering the zone lowers the temperature back toward its set point. As long as the quantity of heat generated in the zone remains at its present level, the hot water valve will stay closed and the mixing dampers will maintain the zone at its temperature set point.

Zone Humidity Process Response

1. An increase in the amount of dry outside air entering the space in response to the opening of the outside air dampers reduces the zone's relative humidity.
2. The humidifier's load increased, causing the signal generated by the reverse acting humidifier controller to increase, modulating the steam valve toward 55% open to balance the mass transfer of steam from the humidifier with the load in the zone.

21.3.2 Load Change 2

The outside air temperature decreases from 27 to 20° and the interior gains remain at the same level as in Load Change 1. The outside air dampers are currently open 30% and the steam humidifier valve is open 55%.

Minimum Position Control Response

1. The decrease in the temperature of the outside air entering the unit produces a corresponding drop in the temperature of the air discharged from the unit.

2. The zone's temperature begins to drop due to the cooler air handling unit discharge air temperature.

3. The output signal to the mixing dampers decreases to reduce the volume of outside air entering the unit, thereby raising the mixed air temperature. The dampers try to close from 30% open to 15% open.

4. The decrease in outside air temperature causes the DDC control panel to calculate a new minimum position set point that will maintain 20% ventilation air in the mixed air stream. The mixing dampers are modulated to maintain the new mixed air temperature set point 61.6° [(0.20 x 20°) + (0.80 x 72°) = 61.6°].

5. The zone temperature stabilizes at 72° with the dampers at minimum position and the heating valve 100% closed.

Zone Humidity Process Response

1. The reduction in the amount of dry outside air entering the space produces a related decrease in the humidifier load.

2. The signal generated by the reverse acting humidifier controller decreases, modulating the steam valve from 55% toward 40% open to balance the mass transfer of steam with the load in the zone.

21.3.3 Load Change 3

The thermal load in the zone decreases, causing the heating valve to modulate open from 0% to 10%. The temperature of hot water supplied to the air handling unit's coil simultaneously decreases due to a set point adjustment made by the building operator.

Zone Temperature Process Response

1. The reduction in the hot water supply temperature reduces the heat transfer capacity of the AHU coil, causing the supply air temperature to decrease.

2. The reduction in the supply air temperature causes the zone's temperature to decrease because the heat entering the zone through the supply duct is less than the heat being transferred from the zone, disrupting the balance between the system's capacity and the zone heat load.

3. The zone's thermostat responds by decreasing its output signal to the hot water valve, thereby modulating the valve toward its wide open position. Opening the hot water valve increases the mass flowrate of water through the coil, compensating for the reduction in the heat content of the control agent caused by the operator's reduction in the boiler's hot water set point.

4. The increase in the heat contained in the air entering the zone produces the necessary increase in the zone's temperature. The proportional + integral controller transfer function returns the control point back to the zone temperature

set point. The integral gain element of the controller algorithm maintains the valve at a higher percentage of open (28%) than it was during its previous steady state period (10%).

21.4

TROUBLESHOOTING SYSTEMS

Troubleshooting a control system is accomplished by evaluating a series of logical steps that will either identify the problem or eliminate it as one of the potential causes. Either outcome works toward effectively pinpointing the actual cause of a system malfunction. The key to successfully troubleshoot any system is to be able to ascertain whether the device initially identified as the problem is the cause and not just responding incorrectly because of the actual problem.

The logical place to start when investigating a control problem is to verify that the process sensor is generating the proper output signal for the measured condition. If a sensor is generating an inaccurate output signal, the rest of the control system responds to the incorrect signal, producing an incorrect response. Therefore, there is no merit in checking the controller's response or determining if the final controlled device is in the correct position if the sensor's output signal has not been verified as being correct. Unfortunately, the first component modified during the course of a control related analysis is usually the controller, the most complex device in the control loop.

Once the sensor's response is verified as being correct, the output of the controller (controller algorithm in a DDC panel) is measured to determined if it correlates with the sensor's output signal. The process set point and the control point are entered into the controller's transfer function to calculate the output signal for current process conditions. After the controller is verified to be operating correctly, the final controlled device can be evaluated to determine if it is properly positioned and free to move throughout the entire range of its actuator stroke without binding. Once the operation of these and any other control loop components are confirmed to be operating correctly, the process control agents are analyzed to see if they are of the proper *quality* (temperature, pressure) and *quantity* (cfm, gpm, lb/hr, etc.) to balance the existing load. If the control agent does not arrive at the proper conditions, the primary equipment's control system must be evaluated to determine the cause of the problem.

The following sections identify some typical problems associated with single zone air handling unit configurations. Some probable causes and steps taken in analyzing the system are listed under each scenario. Each succeeding section will present different problems to increase the reader's exposure to the common practices, procedures and techniques used to analyze HVAC control system problems.

21.5

ANALYSIS OF A SINGLE ZONE, CONSTANT VOLUME AIR HANDLING UNIT

21.5.1 Scenario 1

The zone's temperature is too cold even though the air handling unit is operating in a heating mode.

Possible Causes

1. The control agent is not arriving at the AHU heating coil at the proper quantity or quality.
2. The mixing dampers are open beyond their minimum position, reducing the ability of the heating coil to produce the necessary rise in temperature.
3. The flowrate of the air discharged into the zone has been restricted, inhibiting the proper volume of air (heat) from being transported into the zone.

Verify the

1. thermostat's set point is actually placed at the desired setting.
2. thermostat is properly calibrated. Verify calibration by performing the following procedure:
   a. Measure temperature at the thermostat's location.
   b. Adjust the thermostat's set point so it equals the measured temperature.
   c. Measure the thermostat's output signal.
   d. Calibrate the thermostat's output signal bias until it is equal to the maximum value of the heating valve's spring range. At this value the hot and chilled water valves are closed and the outside air damper is at its minimum position. See Figure 21.2.
   e. Place the thermostat's set point equal to the desired zone temperature and let the loop stabilize.
   f. Adjust the thermostat's proportional gain to achieve an initial quarter amplitude decay response to a step change of 5°.
3. thermostat's control signal is present at the actuators. Measure the output of the thermostat to compare it to the signal arriving at the actuators of the final controlled devices.
4. actuators move freely throughout their entire stroke.
5. quality and quantity of the of the control agent. If the hot water is not sufficiently warm for the present conditions or is being supplied at a reduced flowrate due to a pump or system balance problem, the air handling unit's heating coil will be unable to transfer energy at the correct rate.
6. air flowrate into the zone is at its design specification. If the AHU filters are plugged, coils dirty or the fan belts are slipping, the volume flowrate of conditioned air decreases, creating temperature control problems. Check for closed fire, register or balancing dampers obstructing flow through the duct.

7. position of the valves and mixing dampers to make sure they correlate with the output signal from the thermostat. If any of the final controlled devices are incorrectly positioned, calibrate their electric to pneumatic signal converter or pneumatic actuator positioner performing the following procedure:
    a. Set the controller's output signal to its minimum value and apply that signal to the converter or positioner.
    b. Tune the zero or start adjustment until the converter's (positioner's) output signal begins to reposition the connected valve or damper.
    c. Set the controller's output signal to the maximum value of the connected actuator's input range and apply that signal to the converter (positioner).
    d. Tune the span screw or spring position of the converter (positioner) until its output signal repositions the valve or damper at its maximum stroke.

21.5.2 Scenario 2

The zone's relative humidity is too low.

Possible Causes

1. The humidifier's control valve is closed or not modulating.
2. The humidifier's steam distribution nozzles are plugged.
3. Steam is not available to the humidifier.

Verify the

1. humidity set point is placed at the desired value.
2. humidity sensor and controller are calibrated. Use the same procedure as described for calibrating thermostats in Scenario 1.
3. control signal is present at the steam valve's actuator.
4. position of the steam valve correlates with the measured controller output signal.
5. presence of steam in the duct when the humidifier valve is commanded open. If valve is wide open and no steam is visibly exiting the humidifier grids check to be sure all the isolation valves in the steam piping circuit are open, the correct steam pressure is present at the humidifier grid and the steam injection orifices of the humidifier grid are not plugged.
6. calibration of the EP converter or pneumatic positioner using the procedure described in Scenario 1.

21.5.3 Scenario 3

The zone's temperature is too warm even though the outside air temperature is below 55°.

Possible Causes

1. The hot water valve is open beyond its appropriate position.
2. The mixing dampers are not modulating in the economizer mode.

Verify the

1. thermostat set point is positioned at the desired value.
2. unit sensors (mixed, outside, and return air) and the thermostat are calibrated.
3. economizer high limit control is operating properly and releasing the mixing dampers from their minimum position when the zone calls for cooling and the outside air temperature is below the economizer high limit set point.
4. smooth actuation and proper position of final controlled devices with respect to the thermostat's output signal.
5. calibration of the pneumatic positioners or EP converters.

21.6

EXERCISES

Respond to the following statements, questions and problems completely and accurately, using the system drawing in Figure 21.1.

1. What is the probable response of the zone's temperature control loop if the afternoon temperature increases from 50° to 62° and the internal gains remain constant?
2. What is the probable response of the zone's relative humidity control loop when the outside air temperature increases to 62° and the outside air's relative humidity increases from 35% to 55%?
3. What is the probable response of the zone's temperature control loop as the external heat losses through the walls and windows decrease throughout the afternoon?
4. A complaint of a cold zone in a computer center is investigated. The occupant states the temperature is usually comfortable for the first few hours in the morning but the zone gets gradually cooler as the office becomes more productive. The HVAC system appeared to be operating correctly until last week when the office furniture was rearranged. The zone's thermostat is now directly above the laser printer.

    The zone set point is 73°, the hot water valve, chilled water valve and mixing dampers are properly positioned with respect to the thermostat's output signal. The temperature indication on a desk thermometer shows 69° at the occupant's desk. That value is verified with an accurate thermometer. What are some probable causes of the cold zone problem?

5. A zone loses temperature control as the heat gains in the room increase. The set point of the zone is 74° and the control point is 77°. The thermostat is generating the correct output signal to the EP converter. The EP is generating a 15 psi signal to the final controlled device actuators. The hot water valve is closed, chilled water is available and the outside air damper is at its minimum position. The chilled water valve is closed. What are some probable causes of the high zone temperature problem?

Single Path, Constant Volume, Multiple Zone Air Handling Units With Reheat Coils

22.1

SINGLE PATH AIR HANDLING UNITS WITH REHEAT COILS—MULTIPLE ZONE APPLICATIONS

Single zone, single path air handling units can only satisfy the thermal requirements of one zone. These units maintain the zone's temperature set point by varying the capacity of its heating and cooling coils based upon the output signal of a single thermostat. A single path air handling unit can be modified to simultaneously serve the thermal requirements of multiple zones by installing a terminal reheat coil in each zone. This coil adds heat to the cooler air supplied by the air handling unit, varying its temperature to satisfy the thermal requirements of the zone. The reheat coil's valve is modulated by a thermostat in response to changes in the zone's temperature.

In *multiple* zone applications of a single path air handling unit, all the zones receive air that has been conditioned to the same temperature and relative humidity.

The discharge air temperature is maintained at a value that balances the thermal load of the zone experiencing the greatest *cooling* load. This will be the zone whose control point exceeds its set point by the largest amount. Using the zone with the greatest cooling load to set the supply air's temperature set point causes the air handling unit to supply air to all the remaining zones at a temperature that is too cool to balance their thermal loads. The reheat coils in these zones will add additional heat to the supply air to maintain a balance between the energy discharged into the zone and its thermal load.

22.2

CONTROL STRATEGY DIFFERENCES BETWEEN SINGLE PATH, SINGLE ZONE AIR HANDLING UNIT AND SINGLE PATH, MULTIPLE ZONE AIR HANDLING UNIT APPLICATIONS

Single path, single zone air handling units that are configured to serve multiple zones require additional control loops to simultaneously maintain a balance between the various thermal loads that are occurring in each of the zones.

22.2.1 Zone Temperature Control Differences

Each zone connected to the system requires its own temperature control loop consisting of a thermostat (sensor and controller) and a normally open control valve for its reheat coil. The thermostat modulates the reheat valve whenever the zone is experiencing a heating load. During all other periods, the zone is operating in a cooling mode and the thermostat's output signal keeps the reheat valve closed. This response differs from the single zone temperature control strategy described in Chapter 21. In single zone applications, the air handling unit's heating and cooling coil valves are modulated under all load conditions, not just during periods when the zone is experiencing a heat load.

22.2.2 Supply Air Temperature Control Differences

Since the zone thermostat in a single path, multiple zone application is dedicated to modulating a reheat valve, a supply (discharge) air temperature control loop must be installed on the air handling unit. This control loop modulates the air handling unit's heating and cooling coil valves in response to the temperature of the supply air. In non-dehumidification configurations, the normally open heating valve and the normally closed cooling valve are modulated in sequence using positioners, spring ranges or programming code. In dehumidification applications, each coil's valve must be modulated independently using an additional control loop to provide reheat capability.

22.2.3 Mixed Air Temperature Control Differences

Another difference between single zone and multiple zone applications of a single path air handling unit is found in the manner in which the mixing dampers are controlled. The mixed air temperature has its own control loop instead of being sequenced with the hot and chilled water valves. The dedicated mixed air controller also allows the system to incorporate the minimum position strategy based upon temperature of the outside, return and mixed air streams.

22.2.4 Resetting Temperature Set Points Based Upon a Related Variable

In single path multiple zone applications, the air handling unit's set points are typically raised and lowered based upon another process variable. This strategy improves the operating efficiency and controllability of the process. The automated altering of temperature set points is called *reset*. Whenever a set point changes in the opposite direction to the change in the independent variable, the strategy is called *reverse reset*. When the set point and reset variables change in the same direction, the strategy is called *direct reset*.

Process efficiency is improved by the reset strategy because the primary equipment is able to take advantage of the reduction in load whenever the outside air temperature becomes milder by reducing the temperature difference between the control agent and the air's temperature set points. Resetting the temperature set points in single path applications minimizes the amount of reheat in the zones by keeping the supply air temperature at the *highest* possible cooling set point. Resetting set points also improves the controllability of the process by reducing the differential between the control agent and controlled medium temperatures, encouraging the final controlled devices to modulating within the mid-range of their actuator stroke, where their installed response is relatively linear.

Outside Air Temperature Reset Outside air temperature is one of two reset variables commonly used in reset strategies. In these schemes, the supply and mixed air set points are reset to higher values as the outside air temperature decreases. Conversely, as the outside air temperature increases, the mixed and supply air temperature set points are lowered to provide the necessary level of cooling required to offset the increase in heat gains.

Outside air temperature reset strategy is based upon an open loop relationship between changes external to the process and its load. Open loop reset uses a temperature range (40 to 70°) of an extremely large thermal capacitance (the outdoor air volume) to vary the set point between its reset limits. Hydronic heating systems are typically configured so the hot water's temperature is reverse reset, raising its set point as the outside air temperature decreases. The hot water temperature controller can be calibrated so its set point is reset between 100° and 200° as the outside temperature decreases from 70° to 0°.

Zone Load Reset The thermal load of a zone can also be used as the independent variable of a reset scheme in place of the outside air temperature. In this strategy, the output signals from all (or a representative sample) of the zone thermostats are compared to each other. The greatest thermostat signal identifies the zone with the greatest thermal load. The greatest thermostat signal is used to reset the set points of the air handling unit's temperature processes. Note, it is the comparison between the *thermostat output signals* that is used to reset the air handling unit's set points, not the zone *temperatures*. Since each zone can have its own occupant regulated set point, the temperatures cannot be used to determine which zone is experiencing the greatest load. A zone may have a higher temperature due to a higher thermostat set point and actually be experiencing the smallest load relative to the other zones. For example, a zone with a set point of 70° and a control point of 73° has a greater cooling load than a zone with a set point of 75° and a control point of 75.2°. Therefore, it is the signal from the thermostat that is related to the load in the zone, not its temperature.

Using the thermal loads of the zones to reset the air handling unit's set points is better in some aspects than using the outside air temperature. Using zone loads creates a closed loop reset strategy because the actual conditions in the building are being evaluated. Recall from previous chapters, closing any loop with a feedback path increases the operating efficiency of a process while it also increases the risk of introducing instability into the loop. Therefore, implementing reset based upon the thermal loads of the zones increases the probability for instability in the reset process. This risk is present because a small (5° maximum) change in the zone's temperature is used to reset an AHU set point from its low limit to its high limit. This is equivalent to a high gain response in a proportional controller. In small capacitance process applications, this five degree reset range will in all likelihood be too small to produce a stable reset response. To minimize the possibility of oscillations in a zone load reset application, only the thermostat signals from the larger thermal capacitance zones are used. Calculation periods can also be implemented in DDC based applications that limit the number of times per hour the reset set point will be calculated. When commissioned correctly, closed loop reset strategies are the preferred method of air temperature reset.

Humidity Reset   Reset strategies are also used to vary relative humidity set points of a zone. The rate of condensation on window surfaces is directly related to the outside air temperature. This relationship makes the outside air temperature a logical choice for the independent variable in the reset strategy. A direct reset strategy is used with humidifying processes to lower the zone's humidity set point as the outside air temperature decreases. By reducing the zone's humidity level, condensation on window surfaces and within wall cavities is reduced.

Reset Schedules   One of the positive aspects of linear control systems is that the same basic formula is used to calculate all the responses in the control loop. This formula is also used to develop the reset schedule of a process and to calculate the process set point based upon the current value of the reset variable. As in all proportional relationships, the change in the output response is divided by the change in the input response to determine the amount of change that occurs for each incremental change in the input value. In reset strategies, the output response of the equation is the change in the process set point and the input response is the change in the reset variable (outside air temperature, zone load, etc.). Applying these relationships to a hot water reset application, if the hot water set point is reverse reset from 180 to 110° based upon an outside air temperature change of 0 to 65°, the set point will change $-1.1°$ for each one degree increase in the temperature of the outside air. The negative sign denotes a reverse reset relationship.

$$\frac{-(180 - 110)}{(65 - 0)} = \frac{-70°}{65°}$$

$$= -1.1° \; \textit{change in the hot water set point per } 1° \textit{ increase in outside}$$
$$\textit{air temperature}$$

To determine the set point at a given outside air temperature, multiply the change in the outside air temperature by the reset ratio. If the outside air temperature is 40°, the change in the set point equals:

$$40° \times \frac{-1.1° \; \textit{set point}}{1° \; \textit{OA Temperature}} = -44°$$

Add the change in the set point from the value of the set point when the outside air temperature is 0° to calculate the hot water set point.

$$180° - 44° = 136°$$

These calculations have the same format and calculating procedures used in the transfer function of a sensor. They are used to calibrate controllers that have two inputs, one for the controlled variable and the other for measuring the reset variable. When analyzing a loop that incorporates a reset schedule, the present set point must be calculated using the reset formula before the controller can be evaluated for a proper response.

22.2.5 Tracking Supply and Mixed Air Set Points

In single path air handling unit applications that use separate control loops to modulate their mixing dampers and valves, the supply and mixed air temperature set points are maintained equal to each other. This strategy permits the mixed air control loop to maintain the supply air temperature (single path) as long as possible, minimizing the opening of the hot and chilled water valves. In applications that incorporate reset strategies, the supply and mixed air temperature set points are configured to *track* (follow) each other. Tracking single path temperature set points improves the system's energy utilization efficiency in comparison to strategies that use a fixed set point for the mixed air temperature loop and reset the supply air temperature set point. Tracking allows the mixed air dampers to modulate to achieve the supply air temperature set point whenever possible. This strategy also keeps the hot water and chilled water valves closed as much as possible, minimizing boiler and chiller operation. In applications where the mixed air set point is maintained at a fixed value, the supply temperature control loop is required to modulate one of its valves to reduce the offset that exists as a consequence of the different set points, expending energy that would otherwise not be needed.

In systems with large fans or where the fan motor is located inside the duct, a temperature rise occurs as the airstream passes through the fan blades and over the motor. In these applications, a fixed temperature differential is applied to the tracking set points to optimize process efficiency. Taking this temperature rise into consideration, the mixed air set point is programmed to equal the supply air temperature set point minus a fixed differential of 2 to 5°, dependent upon the amount of heat added by the fan and its motor. As the cooler mixed air passes through the fan, its temperature is raised to the supply air temperature set point, improving the efficiencies associated with the temperature control processes.

22.3

ILLUSTRATION OF A SINGLE PATH, CONSTANT VOLUME, MULTIPLE ZONE AHU WITH REHEAT

This example illustrates the response of a dual fan, single path, constant volume, multiple zone air handling unit with zone reheat coils. The system has the following control loops in its control system:

1. Supply and return fan start/stop.
2. Mixed air temperature control.
3. Supply air temperature control.
4. Zone air temperature control.
5. Low temperature limit protection.

22.3.1 System Drawing

A system drawing for a single path multiple zone AHU is shown in Figure 22.1.

22.3.2 Sequence of Operation

The sequence of operation for the single path, multiple zone air handling unit with zone reheat coils that is illustrated in Figure 22.1 is explained in this section.

Fan Start/Stop

The supply fan is commanded on based upon a user defined occupancy schedule. The return fan is electrically interlocked with the supply fan so they operate in unison. Whenever the fans are off, the mixing dampers along with the hot and chilled water valves are commanded to their normal position.

If either fan trips off due to the presence of a fault condition, the other fan's control circuit is also opened, disabling both fans. The mixing dampers along with the hot and chilled water valves are commanded to their normal position.

Low Temperature Limit

If the low limit temperature device that is located downstream of the hot water coil measures a temperature below 36 °F, the supply fan is commanded off, dis-

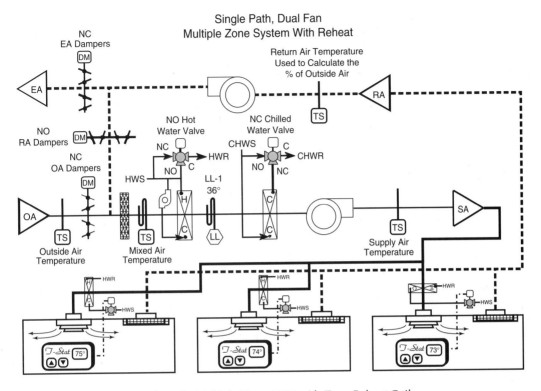

Figure 22.1   Single Path, Multiple Zone AHU with Zone Reheat Coils

abling the return fan. The mixing dampers along with the hot and chilled water valves are commanded to their normal position.

Mixed Air Temperature Control

The normally closed outside and exhaust air dampers along with the normally open return air dampers are modulated to maintain the mixed air temperature set point. The dampers are modulated in a manner that insures a minimum volume of ventilation air enters the air handling unit whenever the building is occupied.

During periods when the supply air is being heated, the mixing dampers are commanded to maintain their minimum position flowrate based upon the temperatures of the outside, return and mixed air streams. The minimum position mixed air set point is calculated every 10 minutes to respond to changes in the outside or return air temperatures.

The economizer cycle begins as soon as the outside air dampers modulate beyond their minimum position. As the mixed air temperature increases above its set point, the direct acting controller response increases the signal to the damper actuators, opening the outside and exhaust dampers beyond their minimum position while proportionately closing the return air dampers.

Whenever the outside air temperature exceeds the economizer high limit set point, the economizer strategy is disabled and the mixing dampers are modulated to maintain their minimum position flowrate based upon temperature. Whenever the temperature difference between the outside and return air streams is less than 5°, the dampers are commanded to their minimum position by sending a 20% output signal to their actuators.

Supply Air Temperature Control

The normally closed chilled water valve and the normally open hot water valve are modulated in sequence to maintain the supply air temperature set point. As the supply air temperature increases, the signal to both valves also increases, modulating the hot water valve closed and the chilled water valve open. A one degree deadband is programmed into the valve response to prohibit the simultaneous heating and cooling of the airstream. The valve response is shown by the graph in Figure 22.2.

Supply and Mixed Air Temperature Set Point Reset

The supply and return air temperature set points are reverse reset based upon the zone experiencing the greatest cooling load. As its cooling load increases, the supply air temperature set point decreases in accordance with the schedule shown in Figure 22.3. The mixed air temperature set point is calculated to be 2° below the supply air temperature set point to compensate for the addition of fan heat to the airstream.

Zone Temperature Control

The normally open hot water reheat valve is modulated to maintain the zone's temperature set point. As the zone temperature decreases, the signal from the direct acting thermostat also decreases, modulating the valve toward its open

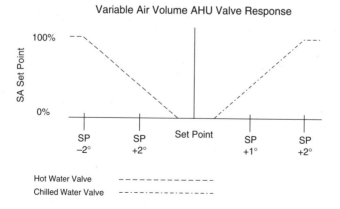

Figure 22.2   Single Path, Multiple Zone AHU Valve Response

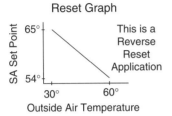

Reset Schedule

Outside Air Temperature	Supply Air Temperature Set Point
30°	65°
60°	54°

Reset Graph

This is a Reverse Reset Application

SA Set Point: 65°, 54°
Outside Air Temperature: 30°, 60°

Figure 22.3   Temperature Reset Schedule and Graph

position. When the air handling unit fans are commanded off, the reheat valve continues to be modulated by the thermostat.

22.3.3 Steady State Process Conditions for the Single Path, Multiple Zone Unit

Table 22.1 lists the current process conditions for the single path, multiple zone unit with zone reheat coils that is shown in Figure 22.1.

Table 22.1   STEADY STATE PROCESS CONDITIONS

Process	Set Point	Control Point	FCD	Position
Outside Air Conditions		67° DB 55% RH		
Return Air Conditions		78° DB 54% RH		
Mixed Air Temperature	Economizer 54.6°	67°	OA & EA RA Damper	100% Open 0% Open
Supply Air Temperature	56.6°	56.6°	CHWS Valve HWS Valve	38% Open 0% Open
Temperatures Zone 1 Zone 2 Zone 3*	73° 74° 75°	72.5° 72.0° 76.1°	Hot Water Reheat Valves	0% Open 5% Open 0% Open

*highest cooling load

22.4

PROCESS RESPONSES TO LOAD CHANGES

The following passages list the probable responses of the air handling unit and zone control loops to the described change in a process load.

22.4.1 Load Change 1

The temperature in Zone 3 increases in response to an increase in its internal heat gains. (Note the changes between the following response and that of a single zone AHU response.)

Supply Air Temperature Process Response

1. The thermostat in Zone 3 increases its output signal in response to the increase in the zone's load. The increase in the thermostat's signal resets the supply air temperature set point from 56.6 to 54.5°.
2. The decrease in the supply air temperature set point increases the error between the new supply air set point (54.5°) and the previous control point (56.6°). The supply air temperature controller responds to the error by increasing its output signal to the chilled water valve, positioning it to approximately 60% open.
3. The increase in chilled water flow through the coil absorbs more heat from the airstream, decreases its temperature to the new set point (54.5°) and the loop stabilizes.

Mixed Air Temperature Process Response

1. The mixed air set point decreases from 54.6 to 52.6° (2° for fan heat) in response to the reset of the supply air set point from 56.6 to 54.5°.
2. The mixed air damper position is unaffected by this change in set point. The dampers remain at 100% ventilation position because the loop is still unable to maintain its new set point with 67° outside air and the outside air temperature remains below the economizer high limit set point.

Zone Temperature Process Response

1. The thermostat in Zone 3 measures the increase in room temperature.
2. The control signal to its reheat valve increases so the hot water valve remains closed.
3. The cooler supply air temperature offsets the heat gain in Zone 3, reducing its temperature to its 75° set point.

4. The colder discharge air entering Zone 2 causes its temperature to drop and the thermostat responds by decreasing its output signal, modulating its reheat valve toward 20% open.

5. The colder air entering Zone 1 also causes its thermostat to reduce its output signal, modulating its reheat valve toward 5% open because the control point now exceeds the 1° degree deadband of the thermostat.

22.4.2 Load Change 2

The outside air temperature increases from 67 to 71° while the thermal loads of the zones remain at the levels achieved during Load Change 1.

Minimum Position Process Response

1. The outside air sensor signals the system that the economizer high limit set point has been exceeded. The mixing dampers are commanded to modulate to their minimum position based upon the outside and return air temperatures because a minimum of 5° exists between the temperatures.

2. A new mixed air temperature set point is calculated every 10 minutes to maintain a 25% minimum flowrate of ventilation air. The mixed air temperature controller algorithm modulates the dampers to maintain the mixed air minimum position set point of 76.25° based upon the calculation: mixed air set point = $(0.25 \times 71°) + (0.75 \times 78°)$.

Zone Temperature Process Response

1. The increases in the outside air temperature caused an increase in the quantity of heat transferred into each of the zones through their walls, windows and roof.

2. A thermal response similar to that in Load Change 1 occurs as the cooling load increases in Zone 3.

3. Although Zone 1 and Zone 2 experience an increase in their cooling load as a result of the increase in the outside air temperature, the temperature of the air entering their zones also decreases after being reset based upon the increased load in Zone 3. The position of the reheat valves in Zone 1 and Zone 2 is a function of the net effect of these two changes.

Supply Air Temperature Process Response

1. An increasing signal from the thermostat in Zone 3 resets the supply air temperature set point. Since the mixing dampers were commanded to minimum position, they cannot be modulated to offset some of the increase requirements of the mechanical cooling system.

2. The decrease in the supply air temperature set point increases the offset between the supply air set point and its control point. The supply air temperature controller increases the output signal to the chilled water valve, positioning it closer to its 100% open position.

3. The increase in chilled water flow through the coil decreases the temperature of the mixed air passing across the cooling coil. The loop stabilizes once the control point equals the new set point.

22.4.3 Load Change 3

The outside air temperature decreases from 67 to 60°

Mixed Air Temperature Process Response

1. The drop in outside air temperature below the economizer high limit set point enables the economizer cycle and the mixing dampers are allowed to modulate open beyond their minimum position.
2. Since the outside air temperature exceeds the mixed air temperature set point (58°), the outside and exhaust dampers modulate to 100% open and the return air dampers close. The air passing through the mixing dampers equals the outside air temperature of 60°.

Supply Air Temperature Process Response

1. The reduction in the cooling load in Zone 3 resets the supply air set point to 60°.
2. The increase in the supply air set point to 60° causes the cooling valve to modulate toward its closed position because the outside air dampers are 100% open and the 60° outside air temperature satisfies most of the load. The chilled water valve is modulated to remove the heat added by the supply fan.

Zone Temperature Process Response

1. Decreases in outside air temperature reduce the heat transferred into the zones through their walls, windows and roof. The reduction in the cooling load of Zone 3 resets the supply and mixed air temperature set points.
2. The loads in the Zones 1 and 2 also decrease. The increase in the supply air temperature that results from the resetting of the temperature set points offsets the lower zone loads and their thermostats may or may not call for additional reheat.

22.5

ANALYSIS OF A SINGLE PATH, MULTIPLE ZONE AIR HANDLING UNIT WITH ZONE REHEAT

22.5.1 Scenario 1

The zone's temperature is too warm even though the air handling unit is operating in a cooling mode.

Possible Causes

1. Chilled water is not available at proper quantity or quality to satisfy the load.
2. The zone's reheat valve is stuck open or leaking by its seat.
3. The supply air flowrate into the zone is restricted.

Verify the

1. thermostat's set point is actually placed at the desired setting.
2. thermostat is properly calibrated. Verify calibration by performing the following procedure:
   a. Measure temperature at the thermostat's location.
   b. Adjust the thermostat's set point so it equals the measured temperature.
   c. Measure the thermostat's output signal.
   d. Calibrate the thermostat's output signal bias until it is equal to the maximum value of the reheat valve's spring range. At this value the reheat valve is closed when the zone temperature equals its set point.
   e. Place the thermostat's set point equal to the desired zone temperature and let the loop stabilize.
   f. Adjust the thermostat's proportional gain to achieve an initial quarter amplitude decay response to a step change of +5°.
3. the reheat valve is correctly positioned. Measure the output of the thermostat to compare it to the signal arriving at the reheat valve's actuator. Lower the thermostat set point to its minimum value. Check the position of the hot water valve, it should be 100% closed. Determine if the valve is leaking through by momentarily closing the isolation ball valves on either side of the coil and monitoring the temperature of the air downstream of the reheat coil. If everything appears in order, analyze the control loops on the air handling unit.
4. supply air flowrate into the zone is at design level. If the AHU filters or coil surfaces are dirty, the volume flowrate of conditioned air will decrease, creating temperature control problems. If the problem is only in one zone, check to see if its reheat coil is dirty or that a fire damper is not closed upstream of the zone's diffusers.
5. calibration and response of the supply air temperature sensor.
   Verify sensor calibration by performing the following procedure for pneumatic sensors:
   a. Measure the control point using an instrument of verifiable accuracy.
   b. Measure the output signal of the sensor.
   c. Using the sensor's transfer function, calculate the sensor's output signal using the value measured with the accurate instrument.
   d. Compare the calculated output signal value with the measured output signal value. If they are close, the sensor is calibrated. If they differ more than 2%, calibrate or replace the sensor.

Verify sensor calibration by performing the following procedure for DDC sensors:

    a. Measure the control point using an instrument of verifiable accuracy.

    b. Read the value of the point displayed on the DDC panel.

    c. Compare the measured value with the displayed value. If they are within 2% of each other, the sensor is calibrated. If they differ more than 2%, and the sensor has a signal converter, calibrate the sensor. If the sensor does not have a signal converter, place a compensation term in the database to correct the error or replace the sensor.

6. calibration of the supply air temperature controller by performing the following procedure:

    a. Initiate a step change in the set point of the process.

    b. Evaluate the controller's response using the procedures outlined in Chapter 12.

    c. If the steady state error is too large, increase the controller's gain.

    d. If the response is unstable, reduce the controller's gain.

    e. If the process is difficult to stabilize, make adjustments to the output signal bias and evaluate the response.

7. calibration of any signal converters or positioners using the manufacturer's instructions or the procedure in Chapter 21.

8. controller's output signal is present at the valve actuators and that the valves are correctly positioned. Verify that the valves can stroke from 0% to 100% without binding.

9. quality and quantity of the chilled water at the air handling unit. If the chilled water is too warm, the chiller control system must be evaluated.

10. calibration of the mixed air temperature controller by performing the following procedure:

    a. Determine if the dampers should be operating in economizer mode or at minimum position. If the outside air temperature exceeds the economizer high limit set point, the dampers should be modulating to maintain minimum position flowrate.

    b. If the dampers are under minimum position control, measure the return, mixed and outside air temperatures to determine the quantity of air being brought in.

    c. If the dampers are in minimum position but not modulating to maintain the correct flowrate of ventilation air, check the formula used to calculate the minimum position set point, or in pneumatic systems, adjust the minimum position regulator until the correct mixed air temperature corresponding to the correct ventilation rate appears.

    d. After validating the reset set points are correct, initiate a step change in the set point of the process to evaluate the controller's response.

    e. Evaluate the controller's response using the procedures outlined in Chapter 12.

    f. If the steady state error is too large, increase the controller's integral gain.

    g. If the response is unstable, reduce the controller's proportional gain.

    h. If the process is difficult to stabilize, make adjustments to the output signal bias and evaluate the response.

    i. Verify that the controller output signal is present at the mixing damper actuators and the dampers are correctly positioned. Calibrate any signal converters or positioners using the manufacturer's instructions or the procedure in Chapter 21.

22.5.2 Scenario 2

The chilled water temperature control loop is calibrated correctly but the supply air temperature control loop cannot maintain the supply air temperature set point.

Possible Causes

1. The chilled water valve is not opening all the way.
2. The chilled water is not arriving at the coil at correct quantity or quality.
3. The latent heat load from outside air is overloading the capacity of the coil.

Verify the

1. controller's output signal is arriving at the chilled water valve's actuator. If it is and the valve is not opening, determine if the actuator's diaphragm is leaking control air causing the actuator to stay in normal position.
2. filters or the chilled water coil are not dirty, limiting air flow through the air handling unit. Correct any abnormal condition.
3. chiller, condensing unit and associated pumps have been enabled and are operating. Check the pump couplings and fan belts. Check the condensing unit for proper cooling operation and clean fins. Check cooling tower spray nozzles for scale buildup.
4. loss of sensible heat capacity may be due to an abnormal increase in latent heat load of the ventilation air. If the mixing dampers are still operating in an economizer cycle, the high limit set point should be lowered, commanding the dampers to minimum position.

22.5.3 Scenario 3

All of the zones are too warm when the outside air temperature is 65°.

Possible Causes

1. There is no chilled water available.
2. There is insufficient air flowrate into zone.

Since all the zones are warm, the problem is most likely the result of an air handling unit or chiller malfunction.

Verify the

1. operation of the chilled water system as described in Scenario 2.
2. supply and mixed air temperatures appear correct based upon the current load.
3. air flow through the distribution system. Although the fans may appear to be operating correctly, there will be insufficient flow through the main duct.
4. both fans are operating, their belts are tight and all fire dampers are open. Verify that they are rotating in the correct direction. If a fan is inoperative due to broken belts, current overload or manually turned off, all the zones will warm up.

22.6

EXERCISES

Respond to the following statements, questions and problems completely and accurately, using the system drawing in Figure 22.1.

1. What is the response of a zone's temperature control loop if the set point of the hot water supplying the reheat coil increases 15° in response to a decrease in the outdoor air temperature but the zone's load remains relatively constant? Explain your answer.
2. What would be the entire system's response if a heat generating appliance was placed below one of the zone's thermostats? Explain your answer.
3. What is the response of the air handling unit and zone temperature control systems when the air handling unit's low temperature limit trips during occupied operation? Explain your answer.
4. Calculate the minimum position set point for the mixed air temperature control loop when the return air temperature is 73.5°, the outside air temperature is 25.2° and 15% of the air entering the zone must be ventilation air.
5. a. How much ventilation air is entering an air handling unit that serves non-smoking offices if the outside air temperature is 42°, return air temperature is 74.3° and the mixed air temperature is 60.4°?
   b. If the ASHRAE requirement for these zones is 18%, what are the energy ramifications of the present minimum position calibration? Explain your answer.
6. A hot water converter system has the following reset schedule. Determine the set point of the hot water control loop when the outside temperature is 30°, 50° and 60°.

OA Temperature	HW Set Point
−10°	200°
70°	120°

7. A zone humidity control loop has the following reset schedule. Determine the set point of the humidifier control loop when the outside temperature is 0°, 20° and 40°.

OA Temperature	Humidity Set Point
−10°	30%
50°	50%

Single Path, Variable Volume, Multiple Zone Air Handling Units With Reheat Coils

23.1

VARIABLE AIR VOLUME SYSTEMS

Many newer building HVAC designs incorporate single path, *variable* volume air handling units in their mechanical systems. These systems have replaced the constant volume units previously used for multiple zone applications. Variable air volume systems vary the *volume* flowrate of conditioned air entering a zone in response to its thermal load. As the cooling load decreases, the flowrate of air entering the zone also decreases because a smaller volume of supply air is all that is needed to maintain the balance with the zone's thermal load. The supply air temperature is typically held between 50 and 60°, making these primarily cooling air handling units.

Variable volume air handling units are more energy efficient than the single path constant volume systems they replace. Varying the flowrate of the supply and return air streams reduces the amount of electrical energy required to operate the system's fans. The supply and return fans in *constant* volume systems are sized to supply the volume of air that will be needed to satisfy the design cooling load of each zone as if it occurred at the same time on design day. Although correct, this procedure over-sizes the fans because each zone will not experience its full load at the same time. This design wastes fan power throughout the entire year. It also substantially increases the reheat requirements of the system because larger quantities of cool air must be warmed before they are discharged into a partially loaded zone. Reset strategies are implemented to improve the thermal efficiency of constant volume systems but these strategies do not improve the fan's energy utilization efficiency.

Variable volume designs responds to the *diversity* of a building's cooling loads. Diversity describes the shifting load characteristic of multiple zone systems. In reality, the cooling load shifts from the east side of the building in the late morning, to the south side in the afternoon, followed by the west side in the evening. A properly designed VAV system will take advantage of this diversity of its load profile by shifting the volume of cool air from partially loaded zones to those with higher loads. This strategy reduces the initial size requirements of the system's fans and the amount of reheat required in the partially loaded zones.

23.1.1 Variable Air Volume – Variable Air Temperature Systems

When first developed, VAV systems were designed to operate with a fixed supply air temperature set point. Thermal load balancing was accomplished through the variable air volume control of a fixed temperature airstream. This strategy was opposite to that used in constant volume systems where the flowrate was kept constant and its temperature varied. Reset strategies were introduced into constant volume systems to improve operational efficiency and controllability of the processes. Current VAV designs also incorporate set point reset. In similarity with constant volume systems, temperature reset is implemented in the supply and mixed air temperature processes to maintain a set point that satisfies the zone with the highest cooling load.

The name variable air volume and temperature (VAVT) was initially used to differentiate between the system configurations that applied reset and those that did not. VAV is now generally used to describe either system. The advantages gained by resetting supply and mixed air set points are the same as those described in the previous chapters. Altering the air temperatures as a function of outside air temperature or zone load reduces the amount of cooling and reheating that must be performed throughout the system. It also improves the controllability of the process by requiring the final controlled devices to modulate near their mid-stroke position. Combining the operational saving generated by the reductions of fan power and reheat requirements along with the improved control of valves and dampers have made VAV systems the current choice of mechanical system in new building designs and retrofit applications.

To convert a single path, constant volume air handling unit a multiple zone system, a means of varying the fan's capacity must be added to the air handling unit and VAV terminal boxes must be installed in each zone. VAV boxes have a control damper located in the primary airstream to alter the volume of air being discharged into the zone. As a zone's cooling load decreases, the damper modulates toward its minimum position to maintain a balance between the energy discharged into the zone and its current load. If a zone's winter design cooling load is calculated to be lower than the energy entering the zone when the VAV box is at its minimum position, a reheat coil can be installed at the outlet of the VAV box. This coil supplies additional heat to the zone during periods of light thermal load. Perimeter baseboard convectors can be used in place of or to supplement a terminal reheat coil to provide the necessary heat.

23.2

ILLUSTRATION OF A SINGLE PATH, VARIABLE VOLUME, MULTIPLE ZONE AHU WITH REHEAT

This example illustrates the response of a dual fan, single path, variable volume, multiple zone air handling unit with VAV box reheat coils. The system has the following control loops in its control system:

1. Supply and return fan start/stop.
2. Supply fan volume control.
3. Return fan volume control.
4. Mixed air temperature control.
5. Supply air temperature control.
6. Zone air temperature control.
7. Low temperature limit protection.

23.2.1 System Drawing

A system drawing for a single path, variable air volume AHU is shown in Figure 23.1.

23.2.2 Sequence of Operation

The sequence of operation for the variable air volume, multiple zone air handling unit with terminal box reheat coils that is illustrated in Figure 23.1 is explained in this section.

   (Note: because of the similarities between multiple zone air handling systems, many of the following sequences of operation are identical to those found in the constant volume, multiple zone unit described in the previous chapter.)

Fan Start/Stop

The supply fan is commanded on and off based upon a user defined occupancy schedule. The return fan is electrically and software interlocked with the supply fan so they operate in unison. Whenever the fans are off, the mixing dampers along

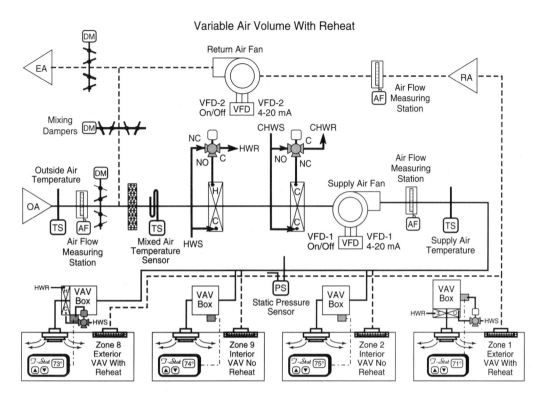

Figure 23.1 Variable Air Volume Air Handling Unit

with the hot and chilled water valves are commanded to their normal position. The modulating signal to both variable frequency drives is also commanded to 0%.

If either fan trips off due to the presence of a fault condition, the other fan's control circuit is also opened, commanding it off. The mixing dampers along with the hot and chilled water valves are commanded to their normal position. The modulating signal to both variable frequency drives is also commanded to 0%.

Low Temperature Limit

If the low limit temperature device that is located downstream of the hot water coil measures a temperature below 36 °F, the supply and return fans are commanded off. The mixing dampers along with the hot and chilled water valves are commanded to their normal position. The modulating signal to both variable frequency drives is also commanded to 0%.

Mixed Air Temperature Control

The normally closed outside and exhaust air dampers along with the normally open return air dampers modulate to maintain the mixed air temperature set point. The mixing dampers are modulated to insure a minimum volume of ventilation air enters the unit whenever the building is occupied.

During periods of when the AHU is in a heating mode (except during warm-up), the mixing dampers are commanded to maintain their minimum position based upon the temperatures of the outside, return and mixed air streams. A new minimum position mixed air set point is calculated every 10 minutes to respond to changes in the outside or return air temperatures.

The economizer cycle begins as soon as the outside air dampers modulate beyond their minimum position. As the mixed air temperature increases above its set point, a direct acting controller response increases the signal to the damper actuators, opening the outside and exhaust dampers beyond their minimum position while proportionately closing the return air dampers.

After the outside air temperature exceeds the economizer high limit set point, the economizer strategy is disabled and the mixing dampers are modulated to maintain their minimum position based upon temperature. Whenever the temperature difference between the outside and return air streams is less than 5°, the dampers are commanded to their minimum position by sending a 20% output signal to their actuators.

Supply Air Temperature Control

The normally closed chilled water valve and the normally open hot water valve are modulated in sequence to maintain the supply air temperature set point. As the supply air temperature increases, the signal to both valves also increases, modulating the hot water valve closed and the chilled water valve open. A one degree deadband is programmed into the valve sequence to prohibit the simultaneous heating and cooling of the air stream. The valve response is shown by the graph in Figure 23.2.

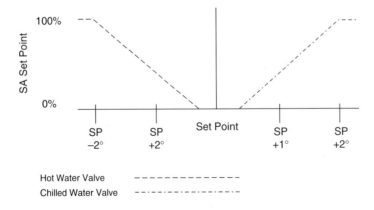

Figure 23.2 Variable Air Volume AHU Valve Response

Supply and Mixed Air Temperature Set Point Reset

The supply and mixed air temperature set points are reverse reset based upon the outside air temperature. As the outside air temperature increases, the temperature set points decrease in accordance with the schedule shown in Figure 23.3.

Supply Air Volume Control

The supply air fan's variable frequency drive is modulated to maintain the supply duct static pressure at a set point of 1.2 inches of water. As the static pressure in the supply duct increases in response to some of its VAV zone boxes throttling toward their minimum position, the reverse acting output signal from the controller decreases, reducing the fan's rotational speed, capacity and power consumption.

Return Air Fan Control

The return fan's variable frequency drive is modulated to maintain the return air volume flowrate equal to the current supply air volume minus the fixed exhaust requirements of the building. In response to an increase in the supply air volume flowrate, a new return air volume set point is calculated. The reverse acting return fan controller increases its signal to the drive, increasing the speed of the return fan's motor, to generate a higher return air flowrate.

Zone Temperature Control

The VAV box damper and reheat valve are modulated in sequence to maintain the zone's temperature at set point. As the cooling load in the zone decreases, the VAV

Reset Schedule

Outside Air Temperature	Supply Air Temperature Set Point
30°	65°
60°	54°

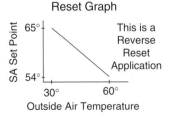

Figure 23.3   Temperature Reset Schedule and Graph

box damper is modulated toward its minimum volume position. If the zone temperature continues to decrease below its set point, the VAV damper remains at its minimum position and the normally open hot water valve is modulated to maintain the zone temperature set point. When the fans are commanded off, the VAV damper and the reheat valve continue to be positioned by the zone's thermostat.

23.2.3 Steady State Process Conditions for the Variable Air Volume, Multiple Zone Unit

Table 23.1 lists the current process conditions for the variable air volume, multiple zone unit with terminal box reheat coils that is shown in Figure 23.1.

Table 23.1   STEADY STATE PROCESS CONDITIONS

Process	Set Point	Control Point	FCD	Position
Outside Air Conditions		75° DB 65% RH		
Return Air Conditions		80° DB 60% RH		
Mixed Air Temperature	20%Min Pos 4000 cfm	4000 cfm	OA & EA RA Damper	15% Open 85% Open
Supply Air Temperature	55.0°	55.0°	CHWS Valve HWS Valve	58% Open 0% Open
Supply Air Static Pressure	1.2 inch Water	1.2 inch Water	Supply Fan VFD	75% of Maximum Flow
Return Air Volume Flowrate	16000 cfm	16000 cfm	Return Fan VFD	80 % of Maximum Flow
Zone 1	74°	73.5°	Damper Reheat Valve	42% Open 0% Open
Zone 2	75°	73.8°	Damper	Min Position
			Reheat Valve	5% Open
Zone 8	73°	73.2°	Damper	52% Open
			Reheat Valve	0% Open
Zone 9	74°	75°	Damper	60% Open
			Reheat Valve	0% Open

23.3

PROCESS RESPONSES TO LOAD CHANGES

The following passages list the probable responses of the air handling unit and zone control loops to the described change in a process load.

23.3.1 Load Change 1

The zone cooling loads decrease as most equipment is turned off and the occupants leave the office to go to lunch.

Zone Temperature Process Response

1. The thermostats respond to the reduction in their zone temperatures as the cooling load decreases.
2. The control signal from the zone thermostats to their VAV box controllers decrease, modulating the control dampers in zones 1, 8 and 9 toward their 30%, 40% and 50% open positions respectively. The damper in zone 2 remains at its minimum position and its reheat valves opens toward 15% to offset the increase in its heating load.

Supply Air Static Pressure Process Response

1. The cooling loads in many of the building's zones decrease. Their thermostats modulate their VAV box dampers to reduced flowrate positions.
2. The decrease in the supply air flowrate requirements produces an increase in static pressure of the air in the supply duct.
3. The supply air static pressure controller decreases its output signal to the supply fan VFD, reducing the frequency of the power supplied to the motor.
4. The reduction in power frequency decreases the motor's rotational speed, lowering the static pressure in the supply duct back to its set point.

Return Air Flowrate Process Response

1. The return air volume control algorithm subtracts the fixed exhaust requirements of the building from the new supply air flowrate to determine the new return air flowrate set point.
2. The decrease in the flowrate set point causes the return air volume controller to reduce the return air fan's drive frequency thereby decreasing the volume of air drawn back from the zones and re-establishing the balance between the supply fan's capacity, fixed building exhaust and the return fan volume.

Supply Air Temperature Process Response

1. The outside air temperature remains constant so the supply air temperature set point also remained the same (55°).
2. The decrease in the supply air volume reduces the air flowrate across the chilled water coil. This reduction in mass flowing across the heat transfer surfaces permits the coil to transfer more heat from each pound of air, producing a reduction in the supply air temperature.

3. The supply air temperature controller responds by reducing its output signal, modulating the chilled water valve from 58% open toward its 45% open position.
4. The decrease in chilled water flow through the coil increases the supply air temperature back to its set point.

Minimum Position Process Response

1. The mixing dampers remain at their minimum position. The dampers modulate to maintain a flowrate of 4000 cfm passing through the outside air dampers as measured by the flow measuring station located in the outside air duct.

23.3.2 Load Change 2

The chilled water temperature increases in response to an increase in the outside air temperature from 75 to 92 °F.

Supply Air Static Pressure Process Response

1. The outside air temperature exceeds the summer design temperature by 2°.
2. The exterior zone VAV boxes are positioned nearly 100% open, the maximum design flowrate as calibrated in the box's volume controller.
3. With many of the boxes calling for their maximum flowrate, the signal to the supply fan VFD commands the fan to 93% of its maximum speed in order to maintain the supply duct static pressure set point.

Return Air Flowrate Process Response

1. The return volume control loop subtracts the fixed exhaust volume of the building from the increase in the supply air flowrate to calculate its new volume set point.
2. The increase in flowrate set point causes the control loop to increase its drive signal to 95%, increasing the volume of air drawn back from the zones.

Mixed Air Temperature Process Response

1. The outside air temperature still exceeds the economizer high limit set point, causing the mixing dampers to continue to modulate to their minimum position. The mixing dampers are modulated to maintain a flowrate of 4000 cfm passing through the outside air dampers as measured by the outside air flow measuring station.

Supply Air Temperature Process Response

1. The outside air temperature exceeds the summer design temperature by 2°.
2. The supply air temperature set point is 54° and the control point is 57°.

3. The chiller and its associated components are operating above the design temperature, reducing their capacity. The chilled water valve is wide open and the loop is operating *out of control.*

Zone Temperature Process Response

1. The thermostats measure the increase in their zone temperature caused by design load heat gains.
2. The increase in the external heat gains in the exterior zones, coupled with the higher supply air temperature cause these zone dampers to modulate to their maximum open position. The interior zones respond to the increase in the supply air temperature by opening their dampers, increasing the flowrate of supply air into the zone.

23.3.3 Load Change 3

The outside air temperature decreases to from 92 to 55° by the end of the second shift.

Zone Temperature Process Response

1. The zone thermostats measure a decrease in temperature that accompanies the decrease in outside air temperature.
2. The decrease in the external heat gains in the *exterior* zones, coupled with the increase in the supply air temperature determines the position of the box dampers and their reheat coil valves.
3. The *interior* zones are not affected by the change in the outside air temperature. These zones respond to the increase in the supply air temperature by modulating their dampers to a more open position, increasing the flowrate into the zone.

Supply Air Static Pressure Process Response

1. The decrease in outside air temperature reduces the heat gains within the exterior zones and their boxes modulate toward their minimum position.
2. The decrease in the VAV box discharge air flowrate causes the static pressure in the supply duct to increase. The response of the supply air static pressure loop is the same as described in Load Change 1. The controller modulates the supply fan's capacity to a reduced value to balance its output with the zone requirements.

Return Air Flowrate Process Response

1. The return air flowrate control loop subtracts the fixed exhaust volume of the building from the reduced supply air flowrate to calculate its new volume set point.

2. The reduction in volume flowrate set point of the return air system causes the return fan control loop to decrease its drive frequency, reducing the volume of air drawn back from the zones to maintain the balance between the supply, fixed exhaust and return air volumes.

Mixed Air Temperature Process Response

a. The decrease in outside air temperature enables the mixing damper economizer cycle. The dampers modulate to maintain the mixed air temperature set point.
b. The decreasing outside air temperature resets the supply and mixed air temperature set points from 54 to 55.8°.
c. The new mixed air temperature set point is 55.8° and the outside air temperature is 55° so the outside air dampers are positioned nearly 100% open.

Supply Air Temperature Process Response

1. The decrease in outside air temperature reverse resets the supply air temperature set point from 54 to 55.8°.
2. The decrease in the supply air volume coupled with the increase in the supply air temperature set point generates a response that closes the chilled water valve. Since the mixed air temperature control loop is operating in an economizer mode and the outside air temperature is less than the supply air temperature set point, all cooling energy is provided by the outside air.

23.4

ANALYSIS OF A SINGLE PATH, VARIABLE VOLUME AIR HANDLING UNIT WITH REHEAT

23.4.1 Scenario 1

There are annoying whistling noises at entrance doors of the building.

Possible Causes

1. Several fixed exhaust fans in the building are not operating.
2. Either the supply or the return fan is not operating correctly.
3. The supply and return fan flowrates are not tracking correctly.

Verify the

1. operation of the fixed exhaust fans. All should be functioning with their dampers open, belts tight and screens unobstructed. If these fans are not operating properly, the supply air fan will over-pressurize the building. The

excess air escapes through exterior doors and windows creating a whistling noise.

2. supply and return fans are operating correctly, fan belts intact and under the proper tension, inlets and outlets to the fan are free of obstructions and blades are clean.

3. operation and calibration of the supply air static pressure sensor by performing the following procedure:

   Verify sensor calibration by performing the following procedure for pneumatic sensors:

   a. Measure the control point using an inclined manometer or other pressure measuring instrument with verifiable accuracy.

   b. Measure the output signal of the sensor.

   c. Using the sensor's transfer function, calculate the sensor's output signal using the value measured with the accurate instrument.

   d. Compare the calculated output signal value with the measured output signal value. If they are close, the sensor is calibrated. If they differ more than 2%, calibrate or replace the sensor.

   Verify sensor calibration by performing the following procedure for DDC sensors:

   a. Measure the control point using an inclined manometer or other pressure measuring instrument with verifiable accuracy.

   b. Read the value of the point displayed on the DDC panel.

   c. Compare the measured value with the displayed value. If they are within 2% of each other, the sensor is calibrated. If they differ more than 2%, calibrate the static pressure sensor's signal converter.

4. calibration of the supply air static pressure controller by performing the following procedure:

   a. Initiate a 0.2 inch step change in the set point of the process.

   b. Evaluate the controller's response using the procedures outlined in Chapter 12.

   c. If the steady state error takes too long to dissipate, increase the controller's integral gain.

   d. If the response is unstable, reduce the controller's proportional gain.

   e. If the process is difficult to stabilize, make adjustments to the output signal bias and evaluate the response.

5. calibration and operation of the air flow measuring stations by performing the following procedure:

   a. Measure the volume of air flow through the air measuring stations when the fan is commanded to 100% flow using a pitot traverse or similar method. Enter these values into the DDC database.

   b. Calibrate the signal converters on the flow measuring stations so their output signal is equal to the upper limit of its range when the fan is operating at full capacity.

   c. Turn the system fans off and zero the signal converters so they output the lower limit value of the converter's output signal when there is no air flowing through the system.

6. calibration and response of the return air flowrate controller by performing the following procedure:
   a. Initiate a step change in the flowrate set point of the process.
   b. Evaluate the controller's response using the procedures outlined in Chapter 12.
   c. If the steady state error takes too long to dissipate, increase the controller's integral gain.
   d. If the response is unstable, reduce the controller's proportional gain.
   e. If the process is difficult to stabilize, make adjustments to the output signal bias and evaluate the response.
7. controllers output signals are present at their fan's final controlled device. Check to see if the VFD (or inlet vanes) modulate from 0% to 100% without problems. Verify the calibration of any signal converters or positioners using the manufacturer's instructions or the procedure in Chapter 21.

If whistling still persists, adjust the fixed exhaust value entered into the DDC controller until the noise goes away. The return air volume should always be less than the supply air volume to maintain a slight positive pressure within the building.

23.4.2 Scenario 2

One of the building's zones is too warm even though its thermostat is calling for full cooling.

Possible Causes

   a. There is no chilled water available from the chiller.
   b. There is insufficient air flow into zone.
   c. The reheat valve is open or leaking by its seat.

Note: since only one zone is warm, the problem is most likely in that zone instead of with any air handling unit processes.

Verify the

1. the correct air flowrate is entering the zone. If reheat coil's surface is plugged, a fire damper is closed or the VAV box control damper is closed, the volume flowrate of cool air will decrease, creating temperature control problems.
2. thermostat's set point is actually placed at the desired setting.
3. thermostat is properly calibrated. Verify calibration by performing the following procedure:
   a. Measure temperature at the thermostat's location.
   b. Adjust the thermostat's set point so it equals the measured temperature.
   c. Measure the thermostat's output signal.
   d. Calibrate the thermostat's output signal bias until it is equal to the minimum value of the VAV damper's operating range. At this value the

damper is at minimum position and the VAV box reheat valve is closed when the zone temperature equals its set point.

    e. Place the thermostat's set point equal to the desired zone temperature and let the loop stabilize.

    f. Adjust the thermostat's proportional gain to achieve an initial quarter amplitude decay response to a step change of +5°.

    g. Measure the output of the thermostat to compare it to the signal arriving at the reheat valve's actuator.

4. calibration and response of the VAV box by performing the following procedure:

    a. Set the thermostat's set point to its maximum value (85°). This causes the box damper to modulate to its minimum flow position.

    b. Measure the velocity pressure of the primary air entering the box. Convert this value into a volume flowrate (cfm) using the chart on the side of the VAV box or manufacturer's specification sheet.

    c. With the minimum signal from the thermostat being applied to the VAV box controller, tighten the set screws that secure the damper shaft to the actuator. Adjust the volume controller's minimum position potentiometer or the minimum flowrate parameter in the DDC program until the air flow through the box equals the design minimum value. The minimum and maximum values are found on the system prints and may also be written on a label attached to the side of the VAV box.

    d. Set the thermostat set point to its minimum value (55°). This modulates the VAV box to its maximum flow position. Repeat the steps listed above for the maximum flow calibration.

5. control signal is present at the VAV box damper and reheat valve's actuator.

6. the position of the hot water valve when the thermostat's output signal is at its minimum value, it should be closed. Determine if the valve is leaking through by momentarily closing the coil's isolation valves and monitoring the temperature of the air downstream of the reheat coil. If the temperature declines, the valve is leaking by it seat and must be repaired or replaced.

23.4.3 Scenario 3

The supply and return fans did not start.

Possible Causes

1. An occupancy or holiday schedule has the fans commanded off.
2. One of the fan drives has tripped off due to electrical problems.
3. The fan's operation proof was not confirmed in the allotted time.
4. The low temperature limit has tripped open.
5. One or both fan disconnect switches are open.
6. Another electrical problem exists.

Verify the

1. occupancy schedule has commanded both fans on.
2. DDC system has not commanded the fans off as the result of a system or programming fault.
3. low temperature limit has not tripped. If it has it must be manually reset. If the low limit has tripped:
   a. Check the quantity and quality of the hot water (steam) supplied to the air handling unit to determine if the preheat or heating valve was not operating correctly thereby causing the low limit to trip or if was a primary equipment problem.
   b. Check the primary and secondary hot water circulating pumps for proper operation. Their motors should be operating and pump couplings intact.
   c. Verify that the mixing dampers are modulating to maintain minimum position and close completely when the fan's are commanded off.
   d. Verify the return dampers are proportionately open and modulating correctly.
4. VFD's did not trip off on an internal or external fault condition. Determine the cause of any fault by reading the drive's diagnostic display and referencing its service manual.

23.5

EXERCISES

Respond to the following statements, questions and problems completely and accurately, using the system drawing in Figure 23.1.

1. What is a probable response of the return volume control loop when the occupants return after lunch hour and turn on all the heat producing equipment within the zones? Explain your answer.
2. What is a probable response of the supply air static pressure loop when the filters in the AHU plug up with dirt, dust or snow? Explain your answer.
3. What is a probable response of the VAV box if the signal line to the box is severed during a remodel of the zone? Explain your answer.
4. What is the related response of the supply and return fan control loops when the problem described in Item 3 occurs? Explain your answer.
5. If a perimeter convection heating system is installed in the exterior zones along with a VAV box reheat coil, what would be likely operating ranges for all three devices, the reheat valve, control damper and perimeter heat valve if they were all operated by the same thermostat signal. Explain your reasoning for the ranges selected and draw a graph of the response.

CHAPTER 24

Dual Path, Constant Volume Air Handling Units

24.1

DUAL PATH AIR HANDLING UNITS

Dual path air handling units have two separate flow passages that air flows through to be conditioned to different set points. One path allows some of the mixed airstream to be heated while air flowing through the other path is cooled. There are two types of *dual* path units used in HVAC applications, *dual duct* units and *multizone* units. Both of these system designs maintain the zone's temperature set point by blending the heated airstream with the cooled airstream to produce a mixture that satisfies the thermal load of the zone.

The warm and cool air streams of dual path units are produced in separate sections of the air handling unit called the *hot deck* and the *cold deck*. The hot deck has a heating coil and its necessary control loops to produce a constant supply of warm air. The cold deck has a cooling coil and its control loops that produce a cold airstream. In cold weather operation, the mixing dampers supply the cooled air to the cold deck so the chiller can be disabled whenever the outside air temperature is lower than the cold deck's temperature set point. Both decks operate simultaneously, producing heated and cooled air throughout the year.

Both dual path units have similar configurations upstream of their supply air distribution ducts. Each system utilizes similar mixed air (economizer), hot deck, cold deck, fan start/stop, humidity and safety control strategies. They only differ in the delivery and mixing configuration used to supply each zone with air at the correct temperature. *Multizone* systems mix the cold and warm air streams at the air handling unit. The blended airstream is delivered to the zone through a single duct. *Dual duct* systems deliver the hot and cold air to each zone using two separate supply ducts. A mixing box is located at the zone to blend the two air streams before the air enters the zone.

The mixing process used to satisfy the zone loads of dual path, multiple zone systems creates some obvious operating inefficiencies. In the hot deck, energy is consumed warming an airstream that is blended with a cooler airstream at the zone so it will not overheat the occupants. These inefficiencies limit the number of new installations using this equipment. Due to the similarities in the operation and analysis of multizone and dual duct air handling units, both will be described together in this chapter. The following sections highlight the basic differences between the two systems.

24.1.1 Multizone Air Handling Unit Systems

The first identifying characteristic of a multizone air handling unit is the location of its zone mixing dampers. Multizone air handling units have their mixing dampers located on the outlet panel of the air handling unit. Figure 24.1 depicts the location of a multizone's mixing dampers.

Each zone has its *own* normally open hot deck damper and normally closed cold deck damper located on the air handling unit. These dampers blend the correct proportions of cold and hot air to produce the supply air temperature needed by

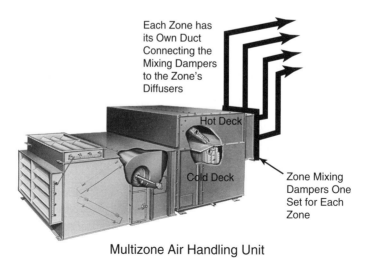

Figure 24.1   Multizone Dampers

the zone. The dampers operate in unison, driven by a single damper actuator that is positioned by the zone's thermostat.

A second characteristic that differentiates a multizone unit from a dual duct unit is the number of supply air ducts that extend from the unit. In multizone applications, each zone receives blended, conditioned air through its own supply duct. This *single* duct extends from the zone's mixing dampers at the air handling unit to the supply air diffusers. Therefore, multizone units are quickly identified by the number of supply ducts on the air handling unit.

The number of zones served by a single multizone unit is a function of the number of ducts that can be terminated on the air handling unit. The total width of the unit increases as the number of zones served by the unit increases. Therefore, it is the air handling unit's width that limits the maximum number of zones served by the unit to about twelve. When more the twelve zones exist in the building, additional air handling units must be installed.

24.1.2 Dual Duct Air Handling Unit Systems

Dual duct air handling units have their mixing dampers located within a terminal mixing box located at each zone. A normally open hot duct damper and normally closed cold duct damper blend the cold and warm air streams in proportion to the zone thermostat's signal.

Dual duct systems are also identified by the number of supply ducts connected to the air handling unit. Only two supply ducts are connected to a dual duct unit, one serving each deck. Smaller branch ducts extend from the main hot and cold distribution ducts to connect with the zone mixing boxes. Because each zone branches off the main hot and cold ducts, there is no dimensional width limitation to the number of zones that can be served by the dual duct unit. Figure 24.2 shows a schematic of a dual duct system.

The supply ducts of dual duct systems are typically different sizes. The hot air ducts are smaller than the cold air ducts because the heated air carries more energy per cfm of air. Consequently, a larger volume of cooled air is required to make the same temperature change in a zone as a smaller volume of hot air can produce.

24.1.3 Improving Operational Efficiencies of Dual Path AHUs

Reset strategies are used to vary the hot and cold deck temperature set points. Both set points are reverse reset using outside air temperature or zone load as the reset variable. Zone load reset can be easily implemented in multizone AHU applications using pnuematic thermostats because the thermostat signal from each zone is already present at the mixing dampers on the air handling unit. The signals can be connected into a hi/lo signal selector. The highest thermostat signal defines the zone with the greatest cooling load so its signal resets the cold deck set point. The lowest thermostat signal defines the zone with the greatest heating load so its signal resets the hot deck set point. Dual duct systems use outside air temperature as the reset variable because the zone's thermostat signals would have to be run to a common location before they could be compared. This increases the cost of implementing the strategy to prohibitive levels in pnuematic applications.

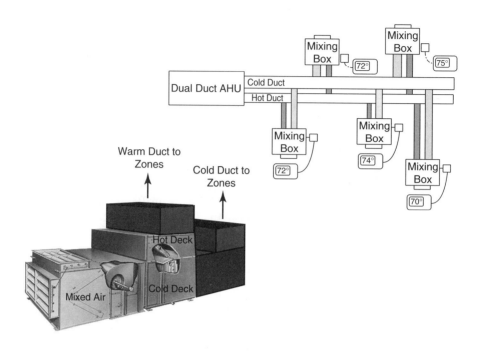

Figure 24.2   Dual Duct Layout

Both types of dual path air handling units can be retrofit to operate as variable air volume systems. In these applications, each deck would have its own fan that varies the flowrate of air in proportion to the total heating and cooling load of the building. The mixing dampers of each system have to have separate hot and cold deck damper actuators in order to modulate the capacity of the fans with respect to the zone loads. With the addition of these changes, the operation of the fans and zone controls become very similar to a single path, VAV system with reheat.

24.2

ILLUSTRATION OF A MULTIZONE AHU

This example illustrates the response of a multizione air handling unit. The system has the following control loops in its control system:

1. Supply and return fan start/stop.   2. Mixed air temperature control.
3. Supply air temperature control.   4. Zone air temperature control.
5. Low temperature limit protection.

24.2.1 System Drawing

A system drawing for a multizone AHU is shown in Figure 24.3.

24.2.2 Sequence of Operation

The sequence of operation for the multizone air handling unit that is illustrated in Figure 24.3 is explained in this section.

Fan Start/Stop

The supply fan is commanded on and off based upon a user defined occupancy schedule. The return fan is electrically and software interlocked with the supply fan so they operate in unison. Whenever the fans are off, the mixing dampers along with the hot and chilled water valves are commanded to their normal position.

If either fan trips off due to the presence of a fault condition, the other fan's control circuit is also opened, commanding it off. The mixing dampers along with the hot and chilled water valves are commanded to their normal position.

Low Temperature Limit

If the low limit temperature device that is located downstream of the hot water coil measures a temperature below 36 °F, the supply and return fans are commanded off. The mixing dampers along with the hot and chilled water valves are commanded to their normal position.

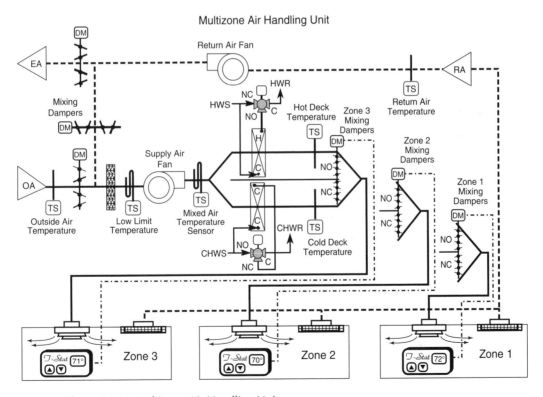

Figure 24.3 Multizone Air Handling Unit

Mixed Air Temperature Control

The normally closed outside and exhaust air dampers along with the normally open return air dampers modulate to maintain the mixed air temperature set point. The mixing dampers are modulated to insure a minimum volume of ventilation air enters the air handling unit whenever the building is occupied.

During cold weather operation, the mechanical cooling system is disabled and the economizer strategy supplies the cold deck with air at the required temperature set point. Whenever the cold deck set point would cause the mixing dampers to modulate below minimum position, the mixing dampers are commanded to maintain their minimum position based upon the temperatures of the outside, return and mixed air streams. A new minimum position mixed air set point is calculated every 10 minutes to respond to changes in the outside or return air temperatures.

The economizer cycle begins as soon as the outside air dampers modulate beyond their minimum position. As the mixed air temperature increases above its set point, a direct acting controller response increases the signal to the damper actuators, opening the outside and exhaust dampers beyond their minimum position while proportionately closing the return air dampers.

After the outside air temperature exceeds the economizer high limit set point, the economizer strategy is disabled and the mixing dampers are modulated to maintain their minimum position based upon temperature. Whenever the temperature difference between the outside and return air streams is less than 5°, the dampers are commanded to their minimum position by sending a 15% output signal to their actuators.

Hot Deck Air Temperature Control

The normally open hot water valve is modulated to maintain the hot deck air temperature set point. As the hot deck air temperature increases, the signal from the direct acting controller also increases, modulating the hot water valve toward its closed position. The hot deck's control valve response is shown in Figure 24.4.

Cold Deck Air Temperature Control

The normally closed chilled water valve is modulated to maintain the cold deck air temperature set point. The chilled water valve is allowed to modulate whenever the mixed air dampers are 100% open or the outside air temperature has exceeded the economizer high limit set point. As the cold deck air temperature increases, the signal from the direct acting cold deck temperature controller also increases, modulating the chilled water valve open toward its 100% open position. The chilled water control valve and mixing damper responses are integrated together as shown in Figure 24.5.

Cold and Hot Deck Air Temperature Set Point Reset

The cold deck set point is reverse reset based upon the zone experiencing the greatest cooling load. As the cooling load in that zone increases, the cold deck's temperature set point decreases.

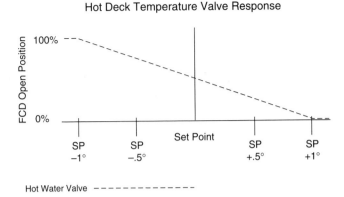

Figure 24.4   Heating Valve Response

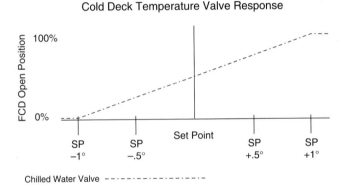

Figure 24.5   Dual Path AHU Cold Deck Valve Response

The hot deck air temperature set point is reverse reset based upon the outside air temperature. As the outside air temperature decreases, the hot deck's temperature set point increases. The set points change in accordance with the schedules shown in Figure 24.6.

Zone Temperature Control

The zone mixing dampers are modulated to maintain the zone's temperature at set point. When the zone calls for heating, the signal from the thermostat to the zone mixing damper actuator decreases, modulating the cold deck damper toward its closed position and the hot deck damper toward its wide open position.

24.2.3 Steady State Process Conditions for the Multizone Air Handling Unit

Table 24.1 lists the current process conditions for the multizone unit that is shown in Figure 24.3.

Hot Deck Reset Schedule

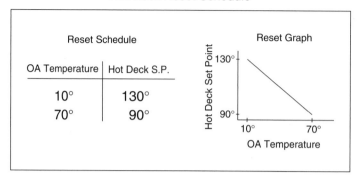

Reset Schedule

OA Temperature	Hot Deck S.P.
10°	130°
70°	90°

Cold Deck Reset Schedule

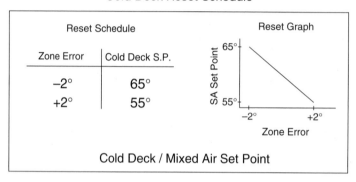

Reset Schedule

Zone Error	Cold Deck S.P.
−2°	65°
+2°	55°

Cold Deck / Mixed Air Set Point

Figure 24.6   Deck Reset Schedules

Table 24.1   STEADY STATE PROCESS CONDITIONS

Process	Set Point	Control Point	FCD	Position
Outside Air Return Air		45° DB 74° DB		
Mixed Air Temperature	53.0°	53.0°	OA & EA RA Damper	72% Open 24% Open
Hot Deck Temperature	106.6°	106.6°	HWS Valve	38% Open
Cold Deck Temperature	53.0°	53.0°	CHWS Valve	0% Open
Zone 1	72°	71.5°	Hot Damper Cold Damper	60% Open 40% Open
Zone 2	70°	70.8°	Hot Damper Cold Damper	55% Open 45% Open
Zone 3	71°	73.°	Hot Damper Cold Damper	0% Open 100% Open

24.3

PROCESS RESPONSES TO LOAD CHANGES

The following passages list the probable responses of the multizone air handling unit and zone control loops to the described change in a process load.

24.3.1 Load Change 1

The outside air temperature decreases from 45 to 35°.

Zone Temperature Process Response

1. The heat losses through the zone's exterior surfaces increase.
2. The increase in heat losses causes the zone temperatures to decrease. The control signal from their thermostats to their zone mixing dampers actuators also decreases. The normally closed cooling dampers modulate toward their closed position and the normally open heating dampers modulate open in proportion to the zone's heating load.

Mixed Air Temperature Process Response

1. The decrease in the zone cooling loads causes the cold deck set point to be reset to 58°. The mixed air temperature set point tracks the change in the cold deck set point so it also changes to 58°.
2. The increase in the mixed air set point causes the air handling unit's mixing dampers to be repositioned. The outside and exhaust air dampers modulate from 72% open to 39% open and the return air dampers modulate proportionately to 61% open.

Cold Deck Temperature Process Response

1. The decrease in the zone cooling load resets the set point of the cold deck to 58°.
2. The chilled water valve remains closed because the mixing dampers operating in their economizer mode can still maintain the required airstream temperature to meet the requirements of the cold deck.

Hot Deck Temperature Process Response

a. The decrease in the outside air temperature resets the hot deck temperature from 106 to 113°.
b. The hot water valve modulates toward 50% open to meet the new set point requirement.

24.3.2 Load Change 2

A malfunction occurs in the boiler plant and the hot water temperature decreases below its required set point (113°). The thermal loads in the zones and outside air temperature remain the same.

Hot Deck Temperature Process Response

1. The hot deck set point remains constant because the outside air temperature did not change.
2. The decrease in the quality of hot water supplied to the hot water coil lowers the heat transfer rate to the mixed airstream passing across the hot water coil. The hot deck temperature sensor measures a corresponding decrease in the hot deck air temperature and the controller modulates the hot water control valve from 50% open toward 100% open.
3. The hot water supply temperature remains below the hot deck temperature set point forcing the hot water valve to 100% open in an attempt to raise the hot deck's air temperature. The hot deck control point continues to decrease because the hot water supply lacks the quality necessary to maintain set point. The loop operates *out of control*.

Zone Temperature Process Response

1. The heat loads within the zones remained constant.
2. The loss in hot deck temperature causes those zones thermostats that were calling for heat to decrease their signals to their zone mixing dampers, modulating their hot air dampers toward 100% open.
3. The zones that were calling for cooling during this event also experience changes in their temperature. The discharge temperature of their zone mixing dampers decreases because of the drop in the hot deck temperature. As cooler mixed air enters the zone, the thermostat signals its mixing dampers to reduce the volume of cold deck air entering the zone in an effort to offset the lower hot deck temperature.

Mixed Air Temperature Process Response

1. The zone with the highest cooling load remains constant so the mixed air set point remains relatively constant.
2. The mixing damper position remains stable because the outside and return air temperatures have not changed. If the hot water temperature does return to normal within a short period of time, the return air temperature will begin to decrease. In response to a decrease in return air temperature, the mixed air temperature controller will modulate the return air damper toward 100% open and the outside air damper toward minimum position in an attempt to maintain the set point.

Cold Deck Temperature Process Response

1. The zone with the highest cooling load remains constant so the cold deck set point remained relatively constant.
2. The volume of air passing across the cooling coil decreases as the mixing boxes throttle their cold deck dampers closed in response to the loss of hot deck temperature.
3. The decrease in volume flowrate across the cooling coil causes the control point of the cold deck to decrease. The direct acting cold deck controller modulates the chilled water valve toward its closed to maintain the set point.

24.3.3 Load Change 3

Instead of load changes #1 and #2 occurring, the system changes from its steady state process conditions to respond to an increase in the outside air temperature from 45 to 75°.

Zone Temperature Process Response

1. The cooling load in the zones increases as heat transfers into the zone from the exterior surfaces.
2. The increase in the zone's heat gain causes its temperatures to increase. The control signal from the thermostat to the zone's mixing damper actuator increases, opening the normally closed cooling damper while proportionately closing the heating damper.
3. The return air temperature increases to 78° in response to the increase in the cooling load within the zones.

Mixed Air Temperature Process Response

1. The outside air temperature increased above the economizer high limit temperature set point. The dampers are commanded to their minimum position.
2. The temperature difference between the outside and return air streams is less than 5°. Therefore, a 15% signal is sent to the mixing dampers.

Cold Deck Temperature Process Response

1. The increase in the zone cooling loads decreases the set point of the cold deck to 54°.
2. The mixing dampers have been commanded to minimum position. The chiller water valve modulates toward 45% open to bring the control point equal to the set point.

Hot Deck Temperature Process Response

1. The increase in the outside air temperature lowers the hot deck temperature from 113° to its minimum value of 90°.
2. The hot water valve modulates to 10% open to maintain a 90° hot deck temperature set point.

24.4

ANALYSIS OF A MULTIZONE AIR HANDLING UNIT

24.4.1 Scenario 1

Many of the zone temperatures are too high when outside air temperature is high.

1. The chilled water temperature is too warm.
2. The cold deck temperature or set point is too high.
3. The hot deck temperature or set point is too high.
4. The mixing dampers are open beyond their minimum position.

Verify the

1. thermostat's set point is actually placed at the desired setting.
2. thermostat is properly calibrated. Verify calibration by performing the following procedure:
   a. Measure temperature at the thermostat's location.
   b. Adjust the thermostat's set point so it equals the measured temperature.
   c. Measure the thermostat's output signal.
   d. Calibrate the thermostat's output signal bias until it is equal to the midpoint of the zone mixing damper actuator spring range. At this value the hot and cold deck dampers will both be 50% open when the zone temperature equals its set point.
   e. Place the thermostat's set point equal to the desired zone temperature and let the loop stabilize.
   f. Adjust the thermostat's proportional gain to achieve an initial quarter amplitude decay response to a step change of +5°.
3. control signal is present at the mixing damper's actuator and the dampers are being correctly positioned. Measure the output of the thermostat to compare it to the signal arriving at the damper actuator. Tighten the hold-down bolts that secure the damper linkage to the damper shaft to prevent slipping. Align the shaft and linkage if necessary.
4. cold deck, hot deck and mixed air control loops are operating with the correct reset set points using the reset schedules located in the sequence of operation and the current conditions.

5. calibration of the hot and cold deck air temperature sensors.
   Verify sensor calibration by performing the following procedure for pneumatic sensors:
   a. Measure the control point using an instrument of verifiable accuracy.
   b. Measure the output signal of the sensor.
   c. Using the sensor's transfer function, calculate the sensor's output signal using the value measured with the accurate instrument.
   d. Compare the calculated output signal value with the measured output signal value. If they are close, the sensor is calibrated. If they differ more than 2%, calibrate or replace the sensor.
   Verify sensor calibration by performing the following procedure for DDC sensors:
   a. Measure the control point using an instrument of verifiable accuracy.
   b. Read the value of the point displayed on the DDC panel.
   c. Compare the measured value with the displayed value. If they are within 2% of each other, the sensor is calibrated. If they differ more than 2%, and the sensor has a signal converter, calibrate the sensor. If the sensor does not have a signal converter, place a compensation term in the database to correct the error or replace the sensor.
6. calibration of the hot and cold deck air temperature controllers by performing the following procedure:
   a. After validating the reset set points are correct, initiate a step change in the set point of the process.
   b. Evaluate the controller's response using the procedures outlined in Chapter 12.
   c. If the steady state error is too large, increase the controller's integral gain.
   d. If the response is unstable, reduce the controller's proportional gain.
   e. If the process is difficult to stabilize, make adjustments to the output signal bias and evaluate the response.
7. calibration of any signal converters or positioners using the manufacturer's instructions or the procedure in Chapter 21.
8. the controller's output signal is present at the valve actuators and that the valves are correctly positioned. Verify that the valves can stroke from 0% to 100% without binding.
9. quality and quantity of the of the hot and chilled water arriving at the air handling unit. If the chilled water is too warm the chiller controls must be evaluated for correct operation. If the hot water is too warm, the boiler or converter control loops must be analyzed for proper operation.
10. calibration of the mixed air temperature controller by performing the following procedure:
    a. Determine if the dampers should be operating in economizer mode or at minimum position. If the outside air temperature exceeds the economizer high limit set point, the dampers should be modulating to maintain minimum position flowrate.

b. If the dampers are under minimum position control, measure the return, mixed and outside air temperatures to determine the quantity of air being brought in.

c. If the dampers are in minimum position but not modulating to maintain the correct flowrate of ventilation air, check the formula used to calculate the minimum position set point, or in pneumatic systems, adjust the minimum position regulator until the correct mixed air temperature corresponding to the correct ventilation rate appears.

d. After validating the reset set points are correct, initiate a step change in the set point of the process to evaluate the controller's response.

e. Evaluate the controller's response using the procedures outlined in Chapter 12.

f. If the steady state error is too large, increase the controller's integral gain.

g. If the response is unstable, reduce the controller's proportional gain.

h. If the process is difficult to stabilize, make adjustments to the output signal bias and evaluate the response.

i. Verify that the controller output signal is present at the mixing damper actuator's and the dampers are correctly positioned. Calibrate any signal converters or positioners using the manufacturer's instructions or the procedure in Chapter 21.

24.4.2 Scenario 2

The low temperature safety limit occasionally trips the air handling unit off when outside air temperature falls below 32°.

Possible Causes

1. The low limit device is out of calibration.
2. The air handling unit's mixing dampers are not modulating correctly.

Verify the

1. temperature at the low limit sensor location to be sure it trips at its design low limit set point of 36°. Adjust trip set point if necessary.
2. mixed air temperature is limited by the cold deck reset schedule to an operating range of 50 to 65°. If the mixed air is falling below 50°, the dampers are not modulating correctly. Calibrate the mixed air temperature control loop using the procedure outlined above.
3. return and outside air damper linkage is responding correctly to changes in the controller's output signal. If the return damper does not move in unison with the outside air damper oscillations may occur that allow pulses of cold air to reach the low limit sensor.

24.4.3 Scenario 3

One zone is too cold during periods of cool weather.

Possible Causes

1. The thermostat is not calibrated.
2. The zone's mixing dampers are not modulating correctly.

Verify the

1. thermostat calibration using procedure detailed above.
2. zone's mixing dampers are responding correctly to changes in the thermostat's output signal. Adjust and tighten linkage if necessary.

24.5

EXERCISES

Respond to the following statements, questions and problems completely and accurately, using the system drawing in Figure 24.3.

1. What is the probable response of the cold deck control loop when the outside air temperature increases by 3° and the zone thermal loads remain the same? Explain your answer.
2. What is the probable response of the mixed air temperature control loop when the outside air temperature increases from 45 to 60° with a corresponding increase in the zones cooling loads? Explain your answer.
3. What is the probable response of the mixing dampers of a multizone system when the internal load of a zone increases at the same time the cold deck set point is decreased? Explain your answer.
4. What is the response of the entire system when the water chiller experiences an increase in its condensing temperature? Explain your answer.
5. What process load changes would cause the mixed air control loop to increase its output signal during economizer mode operation? Explain your answer.
6. Develop a sequence of operation for the dual duct system depicted in Figure 24.7.

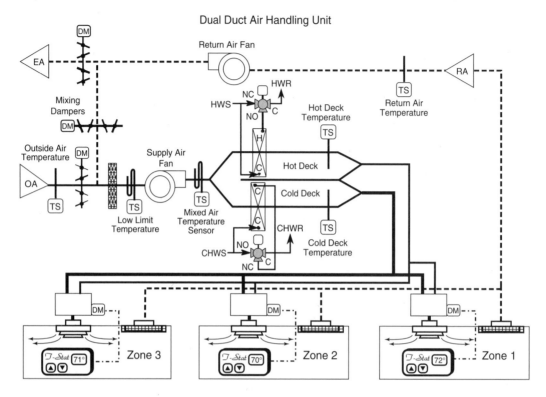

Figure 24.7   Dual Duct Air Handling Unit

Make-Up Air Handling Units

25.1

MAKE-UP AIR HANDLING UNITS

Make-up air handling units condition and deliver 100% outside air into a zone to replace an equal volume of air being exhausted by other equipment within the zone. Many lab and industrial processes use fume hoods, welding booths, spray booths and other equipment that exhausts large quantities of conditioned air out of the building. This air is removed to prevent the build-up of harmful or irritating fumes, gasses or particulates within the work space.

Any air removed from a conditioned space must be replaced with an equal quantity of conditioned air. Make-up units are air handling units used in these applications to provide the conditioned air to *make-up* the air being exhausted from a single zone. The infusion of conditioned air back into the space to replace the exhausted air volume maintains the correct pressure differential between the zone and the adjoining spaces. Negative pressures within a zone increase the infiltration rate of unconditioned air that enters around door and window openings. These uncontrolled flows create drafts, comfort and possible freeze problems during cold weather operation.

Most industrial and fume hood laboratories maintain the static pressure within the zone slightly negative with respect to surrounding spaces. This prevents the process contaminants from leaving the zone by any path other than the exhaust fans. If the area is to be maintained as a clean environment, the make-up air unit will slightly pressurize the zone to prohibit unfiltered air from entering the space. Figure 25.1 shows a picture of a rooftop make-up air unit that is used to maintain environmental comfort, reduce infiltration and to prevent the development of adverse pressure conditions within the space.

Most make-up units used in industrial-based applications are configured as heating only devices. Make-up units can also be provided with a cooling coil and humidity processes if required. Heating make-up air units are installed with out-side air louvers, normally closed dampers, air filters, a heating coil, supply fan and the necessary process control loops. During periods of warmer weather, the unit's only function is to filter the outside air before it is delivered to the zone.

Many installations have the make-up air unit installed within or adjacent to the walls of the zone. In these applications, the units do not require a supply duct to transport the air to the zone. Since these units condition 100% outside air, there are no return ducts or dampers. This lack of ductwork is an identifying characteristic of make-up air units.

25.1.1   Make-up Air Unit Controls

The control loops used for a make-up air unit are similar to those found on a sin-gle zone air handling unit. The operation of the make-up unit's fan is a function of the operating schedule of the exhaust equipment. The fan is interlocked with the process exhaust fans so they all cycle on and off simultaneously.

These are 100% outside air units having no mixing dampers. A single outside air damper is operated by a two-position control signal, opening when the unit's fan is commanded on and closing when it is commanded off. A low temperature limit device protects the heating coil from freezing or allowing unheated outside air from entering the zone. If the low limit senses a temperature below its 36° set point, it commands the fan off and closes the outside air dampers. An alarm should notify the occupants so the exhausting equipment can be cycled off.

Figure 25.1   Make-Up Air Unit

A discharge air temperature control loop regulates the thermal output of the unit. Several different heat sources are commonly used in make-up air applications. The common methods of supplying heat to the zone are: modulating steam valve with integrated face and bypass dampers, a hot water coil with a secondary circulating pump, direct or indirect fired fuel burners and electric heating elements. The hot water and steam coils have the simplest control systems. Fossil fuel burners and electric heating elements require additional safety controls that add complexity to their control systems.

The unit can be supplied with a direct expansion or chilled water coil to cool the outside air before it is discharged to the zone. Of the two, the chilled water coil has the simpler control loop, similar to the cold deck control of a dual path air handling unit. The direct expansion system requires additional electric controls to cycle the system and protect its components from overheating, loss of refrigerant or loss of compressor oil. Sprayed water cooling systems can also be used to provide some temperature relief when the outside air humidity is relatively low throughout the year.

25.2

ILLUSTRATION OF A CONSTANT VOLUME, MAKE-UP AIR UNIT

This example illustrates the response of a make-up air unit having a modulating steam coil with integrated face and bypass dampers. The system has the following control loops in its control system:

1. Supply fan start/stop.
2. Discharge air temperature control.
3. Perimeter fin tube heating control.
4. Low temperature limit protection.

25.2.1 System Drawing

A system drawing for a make-up air unit having a modulating steam coil with integrated face and bypass dampers is shown in Figure 25.2.

25.2.2 Sequence of Operation

The sequence of operation for the make-up air handling unit that is illustrated in Figure 25.2 is explained in this section.

Fan Control

The supply fan is interlocked with the equipment's exhaust fan on/off control. When the exhaust fans are started or stopped, the make-up air unit's fan is also commanded on or off.

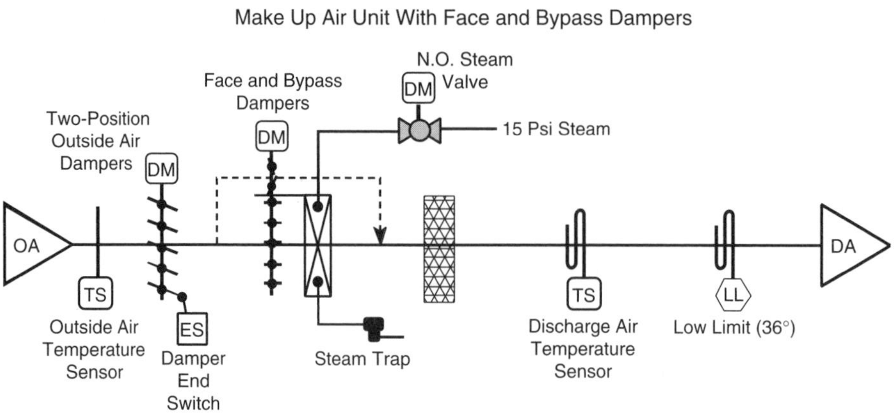

Make Up Air Unit With Face and Bypass Dampers

Figure 25.2 Make-Up Air Handling Unit

Low Temperature Limit

If the low limit temperature device measures a heating coil discharge temperature below 36°, the make-up unit's fan is commanded off, the steam valve, face and bypass dampers and the outside air dampers are commanded to their normal positions. An alarm is annunciated.

Outside Air Damper Control

When the fan exhaust is turned on, an interlock opens the make-up unit's outside air dampers. When the position of the dampers exceed 40% open, a contact closure by an end switch starts the unit's fan.

Discharge Air Temperature Control

Outside Air Temperature Less Than 40°   When the outside air temperature falls below 40°, the steam valve is commanded 100% open and the face and bypass dampers modulate to maintain the discharge air temperature set point.

As the discharge air temperature increases, the direct acting controller's output signal increases, modulating the bypass damper open while proportionately closing the face damper to reduce the temperature of the discharge airstream.

Outside Air Temperature Greater Than 40°   When the outside air temperature rises above 40°, the face and bypass dampers are commanded to their normal position (face open, bypass closed) and the normally open steam valve is modulated to maintain the discharge air temperature set point.

As the discharge air temperature increases, the controller's output signal also increases, modulating the steam valve toward its closed position, reducing the temperature of the air passing through the face damper and coil. The valve and damper response is shown in Figure 25.3.

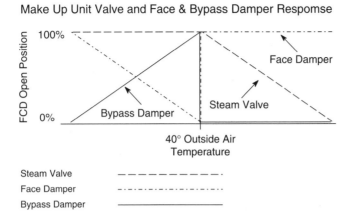

Figure 25.3 Make-Up AHU Valve and Damper Response

Discharge Air Temperature Set Point Reset

The discharge air temperature set point is reverse reset based upon the zone's temperature. As the zone's temperature increases, the discharge temperature set point decreases. The reset schedule is shown in Figure 25.4.

Zone Temperature Control

The thermostat resets the discharge air temperature set point and modulates a normally open perimeter fin tube convector valve. As the temperature in the zone decreases, the signal from the thermostat also decreases, modulating the perimeter heating valve toward it wide open position.

25.2.3   Steady State Process Conditions for the Make-up Air Unit

Table 25.1 lists the current process conditions for the make-up air unit that is shown in Figure 25.2.

Table 25.1   STEADY STATE PROCESS CONDITIONS

Process	Set Point	Control Point	FCD	Position
Outside Air		45°		
Discharge Air Temperature	90°	90°	Steam Valve Bypass Damper	31% Open 100% Open
Zone	71°	71.5°	Perimeter Hot Water Valve	40% Open

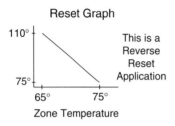

Reset Schedule

Zone Temperature	Discharge Air Temperature Set Point
65°	110°
75°	75°

Reset Graph

110°

75°

65°   75°

Zone Temperature

This is a
Reverse
Reset
Application

Figure 25.4   Discharge Air Set Point Reset Schedule

25.3

PROCESS RESPONSES TO LOAD CHANGES

The following passages list the probable responses of the make-up air unit to the described change in a process load.

25.3.1 Load Change 1

The outside air temperature decreases from 45 to 35°.

Zone Temperature Process Response

1. The heat losses through the exterior surfaces of the zone increase.
2. The increase in heat losses causes the zone's temperature to decline from 71.5 to 70.5°. The thermostat's control signal to the perimeter heat valve decreases and the normally open valve modulates from 40% open toward 50% open.
3. The decrease in zone temperature also resets the make-up air unit's discharge air temperature set point from 89 to 91°.

Discharge Air Temperature Process Response

1. The outside air temperature dropped below 40° so the steam valve is commanded 100% open and the face and bypass dampers modulate to maintain the discharge air set point.

2. The increase in the discharge air temperature set point causes the bypass damper to modulate to 75% open as the face damper moves to 25% open to maintain the discharge air temperature set point.

25.3.2 Load Change 2

The zone's air temperature continues to decrease from 70.5 to 68°.

Zone Temperature Process Response

1. The continued drop in the zone's temperature causes the thermostat's control signal to decrease, modulating the normally open perimeter heat valve from 50% toward 90% open to satisfy the new heating load.
2. The decrease in zone temperature also resets the make-up air unit's discharge air temperature set point from 91 to 100°.

Discharge Air Temperature Process Response

1. The outside air temperature is still below 40° so the steam valve remains wide open and the face and bypass dampers continue to modulate.
2. The increase in the discharge air temperature set point causes the bypass damper to modulate from 75% open to 46% open as the face damper moves to 54% open to maintain the discharge air temperature set point.

25.3.3 Load Change 3

The outside air temperature increases to 71°.

Zone Temperature Process Response

1. The increase in the outside air temperatures raises the zone's temperature to 76°. As the temperature in the zone increases, the thermostat's control signal also increases, causing the normally open perimeter heat valve to modulate closed.
2. The increase in zone temperature also resets the make-up air unit's discharge air temperature set point down to 75°.

Discharge Air Temperature Process Response

1. The outside air temperature is above 40° so the bypass damper is commanded 100% open, closing the face damper. The steam valve modulates to maintain the discharge air temperature set point.
2. The outside air temperature is above the discharge air set point and the steam valve is modulated closed, permitting filtered outside air to enter the zone without any temperature change.

25.4

ANALYSIS OF A MAKE-UP AIR UNIT

25.4.1 Scenario 1

The zone's temperature is too high when outside air temperature is low.

Possible Causes

1. The thermostat's is not operating correctly.
2. The face and bypass dampers or the steam valve is operating incorrectly.
3. One or both valves (steam and perimeter) are leaking past their seat.
4. The reset schedule needs to be adjusted.

Verify the

1. thermostat's set point is actually placed at the desired setting.
2. thermostat is properly calibrated. Verify calibration by performing the following procedure:
   a. Measure temperature at the thermostat's location.
   b. Adjust the thermostat's set point so it equals the measured temperature.
   c. Measure the thermostat's output signal.
   d. Calibrate the thermostat's output signal bias until it is equal to the maximum value of the perimeter heating valve's actuator spring range. At this value the valve will be closed when the zone temperature equals its set point.
   e. Place the thermostat's set point equal to the desired zone temperature and let the loop stabilize.
   f. Adjust the thermostat's proportional gain to achieve an initial quarter amplitude decay response to a step change of +5°.
3. control signal is present at the perimeter heat valve's actuator and the valve is being properly positioned. Measure the output of the thermostat to compare it to the signal arriving at the damper actuator.
4. discharge air temperature control loop is operating with the correct reset set point using the reset schedule located in the sequence of operation and the current zone temperature.
5. calibration of the discharge air temperature sensor.
   Verify sensor calibration by performing the following procedure for pneumatic sensors:
   a. Measure the control point using an instrument of verifiable accuracy.
   b. Measure the output signal of the sensor.
   c. Using the sensor's transfer function, calculate the sensor's output signal using the value measured with the accurate instrument.
   d. Compare the calculated output signal value with the measured output signal value. If they are close, the sensor is calibrated. If they differ more than 2%, calibrate or replace the sensor.
   Verify sensor calibration by performing the following procedure for DDC sensors:

      a. Measure the control point using an instrument of verifiable accuracy.

      b. Read the value of the point displayed on the DDC panel.

      c. Compare the measured value with the displayed value. If they are within 2% of each other, the sensor is calibrated. If they differ more than 2%, and the sensor has a signal converter, calibrate the sensor. If the sensor does not have a signal converter, place a compensation term in the database to correct the error or replace the sensor.

6. calibration of the discharge air temperature controller by performing the following procedure:

      a. After validating the reset set point is correct, initiate a step change in the set point of the process.

      b. Evaluate the controller's response using the procedures outlined in Chapter 12.

      c. If the steady state error is too large, increase the controller's integral gain.

      d. If the response is unstable, reduce the controller's proportional gain.

      e. If the process is difficult to stabilize, make adjustments to the output signal bias and evaluate the response.

7. discharge air temperature controller's output signal is present at the steam valve and the face and bypass damper actuator. If the outside air temperature is below 40°, the signal on the valve actuator should be 0% and the face and bypass dampers receive the modulating signal from the discharge air controller. If the temperature is above 40°, the signal on the face and bypass dampers should be 0% and the valve receives the modulating signal. Be sure the modulating devices are correctly positioned.

8. steam valve is not leaking by after it is commanded closed.

If the control loops are operating correctly and the zone is still too warm, modify the reset schedule to reduce the discharge air temperature.

25.4.2   Scenario 2

The zone's temperature is too low when outside air temperature is low.

Possible Causes

1. The thermostat is not operating correctly.
2. The face and bypass dampers or steam valve is operating incorrectly.
3. One or both valves are leaking by their seat.
4. The reset schedule needs to be adjusted.

Using the same procedures outlined above, analyze the system. If the control loops are operating and the zone is still too cool, change the reset schedule to increase the discharge air temperature. Raise the maximum set point in 5° increments until the zone can maintain a comfortable environment for the occupants.

25.5

EXERCISES

Respond to the following statements, questions and problems completely and accurately, using the system drawing in Figure 25.2.

1. What is the probable cause of people feeling cold around the door or window, but not in the middle of the zone?
2. A make-up air serves the entire load of a zone using heating and cooling coils. The zone's temperature is too warm when the outside air is at its summer design temperatures. What possible problems could cause this condition? Explain your answer.
3. What could cause the discharge air temperature to become too high when the outside air temperature is 35°? Explain your answer.
4. What affect would a plugged steam trap have on the capacity of a steam coil?
5. The outside air temperature is 22° and the zone temperature is at set point (73°). What is the response of a make-up air unit if the exhaust damper sticks closed and both fans continue to operate? Explain your answer.

CHAPTER 26

Steam to Hot Water Converters

26.1

STEAM CONVERTERS

Steam converters are heat exchangers used to produce hot water for building heating systems using steam or high temperature water as the control agent. They are typically used in multiple building facilities that incorporate central boiler plants to produce and distribute steam or high temperature water. The high pressure steam (125 psi) is delivered to each building's mechanical room where its pressure is reduced before it is distributed to the process equipment within that building.

A steam to hot water converter is used to transfer the energy in the steam to the water. Steam converters are shell and tube or plate and frame heat exchangers that allow steam to flow on one side of the heat transfer surface and process water to flow on the other side. As the steam condenses, it gives up its latent heat of vaporization to the water, raising its temperature. By modulating the amount of steam entering the converter, the temperature of the hot water can be maintained at its set point.

26.1.1 Steam Converter Control Strategies

The control system for hot water converters requires few components. A temperature sensor of the appropriate range is located in the process water piping at the outlet of the converter. A controller modulates *two* normally closed, two-way steam valves in *sequence* to maintain the hot water supply at its set point. Two steam valves are used to improve the response of the process. One of the valves is sized to pass one third of the converter's steam capacity and the other passes the remaining 2/3 of the maximum steam flow. The smaller or 1/3 valve has an actuator range that allows it to operate during the lower portion of the controller's output signal range. The larger or 2/3 valve modulates within the upper range of the controller's output signal. A small overlap exists between the actuator ranges to permit a smooth transition between the single and double valve operating modes.

The 1/3, 2/3 valve scheme permits a greater overall turndown ratio to improve process controllability during periods of reduced thermal load. When a single valve is used in place of the two valves, modulation at low loads becomes unstable. Whenever this large valve began to open, excessive quantities of steam enter the converter. The steam quickly heats the water beyond its set point, initiating an immediate response to close the valve. By splitting the valve's maximum capacity over two separate valves, the process can be efficiently controlled at low and high loads. The valve response with overlapping operating ranges is shown in Figure 26.1. Normally closed valves are used so the devices fail closed upon the loss of their control signal. This prevents the overheating and pressurization of the converter during a system malfunction.

Additional control loops are required for converters to protect the building occupants and equipment. A flow switch is inserted in the hot water pipe to disable the converter heating system whenever the flow of water through the converter is not maintained. When the flow switch indicates a loss of flow through the

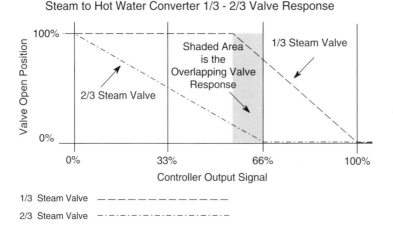

Figure 26.1 Sequenced Steam Valves

converter, the steam valves are commanded closed. This strategy prevents the water located within the converter tubes from boiling within the vessel during a pump coupling failure or other system malfunction. Boiling of the water within a sealed system can cause pipes or seals to burst.

26.1.2 Steam Converter Temperature Reset

The hot water supply temperature is usually reset based upon the outdoor air temperature or system load. Outdoor air is the more common method of resetting hot water because of the close relationship that exists between the heating load and the outside temperature. The higher thermal capacitance of the outside air also reduces the likelihood of the reset process becoming unstable.

Closed loop temperature reset can be performed by monitoring the differential temperature that exists between the hot water supply and return flows. As this differential increases, a rise in the building load is indicated and the set point is raised to reduce the differential temperature. As in all closed loop strategies, the reset may become unstable when the water temperature changes quickly.

26.2

ILLUSTRATION OF A HOT WATER CONVERTER

This example illustrates the response of a steam to hot water converter incorporating a 1/3, 2/3 valve configuration and outside air temperature reset. The system has the following control loops in its control system:

1. System enable/disable.
2. Hot water supply temperature control.
3. Flow verification.

26.2.1 System Drawing

A system drawing for a steam to hot water converter incorporating a 1/3, 2/3 valve configuration and outside air temperature reset is shown in Figure 26.2.

26.2.2 Sequence of Operation

The sequence of operation for the steam to hot water converter that is illustrated in Figure 26.2 is explained in this section.

System Enable/Disable

When the differential pressure switch (installed to read the pressure differential across the hot water pump) indicates that there is water flowing in the system,

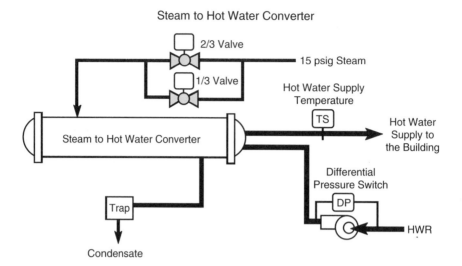

Figure 26.2   Steam to Hot Water Converter System

the converter steam valves are enabled, allowing them to be modulated by the controller. If no hot water flow is detected, the valves are commanded to their normal position (closed), prohibiting the flow of steam into the converter's shell.

Hot Water Supply Temperature Control

The 1/3 and 2/3 steam valves are modulated in sequence to maintain the hot water supply temperature set point. As the temperature of the hot water decreases, the reverse acting signal from the controller increases, modulating the 1/3 valve open. If the temperature continues to decrease, the 2/3 valve begins to modulate open to balance the steam flow with the load. The valve response is shown in Figure 26.1.

Hot Water Supply Temperature Set Point Reset

The hot water supply temperature set point is reverse reset based upon the temperature of the outside air. As the outside air temperature decreases, the hot water set point increases. The reset schedule is shown in Figure 26.3.

26.2.3 Steady State Process Conditions for the Steam to Hot Water Converter

Table 26.1 lists the current process conditions for the variable air volume, multiple zone unit with terminal box reheat coils that is shown in Figure 26.2.

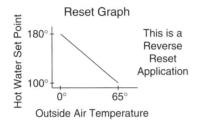

Reset Schedule

Outside Air Temperature	Hot Water Temperature Set Point
0°	180°
65°	100°

Reset Graph

This is a Reverse Reset Application

Figure 26.3   Hot Water Set Point Reset Schedule

TABLE 26.1   STEADY STATE PROCESS CONDITIONS

Process	Set Point	Control Point	FCD	Position
Outside Air		25°		
Hot Water Supply Temperature	149°	149°	1/3 Steam Valve 2/3 Steam Valve	100% Open 22% Open

26.3

PROCESS RESPONSES TO LOAD CHANGES

The following passages list the probable responses of the steam to hot water converter to the described change in a process load.

26.3.1 Load Change 1

The outside air temperature decreases from 25 to 5°.

Hot Water Temperature Process Response

1. The decrease in the outside air temperature causes the hot water set point to be reset from 149 to 174°.
2. The control point is now much lower than the new set point, causing the controller to respond by increasing its output signal. The 1/3 hot water valve remains 100% open and the 2/3 valve modulates toward 84% open to balance the new load.

26.3.2 Load Change 2

The large change in the hot water set point produced a controller response that caused the hot water supply temperature to increases to 180° although the outside air temperature remained at 5°.

Hot Water Temperature Process Response

1. The outside air temperature remained constant so the hot water supply temperature set point remains at 174°.
2. The signal from the controller decreases because the control point is higher than the set point. The 1/3 hot water valve remains 100% open and the 2/3 valve modulates toward 65% open to reduce the hot water temperature and balance the load at 174°.

26.3.3 Load Change 3

The outside air temperature increases to 60°.

Hot Water Temperature Process Response

1. The increase in the outside air temperature resets the hot water supply temperature set point to 106°
2. The signal from the controller decreases, modulating the 2/3 valve closed and the 1/3 valve to 15% open.

26.4

ANALYSIS OF A STEAM TO HOT WATER CONVERTER

26.4.1 Scenario 1

The building zones are too cold and their hot water valves are wide open.

Possible Causes

1. The control signal to converter steam valves has been interrupted.
2. The hot water pump is not operating.
3. The diaphragm in steam valve actuator is leaking.
4. The converter controls are not calibrated.
5. The reset schedule is inappropriate for this application.

Verify the

1. control signals are present to both valve actuators and the valves are free to operate. If the signals are not present, determine if the circulating pump is operating and the flow switch contacts indicate flow is present. If the flow

switch fails to change contact position when the pump starts, the safety loop may be keeping the valves in their normal position.

2. actuators can position the steam valves from 0% to 100% using a hand powered pneumatic pressurization bulb. If electric motors are used, apply the correct control signal to the input of the actuator and position the valve from 0% to 100%. If the control signal is present and the valves move without binding, check the calibration of the controls.

3. hot water temperature control loop is operating with the correct reset set point using the reset schedule located in the sequence of operation and the current outside air temperature.

4. calibration of the hot water temperature sensor.

   Verify sensor calibration by performing the following procedure for pneumatic sensors:
   a. Measure the control point using an instrument of verifiable accuracy.
   b. Measure the output signal of the sensor.
   c. Using the sensor's transfer function, calculate the sensor's output signal using the value measured with the accurate instrument.
   d. Compare the calculated output signal value with the measured output signal value. If they are close, the sensor is calibrated. If they differ more than 2%, calibrate or replace the sensor.

   Verify sensor calibration by performing the following procedure for DDC sensors:
   a. Measure the control point using an instrument of verifiable accuracy.
   b. Read the value of the point displayed on the DDC panel.
   c. Compare the measured value with the displayed value. If they are within 2% of each other, the sensor is calibrated. If they differ more than 2%, and the sensor has a signal converter, calibrate the sensor. If the sensor does not have a signal converter, place a compensation term in the database to correct the error or replace the sensor.

5. calibration of the hot water temperature controller by performing the following procedure:
   a. After validating the reset set point is correct, initiate a step change in the set point of the process.
   b. Evaluate the controller's response using the procedures outlined in Chapter 12.
   c. If the steady state error is too large, increase the controller's integral gain.
   d. If the response is unstable, reduce the controller's proportional gain.
   e. If the process is difficult to stabilize, make adjustments to the output signal bias and evaluate the response.
   f. Calibrate the valve's actuator positioners using the manufacturer's instructions or the procedure in Chapter 21.

If the control loops are operating correctly and the water is still too cool, change the reset schedule to increase the hot water temperature. Raise the maximum set point of the reset schedule in 5° increments until the hot water can satisfy the load in the zones. Monitor the return hot water temperature. When the temperature difference between the supply and return water is about 20°, the set point is adequate. When the differential temperature is less than 20°, the supply water temperature is too high. When the differential temperature is more than 20°, the supply water temperature is too low.

26.4.2 Scenario 2

The zones are too warm and their temperature control loops are becoming unstable.

Possible Causes

1. The converter controls are not correctly calibrated.
2. The reset schedule is inappropriate for this application.

Verify the

1. reset set point is correct and that the control loops are operating correctly using the procedures outlined in Scenario 1.

If the water is still too warm, change the reset schedule to decrease the hot water temperature. Lower the maximum set point in 5° increments until the hot water can satisfy the load in the zones and permit their valves to modulate toward their 50% stroke position. Monitor the return hot water temperature. When the temperature difference between the supply and return water is about 20°, the set point is adequate.

26.5

EXERCISES

Respond to the following statements, questions and problems completely and accurately, using the system drawing in Figure 26.2.

1. List the possible causes of an entire building being too cold when the outside air temperature is below the winter design dry bulb temperature.
2. What is the probable response of the converter control system if the hot water pump coupling fails, stopping flow of water through the heating system? Explain your answer.
3. What would the response of the system be if the pneumatic sensor tube was resting on a the steam pipe and a small hole was melted in the side of the tube? Explain your answer.
4. What could cause the hot water supply temperature to drop below its set point even though both steam valves were wide open? Explain your answer.
5. What is the response of the converter that serves several air handling units when the AHUs were operating in economizer mode and commanded to minimum position due to a reduction in the outside air temperature? Explain your answer.

CHAPTER 27

Fume Hood
Control Systems

In teaching and research facilities there can be many zones, each of which may have a number of fume hoods. The environmental conditions within these zones must be kept constant to minimize their effects on the outcome of an experiment or research. A variable volume air handling unit is typically used to condition the zone. These air handling units operate continuously to maintain the thermal, moisture and pressure conditions within the zone at their set points. Any conditioned air that enters a zone that has a fume hood is not recirculated through the air handling unit. Instead, it is exhausted to the outside of the building along with the air being exhausted from the fume hood enclosures. This prevents the transportation of laboratory agents to other zones within the building. When all of the zones in a building have fume hoods, the air handling unit conditions 100% outside air which passes through the building one time before it is exhausted.

The increasing costs associated with conditioning air that passes through a zone once and is exhausted from the building, coupled with the need to operate lab fume hoods twenty four hours a day, every day of the year, has made variable air volume units the logical building system configuration in these applications. Fume hood control technology has been integrated into the building management control system to operate the facility in a manner that minimizes its operational costs. DDC control panels and variable air volume boxes are used in a configuration that maintains safety and minimizes energy consumption in these applications.

27.1

FUME HOOD SYSTEMS

A fume hood is a ventilated, enclosed work space used to capture and exhaust fumes, gasses and particulates generated by chemicals and other reactions that occur within the hood. Its enclosure is fabricated from a metal box that is typically two to eight feet in length, three feet deep and four feet high. It usually sets upon a counter top or storage cabinet to elevate the work surface to a comfortable height. The sides, baffles, top and work surface are assembled with fire and chemical resistant materials to withstand the harsh environment that exists within the enclosure. There are connections for water, natural gas, compressed air, vacuum, electricity and other utilities installed inside the enclosure to provide the support systems needed to perform an experiment. Figure 27.1 depicts a side view schematic drawing of a fume hood.

Mounted in the front panel of a fume hood is a vertical sliding window made of tempered glass called the *sash*. The sash opens so the user can access the material in the hood or set up an experiment. Mechanical stops are typically used to limit the opening height of the sash to eighteen inches. This limit provides a measure of protection to the user in the event that an uncontrolled reaction occurs when someone standing in front of the hood. The mechanical stops can be overridden so the sash can be opened fully to simplify the set up of an experiment. Once an experiment is underway, the user positions the sash to a level between eighteen inches and closed. The height of the sash has a direct effect on the amount of conditioned air entering the fume hood from the zone.

The fume hood is connected to an exhaust system by a galvanized or stainless steel duct. The exhaust fan draws air, gasses, fumes and particulates out of the enclosure and exhausts them outside of the building through roof mounted stacks. In addition to the sash opening, the exhaust fan draws air from a number of openings designed into the panels that make up the hood's enclosure. Internal baffles direct the flow of gas through the hood, drawing the heavier fumes and gasses from the working surface and the lighter fumes from the top of the hood. When the sash is completely closed, openings near its bottom edge of the work surface allow air to continue to enter the hood to maintain a minimum flowrate through the enclosure at all times.

27.1.1 Fume Hood Face Velocity Control

To operate effectively, the velocity of the air entering the sash opening must be constrained with a range of 80 to 100 feet per minute. Maintaining the proper face velocity minimizes the chances of fumes, gasses or particulates escaping out of the open sash into the zone. When the velocity of the air entering the sash opening is below 80 ft/min, the heavier fumes spill out of the opening, endangering the user and contaminating the zone. When the face velocity exceeds 120 ft/min, turbu-

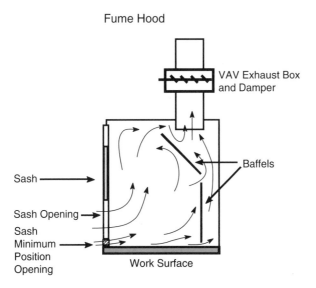

Figure 27.1   Components of a Laboratory Fume Hood

lence and eddy currents develop at the edges of the sash opening, forcing fumes out of the enclosure and into the face of the user. The actual velocity set point is a function of the chemicals or research being performed in the hood.

To maintain the face velocity of a hood at a desired set point, an automatic control system is required. The fume hood control system fume hood maintains the face velocity through its open sash at a *set point* between 80 and 100 feet per minute, regardless of the position or movement of the sash. To control the face velocity, the control point must be either measured directly or calculated. Face velocity can be easily calculated by dividing the volume flowrate of the air being exhausted from the hood by the current area of the sash opening. The relationship between the position of the sash and face velocity through the opening is linear, simplifying the calculation. The width of the sash is a constant, dependent upon the size of the hood. The height of the sash opening varies whenever the sash is moved. Therefore, sash height is the measured variable of the loop. The face velocity is calculated by entering the current value for the sash height into the formula:

$$\frac{cfm}{(height\ of\ sash) \times (width\ of\ sash)} = face\ velocity$$

The sash height is measured with a linear or rotating potentiometer. The height is converted into a variable voltage signal that the DDC panel converts into a measurement of the height in units of feet.

The face velocity can be indirectly measured with a thermal anemometer sensor. The anemometer is mounted in a one inch diameter tube that is inserted through a hole drilled into the side of the hood enclosure. The sensor measures the velocity of the air passing through the tube into the hood due to the differential pressure created by the hood's exhaust fan. As the sash changes position, a linear change in the sensor's output signal is sent to the controller. The DDC panel converts this signal into a sash height.

The face velocity control loop responds to changes in the sash height by varying the amount of air being exhausted from the hood. As the sash is opened, more air must be exhausted from the hood to maintain the velocity set point of the air entering the sash opening. The most common means of varying the exhaust volume is through the use of a VAV box in the exhaust duct of the fume hood. The VAV exhaust box is located between the exhaust fan and the hood. The exhaust fan operates with a VAV control scheme that maintains a constant negative pressure in the exhaust system. As a sash opens, the VAV exhaust box damper opens to increase the volume of air entering the hood, thereby maintaining the velocity set point. The actuation of the VAV box exhaust damper is always performed using pneumatic piston actuators. These are the only actuators that provide the quick response needed to vary the exhaust flow rate when the hood is suddenly opened or closed. Electronic motors that use gear drives have response times that are too long to maintain the proper face velocity. When the actuator's response is too slow, the fumes will exit the sash before the flowrate through the box exhaust damper modulates to its new position. In all applications of VAV exhaust control, the exhaust air passage must remain open upon loss of control signal to insure the fume hood will still be able to contain and exhaust its enclosure.

There are different types of exhaust VAV boxes commonly used to maintain the fume hood's face velocity. One type is a specially designed VAV exhaust valve. This pressure independent air valve has a fixed linear relationship between its actuator stroke and the flowrate through the damper. The fume hood control system calculates the flowrate based upon the actuator's position and the flow characteristics of the air valve. Figure 27.2 shows an air valve VAV damper.

Another type of exhaust box uses pneumatically expanded stainless steel airfoil damper blades to vary the flow through the fume hood. As control air enters the blade's bladder, the airfoil expands, restricting airflow through the box. A third

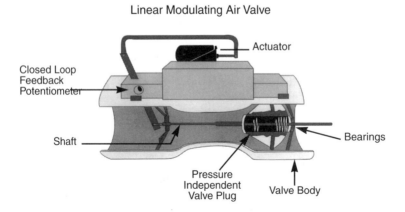

Figure 27.2 VAV Air Valve

configuration uses a single blade damper in the exhaust duct similar to typical supply air VAV boxes. All of these boxes use flow rings, hot wire anemometers or some other measuring means to provide feedback on the actual exhaust flowrate back to their controller. The choice of exhaust damper is based on user request and engineer familiarity.

In most applications, the hoods operate continuously to prevent fumes from accumulating in the enclosed space of the hood. When the sash is completely closed, the hood's exhaust damper modulates to its minimum position to maintain a minimum flow through the fume hood enclosure. Conditioned zone air enters the hood through openings under the sill to maintain flow through the hood when the sash is closed.

27.1.2 Balancing Air Flowrates Within a Zone

A VAV supply air box must be installed in the zone to maintain the correct pressure within a zone incorporating a fume hood employing a VAV exhaust box. The VAV supply air box modulates in response to changes in the exhaust air flowrate through the fume hoods. Whenever an exhaust damper opens in response to an increase in sash height, the increase in air leaving the zone must be made up by the VAV supply box. If the air leaving the fume hood were not replaced, the zone's pressure would drop and there would not be sufficient airflow through the sash opening to maintain its face velocity. Therefore, as the exhaust damper on a fume hood opens, the primary air damper of the supply VAV box opens. Conversely, as the exhaust damper closes in response to the lowering of a sash, the supply box damper modulates toward minimum position.

27.1.3 Zone Temperature Control

The temperature control strategies used for zones with fume hoods is similar to those found in a typical VAV with reheat application. A thermostat modulates the supply VAV box's primary air damper and reheat valve to satisfy the zone's thermal load. Whenever the temperature begins to drop, the supply VAV box modulates its damper toward its minimum position. The minimum volume position is equal to the current volume of air being exhausted by the fume hoods minus a fixed differential of 100 cfm. If the zone temperature continues to drop after the supply damper is at its minimum position, the reheat valve modulates to maintain the zone temperature.

The supply VAV box used in a fume hood applications has two functions that must be integrated together to maintain the safety and comfort of the lab personnel. It must modulate the flowrate of the primary air based upon the exhaust flowrate of the fume hoods and it must modulate the flowrate of the primary air to maintain the proper temperature in the zone. Of these two functions, maintaining the flowrate of primary air to make up the air exhausted by the fume hoods takes precedence over temperature control.

27.1.4 Zone Pressure Control

The static pressure of laboratories is typically maintained at a negative value with respect to its adjacent hallway and zones. This prevents the migration of harmful contaminants into the other spaces within the building. To maintain the negative pressure within the zone, an additional control loop and exhaust VAV box are required in the zone. The new exhaust VAV box is called the *general exhaust box*. Its function is to regulate the amount of air leaving the zone that does not pass through a fume hood.

The general exhaust VAV box modulates open to maintain a negative pressure within the zone by exhausting the supply air that does not leave through the fume hood exhaust VAV boxes. Its controller modulates to maintain the volume flowrate of air leaving the zone equal to the volume of air entering the zone through the supply VAV box minus the volume being exhausted by the fume hoods minus a fixed quantity of 100 cfm as shown in the following formula:

$$Supply\ Volume - Exhaust_{Fume\ Hood} - 100\ cfm = Exhaust_{General}$$

A volume control loop modulates the position of the general exhaust damper to maintain the flowrate calculated by this formula. By exhausting more air from a zone than is entering through the supply VAV box, the zone is maintained in a slight negative pressure with respect to its surrounding zones.

The general exhaust box also responds to changes in the zone's temperature. Whenever the cooling load increases, the supply VAV box increases its flowrate to offset the increasing heat gains. If the hood's sash position remains constant, the extra volume of air supplied into the room is relieved by the general exhaust VAV box to maintain the 100 cfm differential between the supply and exhausted air volumes. Conversely, as the zone's cooling load decreases, the supply VAV box throttles its damper toward its minimum position. The minimum volume position is equal to the current volume of air being exhausted by the fume hoods minus a fixed differential of 100 cfm. If the amount of air entering the zone can be completely exhausted by the fume hoods, the general exhaust box will modulate closed.

27.1.5 Control System Summary

In newer lab facilities, a 100% outside air VAV air handling unit is used to condition the air supplied to the zones. The configuration of the VAV boxes used in the zones is based upon the presence of a fume hood in the space. Zones that do not have a fume hood require two VAV boxes, a supply VAV box to vary the primary air from the air handling unit and an exhaust VAV box to maintain the proper static pressure differential between the zone and its adjoining spaces. Zones that have a fume hood require three VAV boxes, one supply box and *two* exhaust boxes. These exhaust boxes work together to regulate the amount air leaving the fume hood and the zone. The response of all three boxes is integrated together using strategies that vary the flow entering and leaving the zone based upon the fume hood's sash position, the pressure differential between the zone and the room temperature.

The complexity of fume hood strategies is best performed by DDC systems. DDC systems are well suited for the proper loop coordination while maintaining safety, face velocity, zone temperature, pressure and process efficiency. The microprocessor can easily evaluate the sensor data of sash position, hood exhaust volume, supply volume, zone exhaust volume, zone temperature and differential pressure in order to correctly position all three VAV boxes.

27.2

ILLUSTRATION OF A FUME HOOD CONTROL SYSTEM

This example illustrates the response of a VAV fume hood control system incorporating supply, general exhaust and fume hood exhaust VAV boxes to maintain the face velocity of the hoods along with the temperature and pressure requirements of the zone. The system has the following control loops in its fume hood control system:

1. Fume hood face velocity control.
2. Zone static pressure control.
3. Zone air temperature control.

27.2.1 System Drawing

A system drawing for a VAV fume hood control system incorporating supply, general exhaust and fume hood exhaust VAV boxes to maintain the face velocity of the hoods along with the temperature and pressure requirements of the zone is shown in Figure 27.3.

27.2.2 Sequence of Operation

The sequence of operation for the VAV fume hood control system incorporating supply, general exhaust and fume hood exhaust VAV boxes to maintain the face velocity of the hoods along with the temperature and pressure requirements of the zone that is illustrated in Figure 27.3 is explained in this section.

(Note: the air handling unit sequence of operation is very similar to the one described in Chapter 23. The following descriptions only describe the operation of the fume hood and the interactions of the VAV boxes.)

Supply Air Volume Control

The supply air fan's variable frequency drive is modulated to maintain the supply duct static pressure at a set point of 1.8 inches of water. As the static pressure in the supply duct increases, in response to some of the supply VAV boxes throttling toward their minimum position, the reverse acting output signal from controller

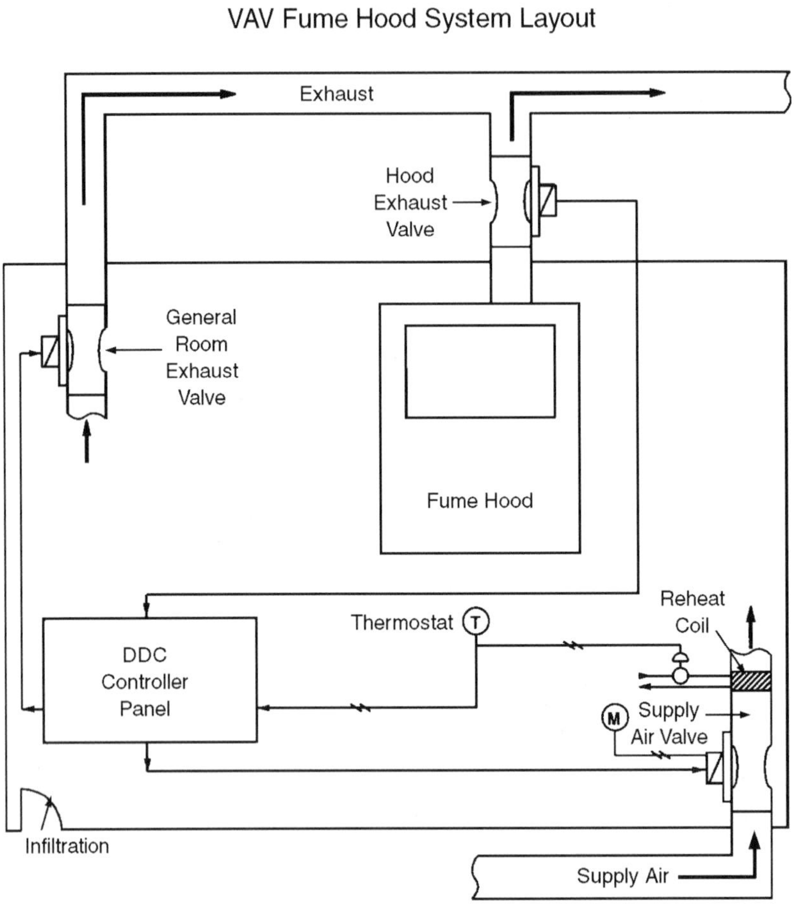

VAV Fume Hood System Layout

Figure 27.3 Laboratory Fume Hood Layout

decreases, reducing the fan's rotational speed and capacity as it reduces the motor's power consumption.

Exhaust Fan Control

The exhaust fan's variable frequency drive is modulated to maintain the fume hood and general exhaust duct system's static pressure at a set point of −2.0 inches of water. As the static pressure in the exhaust duct increases in response to fume hood or general exhaust dampers throttling toward their maximum open position, the output signal from the controller increases, modulating the fan motor to a faster speed, increasing its capacity. If the exhaust fan is operating at its lowest frequency and the negative static pressure in the exhaust duct decreases below its −2.0 inches of water set point, the outside air inlet damper installed in the exhaust duct system modulates open to maintain the set point.

Fume Hood Face Velocity Control

The fume hood's exhaust damper is modulated to maintain a face velocity of 100 feet per minute through the sash opening. A position sensor mounted to the sash is used to calculate the face velocity. As the sash height increases, the signal from the face velocity controller increases, opening the fume hood exhaust damper and increasing the volume flowrate through the hood to maintain the face velocity set point.

A control signal based upon the face velocity controller's response modulates the zone's supply VAV box damper actuator open to supply the additional exhaust requirements of the hood.

Zone Pressure Control

The zone's general exhaust VAV box modulates to maintain a fixed 100 cfm volume flowrate differential between the air entering the zone through the supply VAV box and the total quantity of air being exhausted through the fume hood and general exhaust boxes. When the mathematical difference between the supply volume flowrate and exhaust volume flowrate increases, indicating more air is being exhausted than supplied, the signal to the general exhaust damper decreases, modulating it toward its closed position.

Zone Temperature Control

The supply VAV box damper is modulated to maintain the zone's temperature at set point. As the cooling load in the zone decreases, the damper is modulated toward its minimum volume position. The minimum volume position is equal to the current volume of air being exhausted by the fume hoods minus a fixed differential of 100 cfm. The steady state position of the supply VAV box damper must supply sufficient air to the fume hoods to maintain their face velocity requirements. If the zone temperature continues to decrease due to thermal load changes or the fume hood exhaust requirements, the normally open hot water reheat valve is modulated to maintain the zone temperature set point.

27.3

PROCESS RESPONSES TO LOAD CHANGES

The following passages list the probable responses of the fume hood control system to the described change in a process load.

27.3.1 Load Change 1

A fume hood's sash is opened from 0% to 100% (18″) while the thermal load of the zone remains constant.

Fume Hood Face Velocity Response

1. The position sensor on the fume hood's sash measures the change in the sash height and calculates the corresponding change in the exhaust flowrate required to maintain its face velocity set point.
2. The control signal to the fume hood exhaust damper increases, modulating the damper from its minimum position toward its maximum flow position to maintain the hood's face velocity set point.
3. The control signal to the zone's supply VAV box damper also increases, modulating the primary air damper open to supply the additional air that will be drawn through the hood's exhaust damper.

Zone Temperature Process Response

1. The thermostat measures the decrease in the zone temperature in response to the increase in the volume flowrate of primary air entering the zone.
2. The control signal to the supply VAV damper remains constant to fulfill the requirements of the fume hood. The supply box reheat coil's valve modulates to maintain the zone temperature set point.

Zone Pressure Process Response

1. The increase in supply volume is 100 cfm less than the fume hood exhaust flow so the general exhaust damper remains closed. The fume hood exhaust satisfies the differential pressure requirements of the zone.

Supply Air Static Pressure Process Response

1. The fume hood's load increased. The signal from its face velocity controller caused its supply VAV box damper to modulate to higher flowrate position.
2. The increase in the supply air flowrate requirements causes the static pressure in the supply duct to decrease.
3. The supply air static pressure loop increases its signal to the supply fan VFD, increasing the frequency supplied to the fan motor.
4. The increase in motor frequency increases the motor's rotational speed, increasing the static pressure in the supply duct back to its set point.

Exhaust Air Duct Static Pressure Process Response

1. The increase in the fume hood volume flowrate load causes the static pressure in the exhaust duct to increase from -2.0 toward -1.5 inches of water.
2. The exhaust air static pressure loop increases its signal to the exhaust fan VFD, increasing the frequency supplied to the fan motor.
3. The increase in motor frequency increases the motors rotational speed, decreasing the static pressure in the supply duct back to its -2.0 inches of water set point.

27.3.2 Load Change 2

The fume hood sash remains 100% opened at 18″ and the cooling load of the zone increases.

Fume Hood Process Response

1. The position sensor measures no change in the sash height so the control signal to the hood exhaust damper remains the same.
2. The control signal to the supply VAV box damper is also unaffected by the face velocity control loop because there was no change in the sash position.

Zone Temperature Process Response

1. The thermostat measures the increase in the zone's temperature.
2. The control signal to supply box reheat coil valve increases, modulating the normally open valve closed. The signal to the VAV primary air damper also increases to increase the flow of primary air into the zone to maintain the zone temperature set point.

Zone Pressure Process Response

1. The increase in supply air flowrate surpasses the volume being exhausted through the fume hood exhaust damper. The signal to the general exhaust damper increases, opening the damper to maintain the zone pressure slightly negative with respect to the pressure in the hallway.

Supply Air Pressure Process Response

1. The zone's cooling load increased, causing the zone's primary air damper to modulate open to a higher flowrate position.
2. The increase in the supply air flowrate requirements causes the static pressure in the supply duct to decrease.
3. The supply air static pressure loop increases its signal to the supply fan VFD, increasing the frequency supplied to the fan motor, increasing the static pressure in the supply duct back to its set point.

Exhaust Air Pressure Process Response

1. The modulation of the general exhaust box causes the static pressure in the exhaust duct to increase from -2.0 toward -1.7 inches of water because the fume hood and general VAV boxes use the same exhaust fan and duct.
2. The exhaust air static pressure loop increases its signal to the exhaust fan VFD, increasing the frequency supplied to the fan motor.
3. The increase in motor frequency increases the motors rotational speed, decreasing the static pressure in the supply duct back to its -2.0 inches of water set point.

27.3.3 Load Change 3

All of the fume hood sashes are closed at the end of the day and the cooling load in the zones decreases.

Fume Hood Process Response

1. The position sensors on all the fume hood sashes measure their change to a closed position. The only air passing through the fume hood comes through the minimum flow openings at the bottom of each sash.
2. The face velocity control loops signal their fume hood exhaust dampers to modulate to their minimum position.
3. The control signals to the supply VAV box dampers in all the labs also decrease in response to the reduction in fume hood exhaust requirements.

Zone Temperature Process Response

1. The thermostats measure the decrease in their zone temperature that occurs as the lab personnel turn off equipment and leave for the day.
2. The control signal to the supply boxes decrease, modulating the primary air dampers toward their minimum position.
3. If a zone's temperature continues to decrease, its reheat valve modulates to maintain the zone temperature set point.

Zone Pressure Process Response

a. The reduction of the supply air volume to its minimum position, coupled with the decrease in the fume hood exhaust requirements causes the general exhaust damper to modulate closed.

Supply Air Pressure Process Response

1. The system's cooling load decreases. The supply VAV box dampers have modulated to their minimum position.
2. The decrease in the supply air flowrate requirements causes the static pressure in the supply duct to increase.
3. The supply air static pressure loop decreases its signal to the supply fan VFD, decreasing the frequency to the fan motor thereby decreasing the static pressure in the supply duct back to its set point.

Exhaust Air Pressure Process Response

1. The decrease in the zone's exhaust loads causes the static pressure in the exhaust duct to decrease from -2.0 toward -2.7 inches of water.
2. The exhaust air static pressure loop decreases its signal to the exhaust fan VFD to minimum position.

3. The decrease in motor frequency decreases the motors rotational speed, increasing the static pressure in the supply duct back to −2.2 inches of water.
4. The outside air damper located on the exhaust duct modulates to maintain the exhaust duct static pressure at a set point of −2.0 inches of water.

27.4

ANALYSIS OF A FUME HOOD CONTROL SYSTEM

27.4.1 Scenario 1

There is excessive turbulence in front of one fume hood's sash.

Possible Causes

1. The fume hood's exhaust damper open too much.
2. The fume hood's control loop is out of calibration.

Verify the

1. operation of the face velocity control loop by performing the following procedure:
   a. Measure the face velocity at the open sash with a thermal anemometer.
   b. Compare the actual face velocity with the value displayed by the DDC control system.
   c. If the actual and DDC value of the face velocity differ more than 5%, calibrate the sensor following the manufacturers instructions.
2. positioning of the hood's exhaust damper to determine if it correlates with the controller's output signal.
   a. Measure the volume of air flowing through the exhaust duct. Take appropriate precautions if the hood has been used with hazardous substances.
   b. Using the measured face velocity along with the area of the sash opening, calculate the volume flow rate in cubic feet per minute and compare the actual exhaust volume with the calculated value.
   c. If the measured and calculated values differ significantly, determine that the actuator is positioning the damper correctly.
   d. If the damper is being positioned correctly, the control signal from the system face velocity controller needs calibration. The calibration of the loop components differs for open and closed loop strategies. Consult the manufacturers instructions for proper calibration procedures.

27.4.2 Scenario 2

One of the zones is too warm. The static pressure in that zone is being maintained at its set point and the face velocity of the fume hoods are also being maintained at set point as their sashes are repositioned from 0% to 100% open.

Possible Causes

1. The supply VAV box reheat control valve is stuck open or leaking past its seat when closed.
2. The thermostat needs calibration.

Since only one zone is warm, the problem is confined to that zone instead of with the air handling unit processes. Since the static pressure in the zone is being maintained at its set point throughout the repositioning of the sash, the supply VAV box must be modulating correctly. Therefore, the problem must be in the zone temperature control loop.

Verify the

1. thermostat's set point is actually placed at the desired setting.
2. thermostat is properly calibrated. Verify calibration by performing the following procedure:
   a. Measure temperature at the thermostat's location.
   b. Adjust the thermostat's set point so it equals the measured temperature.
   c. Measure the thermostat's output signal.
   d. Verify that the thermostat's signal is equal to the midpoint value of its output range. Make the necessary adjustments on the thermostat or in the DDC database.
   e. Slide the adjustment to both limits of its throttling range to verify that it is generating a linear output signal over a 2 to 4° throttling range. If the thermostat's response is erratic it must be replaced.
   e. Place the thermostat's set point equal to the desired zone temperature and let the loop stabilize.
3. operation of the VAV boxes:
   a. Close all the fume hood sashes in the zone and adjust the thermostat set point to 85°. This causes the supply VAV box to modulate to its minimum flow position. The general exhaust VAV box should also modulate itself closed.
   b. Using the flow measuring device, measure the flow rate supplied to the zone through the supply VAV box.
   c. Compare the measured supply volume flowrate to air volume being exhausted through the fume hoods.
   d. Adjust the supply VAV box volume controller's minimum position potentiometer or regulator until the air flow through the box equals the fume hood total minus 100 cfm.
   e. Measure the general exhaust flow and calibrate its minimum position potentiometer so the flow equals zero.
   f. Set the thermostat set point to its minimum value (55°). This causes the supply VAV box to open to its maximum flow position to satisfy the zone temperature. Adjust the supply VAV box volume controller's maximum flow position potentiometer until the air flow through the box equals the design flowrate stated on the mechanical system the prints.

    g. Position the fume hood sash to 100% open and leave the thermostat set point at 55°. Adjust the general exhaust VAV box volume controller's maximum flow position potentiometer until the air flow through the box equals the supply design flowrate minus the fume hood's maximum volume exhaust minus 100 cfm.

4. control signal is present at the supply VAV box reheat valve's actuator and the valve is correctly positioned.

5. valve is not leaking through by momentarily closing the isolation ball valves and monitoring the temperature of the air downstream of the reheat coil.

27.4.3 Scenario 3

A zone's temperature and its fume hood's face velocity are being maintained at set point but the static pressure in the zone is positive with respect to the adjoining spaces.

Possible Causes

1. The general exhaust box needs calibration.
2. The general exhaust box actuator or damper not functioning.

Since only one zone's static pressure control point is out of range, the problem is confined to its general exhaust damper.

Verify the

1. calibration of the exhaust box using the procedure detailed in the above section.
2. control signal is present at the general exhaust box actuator and the damper move freely from 0% to 100% of their stroke.
3. damper linkage is not slipping on the damper shaft.

27.5

EXERCISES

Respond to the following statements, questions and problems completely and accurately, using the system drawing in Figure 27.3.

1. What are the possible causes of high fume hood face velocities that blow out a gas burner's flame when the sash is only open five inches? Explain your answer.
2. What is the response of the zone control systems if the actuator on the supply VAV box breaks its linkage causing the damper to open to 100%? Explain your answer.

3. What is the response of the control loops in the zone if the boiler malfunctions during cold weather and the reheat water temperature drops from 150° to 70°? Explain your answer.

4. What effect does keeping the fume hood sashes open when they are not in use have on the cost of each process serving the zone? Explain your answer.

5. What is the response of the system control loops when the dirty filters in the air handling unit are replaced with clean ones? Explain your answer.

Glossary

Accuracy

The accuracy of a sensor indicates the maximum difference between the actual value of the measured variable and the value indicated by the sensor.

Actuator

The portion of the final controlled device that converts the control signal from the loop controller into linear or rotational motion that is used to open and close the flow control device.

ADC

Analog to digital converter. An electronic device used to interface analog sensor signals with a digital controller. The ADC converts the magnitude of the input signal into a binary word.

AHU

Air Handling Unit. Mechanical equipment used to heat, cool humidify, dehumidify and/or pressurize an airstream before it is distributed to a zone.

ALU

Arithmetic Logic Unit. A section of a microprocessor that performs mathematic and logic routines.

Analog Control

A device or control loop that generates and responds to modulating signals. The data in analog signals is represented by a continuous varying magnitude.

ASHRAE

American Society of Heating, Refrigeration and Air Conditioning Engineers. A professional society that performs research and develops standards for the air conditioning and refrigeration industry.

Automatic Control

The automated application of a regulating influence upon control system components by another control device in response to a change in a measured variable.

BAS

Building Automation System. A computer based control system that integrates HVAC, security, access and fire/safety functions into a centrally accessed system.

Bias

The output signal of a control device when the change in its input signal is equal to zero.

Binary Control

A device or control loop that can only recognize or respond to the change of state of a signal from one value to another IE, on/off, high/low, yes/no, 1/0 etc. Binary devices are also called two-position devices.

btu

British Thermal Unit. A measure of thermal energy equal to the amount of heat required to raise one pound of pure water 1 °F.

Cavitation

The violent implosion due to a rapid change of state of a vapor to a liquid that results from a change in the process pressure.

cfm

Cubic foot per minute. A measure of the volume flowrate of a fluid.

Close Loop Configurations

Closed loop control configurations generate a response that can maintain the controlled variable at its set point. These configurations incorporate a feedback path between the control point of the process and the controller's input.

Control

The application of a regulating influence upon a device to make it perform in a desired manner.

Control Agent

The fluid that transports mass or energy to of from the process in response to changes in the output signal of the controller, IE hot water, chilled water, refrigerant, etc.

Control Differential

The characteristic of the two-position transfer function that creates a delay between changes in the output signal to reduce the possibility of the output signal repeatedly cycling on and off whenever the control point is near the set point.

Control Loop

A group of interconnected control devices that operate collectively to maintain a single process at its desired set point.**Control Point**
The actual magnitude of a measured variable.

Controlled Medium

The controlled medium is the fluid that absorbs or releases the mass or energy transferred to or from the process.

Controlled Variable

The controlled variable is the property of a controlled medium that a control loop is maintaining at its set point.

Controller

The component in the control loop that calculates the process error and generates an output signal that repositions the final controlled device to maintain a balance between the mass or energy being transferred to the process and its current load.

Converter

A device that converts a control signal from one type to another IE, pneumatic to current, voltage to current etc.

DAC

Digital to Analog converter. An electronic device used to interface digital computer based systems to the analog world. The DAC converts the magnitude of a binary word into an analog output signal.

Damping

Damping is any effect that reduces or impedes a reaction.

DDC

Distributed (Direct) Digital Control System. A microprocessor based control system that performs modulating control, energy management strategies, custom system programming, generates graphic displays, dials out in emergency situations etc.

Derivative Mode

The portion of the controller transfer function that changes its output signal in response to the rate of change in the process error. This mode is responsible for reducing the effects of fast process changes on the stability of the control loop by responding to the anticipated change in the process based upon its current rate of change. Also called Rate Mode.

Direct Acting

The response of a control loop or device which generates a change in its output signal that is in the same direction as the change in its input signal.

DX Coils

Direct expansion refrigeration coils. Heat transfer that use liquid refrigerant to absorb heat from the fluid flowing across its surfaces.

Error

The mathematical difference between the control point and the set point of a process. Error indicates whether the control point is above or below the set point.

FCD

Final controlled device. The component in a control loop that receives a signal from the controller which commands a change in the flow of the control agent.

Flashing

The rapid change of state of a liquid to a vapor due to a change in the process pressure.

Final Controlled Device

The component in a control loop that varies the flow of mass or energy into the process. Valves and dampers are common HVAC modulating FCDs, electric relays, contactors and motor starters are common two-position FCDs.

Floating Controller Mode

Floating controller mode generates a three state output signal. This signal positions a three state actuator that modulates the flow of the control agent into the process.

Flowchart

A symbolic representations of the sequence of operation of a control system used to write and troubleshoot DDC application software programs.

Flow Coefficient

The flow coefficient is a parameter used to standardize the sizing of valves. The abbreviation for Flow Coefficient is C_v.

Full Scale Range

The operating range of the sensor. The full scale range defines the input signal's operating range of the sensor where its response remains with its design specifications.

General Form of the Linear Transfer Function Equation

The general form of a transfer function is an equation that is the foundation for developing transfer functions for any linear control device, reset schedule or process. Most of the equations used by a control technician are based upon the following equation:

$Output = \Delta Input \times Proportional\ Gain + Bias$

HVAC/R

Heating, ventilating, air-conditioning and refrigeration.

IAQ

Indoor air quality. An indication of the quality of the environmental conditions in a building.

Inherent Flow Characteristics

The response of a flow control device (Δ Flow / Δ Stroke) when the pressure drop across the device is kept constant. This is the characteristic assigned to a device by the manufacturer.

Installed Flow Characteristics

The actual response of a flow control device (Δ Flow / Δ Stroke) when the device is installed in a system and its pressure drop varies as its actuator moves.

Integral Mode

The portion of the controller transfer function that changes its output signal in response to the size and length of time a process error exists. This mode is responsible for reducing the process error to zero before the control loop stabilizes. Also called Reset Mode.

Lag Time

A time delay that occurs as a natural response of dynamic systems. A time lag is an period that occurs between the initiation of an action and the appearance of its desired outcome. Lags can be classified as measuring, transmitting, transporting and mixing lags.

mA

Milliamps. A measure of electrical current equal to one-thousandth of an amp.

Manual Control

The application of a regulating influence by a human being. In HVAC control processes, the response of a manual control loop remains in its current position until changed by an operator.

NC

Normally Closed. The contacts in a relay, blades of a damper or plug in a valve are closed when there is no signal being applied to its actuator.

Normal Position

The position (open, closed) of a final controlled device when there is no signal present to its actuator.

NO

Normally Open. The contacts in a relay, blades of a damper or plug in a valve are open when there is no signal being applied to its actuator.

NTC

Negative temperature coefficient. A characteristic of a temperature sensor that experiences a decrease in its output signal as its measured temperature increases.

Offset

The absolute (non-signed) difference between the set point and control point of a process, also known as drift and deviation. The offset does not tell whether the control point is above or below the set point. See Error.

Open Loop Response Characteristics

All open loop configurations generate a binary or two-position response to changes in the measured variable. This configuration is used in applications to prevent the control point from exceeding a predetermined design limit.

Operating Differential

The range of the process control point that results from the combined effects of the control differential and process time lags. The operating differential is always greater than the control differential.

Positioner

A pneumatic device used to properly position pneumatic actuators by applying main air pressure to the actuator until it moves to its proper location as measured by a lever and feedback spring mounted to the FCD stem.

Process

The intended function of a control loop IE, maintaining a specific temperature, pressure or humidity set point. A mixed air temperature control loop maintains the mixed air temperature process at set point.

Proportional Band

The percent of the sensor's input span that is used to drive the final controlled device from 0% to 100% open (or closed). It is always expressed as a percentage.

Proportional Band = (Throttling Range ÷ Sensor Span) × 100.

Proportional Gain

The amount change in the magnitude of the output signal that occurs in response to a change in the input signal. Also called Sensitivity.

Proportional Mode

The response of a controller that changes its output signal in proportion with the change in the process error. For example, if the error changed 7% the output signal also changes 7%. Proportional mode positions the final controlled device so that a balance is created between the process load and the mass or energy transferred by the control agent.

psi

Pounds per square inch. A measure of pressure.

PTC

Positive temperature coefficient. A characteristic of a temperature sensor that experiences an increase in its output signal as its measured temperature increases.

Rangeability

Characteristic of a flow modulating device that is equal to its maximum controllable flow divided by its minimum controllable flow.

Rate Mode

See Derivative Mode

Reset

The control strategy that resets a process set point based upon the magnitude of another variable. Commonly used for temperature processes. If the set point increases when the measured variable decreases, the strategy is called reverse reset. If the set point increases when the measured variable increases, the strategy is called direct reset.

Reset Mode

See Integral Mode

Resolution

The resolution is the smallest change in the measured variable that is required to produce a detectable change in a control device's output signal.

Response

The behavior of a process or control loop that results from a change in an external or internal variable.

Reverse Acting

The response of a control loop or device which generates a change in its output signal that is in the opposite direction as the change in its input signal.

rpm

Revolutions per minute. A measure of the rotation speed of a rotating device.

RAM

Random access memory. Volatile electronic memory used to store data that is erased when the power supply is disconnected. It is used for storing DDC application programs and databases.

ROM

Read only memory. Non-volatile electronic memory used to store data that cannot be erased when the power supply is disconnected.

RTD

Resistance temperature device (detector). A temperature sensing device that changes its output resistance in a predictable fashion whenever its temperature changes.

Sensors

Control devices that measure the control point of a process. They generate an output signal related to the magnitude of the control point. Pneumatic sensors are often called transmitters.

Sequence of Operation

A concise summary of the operational attributes of mechanical equipment and its control loops.

Set Point

The desired magnitude of a process variable.

Signal Span

The difference between the maximum and minimum values of a control device's input or output signal's operating range.

Steady State Error

The offset between the control point and the set point that remains after the controller output signal has stabilized.

Thermal Resistance

Thermal resistance is a measure of the ability of a material to oppose the flow of heat when a temperature difference exists between its surfaces.

Thermal Capacitance

The thermal capacitance of a process indicates the amount of heat that must be transferred to create a one degree change in the temperature of the controlled variable.

Throttling Range

The change in the control point required to drive the final controlled device from 0% to 100% open (or closed). The throttling range always has the same units as the controlled variable.

Time Constant

The time it takes for a control device or process to change 63% of its total change in response to a change in its input variable.

Transfer Function

The mathematical formula that defines how the change in a device's input signal is generated into a related change in its output signal. The transfer function defines the size, direction and timing characteristics of the change in the output signal.

Timed Two-Position Mode

Timed two-position mode incorporates a mean of reducing the operating differential in electric, two-position temperature control applications that use a thermostat.

Turndown ratio

The ratio of maximum required flowrate to a device's minimum controllable flowrate.

Two-Position Mode

The controller mode that generates a binary output response whenever the measured variable exceeds its set point.

Variable Air Volume AHU

A mechanical system that maintains zone temperature by varying the amount of conditioned air that enters the zone in response to the zone's temperature.

VAV

Variable Air Volume. See variable air volume AHU.

VAVT

Variable air volume, variable temperature. A control strategy that incorporates temperature reset of the supply air of a VAV unit. See variable air volume AHU.

VFD

Variable frequency drive. An electronic device used to vary the speed of an induction motor by changing the frequency of its power source.

VSD

Variable speed drive. See VFD.

Zone

A zone is an area within a building where a controlled variable is maintained at the same set point. All the areas (rooms) within a zone have similar thermal and operational loads.

Index

Page references for figures appear in italics